Environmental Risk Assessment of Genetically Modified Organisms

Volume 3. Methodologies for Transgenic Fish

Environmental Risk Assessment of Genetically Modified Organisms Series

Titles available

Volume 1. A Case Study of Bt Maize in Kenya
Edited by A. Hilbeck and D.A. Andow

Volume 2. Methodologies for Assessing Bt Cotton in Brazil
Edited by A. Hilbeck, D.A. Andow and E.M.G. Fontes

Volume 3. Methodologies for Transgenic Fish
Edited by A.R. Kapuscinski, S. Li, K.R. Hayes and G. Dana

Environmental Risk Assessment of Genetically Modified Organisms

Volume 3. Methodologies for Transgenic Fish

Edited by

Anne R. Kapuscinski
Department of Fisheries, Wildlife and Conservation Biology
University of Minnesota
St Paul, Minnesota, USA

Keith R. Hayes
Australian Commonwealth Scientific and Industrial Research Organisation
Division of Marine and Atmospheric Research, Hobart, Tasmania, Australia

Sifa Li
Aquatic Genetic Resources Laboratory
Shanghai Fisheries University
Shanghai, China

Genya Dana
Conservation Biology Graduate Program
University of Minnesota
St Paul, Minnesota, USA

Guest Series Editor

Eric M. Hallerman
Department of Fisheries and Wildlife Sciences
Virginia Polytechnic Institute and State University
Blacksburg, Virginia, USA

Series Editor

Peter J. Schei
Fridtj of Nansen Institute
Lysaker, Norway

CABI is a trading name of CAB International

CABI Head Office
Nosworthy Way
Wallingford
Oxfordshire OX10 8DE
UK

Tel: +44 (0)1491 832111
Fax: +44 (0)1491 833508
E-mail: cabi@cabi.org
Website: www.cabi.org

CABI North American Office
875 Massachusetts Avenue
7th Floor
Cambridge, MA 02139
USA

Tel: +1 617 395 4056
Fax: +1 617 354 6875
E-mail: cabi-nao@cabi.org

A catalogue record for this book is available from the British Library, London, UK.

A catalogue record for this book is available from the Library of Congress, Washington, DC.

ISBN-13: 9781 84593 2961

Typeset by SPi, Pondicherry, India.
Printed and bound in the UK by Biddles Press, King's Lynn.

Contents

Contributors

C. Bambaradeniya, *Coordinator, Asia Regional Species Programme, IUCN – The World Conservation Union 53, Horton Place, Colombo 7, Sri Lanka. E-mail: cnb@iucnsl.org*

Z. Basiao, *Associate Professor, Institute of Biology, College of Science, University of the Philippines, Diliman, Quezon City, The Philippines. E-mail: zbasiao@yahoo.com*

M.A. Burgman, *Professor, Environmental Science, Director, Australian Centre of Excellence for Risk Analysis (ACERA), The School of Botany, The University of Melbourne, Parkville 3010, Victoria, Australia. E-mail: markab@unimelb.edu.au*

A.M. Cooper, *PhD Candidate, Department of Fisheries, Wildlife and Conservation Biology, University of Minnesota, 200 Hodson Hall, 1980 Folwell Ave., St Paul, MN 55108, USA. E-mail: coop0162@umn.edu*

G. Dana, *PhD Student, Conservation Biology Graduate Program, University of Minnesota, 200 Hodson Hall, 1980 Folwell Ave., St Paul, MN 55108, USA. E-mail: dana0010@umn.edu*

R.H. Devlin, *Department of Fisheries and Oceans, The University of British Columbia, 4160 Marine Dr., West Vancouver, British Columbia, Canada V7V 1N6. E-mail: devlinr@pac.dfo-mpo.gc.ca*

M. Dey, *Economist, Policy and Impact Assessment Program, WorldFish Centre, PO Box 500, GPO 10670, Penang, Malaysia. E-mail: m.dey@cgiar.org*

M.R.R. Eguia, *Aquaculture Research Specialist, SEAFDEC Aquaculture Department, Binangonan Freshwater Station, Binangonan, 1940 Rizal, The Philippines. E-mail: mreguia@aqd.seafdec.org.ph*

M.P. Estrada, *Head of Aquatic Biotechnology Projects, Centre for Genetic Engineering and Biotechnology, Animal Biotechnology Division, PO Box 6162, Havana 10600, Cuba. E-mail: Mario.pablo@cigb.edu.cu*

I.A. Fleming, *Professor and Director, Ocean Sciences Centre, Memorial University of Newfoundland, St John's, Newfoundland, Canada, A1C 5S7. E-mail: ifleming@mun.ca*

D. Fonticiella, *Pescavilla Fisheries and Aquaculture Ent., Doble via Esq. a Novena Escambray, Santa Clara V.C. 50100, Cuba. E-mail: direccion@pescavilla.vcl.cu*

C. Fu, *Associate Professor, Institute of Biodiversity Science, School of Life Science, Fudan University, Handan Road 220, Shanghai 200433, China. E-mail: czfu@fudan.edu.cn*

J. Gallardo, *Associate Professor, Escuela de Ciencias del Mar, Pontificia Universidad Católica de Valparaíso, Avda. Altamirano 1480, Valparaíso, Chile. E-mail: jose.gallardo@ucv.cl*

Z. Gong, *Department of Biological Sciences, National University of Singapore, Science Drive 4, Singapore 117549. E-mail: dbsgzy@nus.edu.sg*

J.J. Hard, *Conservation Biology Division, Northwest Fisheries Science Center, National Marine Fisheries Service, 2725 Montlake Blvd East, Seattle, Washington, DC 98112, USA. E-mail: Jeff.Hard@noaa.gov*

K.R. Hayes, *Risk Analyst, CSIRO Division of Marine and Atmospheric Research, GPO Box 1538, Hobart, Tasmania 7001, Australia. E-mail: keith.hayes@csiro.au*

M. Lorenzo Hernandez, *Biotechnology Risk Analyst, National Centre for Biological Safety, Ministry of Science, Technology and Environment, Calle 28 # 502 e/t 5ta. y 7ma. Miramar, Playa Ciudad de la Habana, Cuba. E-mail: mlhdez@infomed.sld.cu; cnsb_99@yahoo.com*

J.I. Johnsson, *Professor, Department of Zoology, University of Goteborg, PO Box 463, 405 30 Goteborg, Sweden. E-mail: jorgen.johnsson@zool.gu.se*

W. Kamonrat, *Department of Fisheries, Government of Thailand, Chatuchak, Bangkok 10900, Thailand. E-mail: wongpatk@fisheries.go.th;* odd_kamonrat@yahoo.com

A.R. Kapuscinski, *Professor, Department of Fisheries, Wildlife and Conservation Biology, University of Minnesota, 200 Hodson Hall, 1980 Folwell Ave., St Paul, MN 55108, USA. E-mail: kapus001@umn.edu*

S. Kunawasen, *Biosafety Program-BIOTEC, National Center for Genetic Engineering and Biotechnology, 113 Thailand Science Park, Phaholyothin Rd, Klong Luang, Pathumthani 12120, Thailand. E-mail: sukun@biotec.or.th; sukunpj@gmail.com*

W. Leelapatra, *Director, Aquatic Animal Genetics Research and Development Institute, 39 Moo 1, Klong 5, Klong Luang, Pathumtani 12120, Thailand. E-mail: walekfc@hotmail.com; walekfc@gmail.com*

S. Li, *Professor and Director, Aquatic Genetic Resources Laboratory, Shanghai Fisheries University, 334 Jun Gong Road, Shanghai 200090, China. E-mail: sfli@shfu.edu.cn*

N. Maclean, *Professor of Genetics (Emeritus), University of Southampton, School of Biological Sciences, Bassett Crescent East, Highfield, Southampton, Hampshire SO16 7PX, UK. E-mail: nm4@soton.ac.uk*

G.C. Mair, *Senior Lecturer, School of Biological Sciences, Flinders University, GPO Box 2100, Adelaide, SA 5001, Australia. E-mail: graham.mair@flinders.edu.au*

R. Martinez, *Center for Genetic Engineering and Biotechnology, Ave 31 y 158 y 190 Cubanacan Playa Apto 6162, Habana 10600, Cuba. E-mail: rebeca.martinez@cigb.edu.cu*

W. Mwanja, *Principal Fisheries Officer- Aquaculture, Ministry of Agriculture, Animal Industry and Fisheries, Department of Fisheries Resources, Plot 29 Luggard Avenue, PO Box 4 Entebbe, Uganda. E-mail: wwmwanja@yahoo.com*

Y.K. Nam, *Department of Aquaculture, Pukyong National University, Nam-gu 599–1, Busan 608–737, South Korea. E-mail: yoonknam@pknu.ac.kr*

R. Neira, *Professor, Department de Producción Animal, Universidad de Chile, Casilla 1004 Santiago, Chile. E-mail: rneira@uchile.cl*

K.C. Nelson, *Associate Professor, Dept of Fisheries, Wildlife and Conservation Biology, University of Minnesota, Room 115 Green Hall, 1530 N. Cleveland Ave. St Paul, MN 55108, USA. E-mail: kcn@umn.edu*

W.O. Ojwang, At time of writing: *Boston University, Department of Biology.* At time of publication: *Kenya Marine and Fisheries Research Institute, Kisumu Research Centre, PO Box 1881, Kisumu, Kenya. E-mail: Ojwang@hotmail.com*

O. Omitogun, *Professor, Department of Animal Science, Faculty of Agriculture, Obafemi Awolowo University, Ile-Ife, Nigeria. E-mail:ogomitogun@hotmail.com*

T.J. Pandian, *Emeritus Professor of Biology, School of Biological Sciences, Madurai Kamaraj University, Madurai 625 021, India. E-mail: tjpandian@eth.net*

K.M. Paulson, *PhD Candidate, Conservation Biology Graduate Program, University of Minnesota, 200 Hodson Hall, 1980 Folwell Ave., St Paul, MN 55108, USA. E-mail: kmp@umn.edu*

A. Ponniah, At time of writing: *WorldFish Center, Penang, Malaysia.* At time of publication: *Director, Central Institute of Brackishwater Aquaculture, 75, Santhome High Road, Raja Annamalipuram, Chennai 600028, India. E-mail: agponniah@gmail.com*

B.D. Ratner, *Regional Director, Greater Mekong, The WorldFish Center, PO Box 1135 (Wat Phnom), Phnom Penh, Cambodia. E-mail: b.ratner@cgiar.org*

H.M. Regan, *Biology Department, San Diego State University, 5500 Campanile Dr, San Diego, CA 92182–4614, USA. E-mail: hregan@sciences.sdsu.edu*

W. Senanan, *Department of Aquatic Science, Burapha University, Bangsaen, Chonburi 20131, Thailand. E-mail: wansuk@buu.ac.th; wansuks2@yahoo.com*

I.I. Solar, *Marine Biologist, S.A.G.I. Research, Fisheries and Aquaculture Consultants, Reproductive Physiology and Genetic Improvement, Casilla 1017, Suc. Valparaíso-1, Valparaíso, Chile. E-mail: isolar@terra.cl*

L.F. Sundström, *Department of Fisheries and Oceans, The University of British Columbia, 4160 Marine Dr, West Vancouver, British Columbia, V7V 1N6 Canada. E-mail: sundstromf@pac.dfo-mpo.gc.ca*

M.I. Toledo, *Ingeniero Pesquero, Laboratorio de Cultivo de Peces y de Alimentación para la Acuicultura, Escuela de Ciencias del Mar, Pontificia Universidad Catolica de Valparaiso, Av. Altamirano 1480 Valparaíso, Chile. E-mail: itoledo@ucv.cl*

J. Trisak, *Department of Fishery Management, Faculty of Fisheries, Kasetsart University, Chatuchak, Bangkok 10900, Thailand. E-mail: jiraporn.t@ku.ac.th*

A. Alcivar-Warren, *Associate Professor and Head, Environmental and Comparative Genomics Section, Department of Environmental and Population Health, Cummings School of Veterinary Medicine at Tufts University, 200 Westboro Road, North Grafton, MA 01536, USA. E-mail: acacia.warren@tufts.edu; aalcivar@yahoo.com*

M. Zakaraia-Ismail, *Environmental Science Division, Institute of Biological Sciences, University of Malaya, Malaysia. E-mail: zakaraia@um.edu.my*

Series Foreword

The advent of genetically modified organisms (GMOs) offers new options for meeting food, agriculture and aquaculture needs in developing countries, but some of these uses of GMOs can also affect biodiversity and natural ecosystems. These potential environmental risks and benefits need to be taken into account when making decisions about the use of GMOs. International trade and the unintentional trans-boundary spread of GMOs can also pose environmental risks depending on the national and regional contexts.

The complex interactions that can occur between GMOs and the environment heighten the need to strengthen worldwide scientific and technical capacity for assessing and managing environmental risks of GMOs.

The Scientific and Technical Advisory Panel (STAP) of the Global Environment Facility (GEF) provides strategic scientific and technical advice on GEF policies, operational strategies and programmes in a number of focal areas, including biodiversity. Its mandate covers *inter alia* providing a forum for integrating expertise on science and technology, and synthesizing, promoting and galvanizing state-of-the-art contributions from the scientific community. The GEF, established in 1991, helps developing countries fund projects and programmes that protect the global environment. The GEF grants support projects related to biodiversity, climate change, international waters, land degradation, the ozone layer and persistent organic pollutants.

Global environmental management of GMOs and the strengthening of scientific and technical capacity[1] for biosafety will require building policy and legislative

[1] By 'scientific and technical capacity' we mean 'the ability to generate, procure and apply science and technology to identify and solve a problem or problems' including 'the generation and use of new knowledge and information as well as techniques to solve problems'. (Mugabe, J. 2000. Capacity Development Initiative, Scientific and Technical Capacity Development, Needs and Priorities. GEF-UNDP Strategic Partnership, October 2000.)

biosafety frameworks. The latter is especially urgent for developing countries, as the Cartagena Protocol on Biosafety of the Convention on Biological Diversity makes clear. The World Summit on Sustainable Development has also identified the importance of improved knowledge transfer to developing countries on biotechnology. This point was stressed in recent international forums such as the Norway/UN Conference on Technology Transfer and Capacity Building, and the capacity-building decisions of the first meeting of the parties to the Cartagena Protocol on Biosafety.

The STAP has collaborated with a number of international scientific networks to produce a series of books on scientific and technical aspects of environmental risk assessment of GMOs. This complements the projects being undertaken by the United Nations Environment Programme and the GEF to help developing countries design and implement national biosafety frameworks.

The purpose of this series is to provide scientifically peer-reviewed tools that can help developing countries strengthen their own scientific and technical capacity in the biosafety of GMOs. Each book in the series examines a different kind of GMO in the context of developing countries. The workshops and writing teams used to produce each book are also capacity-building activities in themselves because they bring together scientists from developing and developed countries to analyse and integrate the relevant science and technology into the book. This third book, on methodologies for risk assessment and management of transgenic fish, was written by 44 authors from 19 countries, with an emphasis on developing countries which have transgenic fish research programmes or aquaculture systems that might use transgenic fish in the future. The first book, a case study of Bt maize in Kenya, was published in 2004, and the second book, on methodologies for assessing Bt cotton in Brazil, in 2006. Each book provides methods and relevant scientific information for risk assessment, rather than drawing conclusions. Relevant organizations in each country will therefore need to conduct their own scientific risk assessments in order to inform their own biosafety decisions.

This book is the outcome of a scientific partnership between the STAP and the WorldFish Center of the Consultative Group on International Agriculture Research (CGIAR). The book also benefited from scientific and strategic advice from relevant experts at the Secretariat of the Convention on Biological Diversity, the Food and Agriculture Organization of the United Nations and other organizations listed in the acknowledgements. Preparation of the book began with a book-drafting workshop led by the STAP, which conducted an independent, international and anonymous scientific peer review. To assure independence of the peer review, a guest series editor (E.M. Hallerman) replaced one of the series editors (A.R. Kapuscinski), who served as a subject matter editor for this volume.

We hope that this book will help governments, scientists, potential users of GMOs and civil society organizations in developing countries involved with aquaculture and other countries of the world to strengthen their understanding of the scientific knowledge and methods that are available for conducting environmental risk assessments of GMOs. We encourage readers to draw their own insights in order to help them devise and conduct robust environmental risk assessments for their own countries.

Yolanda Kakabadse
Chair, Scientific and Technical Advisory Panel, Global Environment Facility
Quito, Ecuador

Anne R. Kapuscinski
Member, Scientific and Technical Advisory Panel, Global Environment Facility
St. Paul, Minnesota, USA

Peter J. Schei
Member, Scientific and Technical Advisory Panel, Global Environment Facility
Trondheim, Norway

6 April 2006

Acknowledgements

We are indebted to many organizations and individuals around the world who made this book possible. Primary funding for planning and convening a week-long workshop, on which this book is based, came from the Global Environment Facility, Scientific and Technical Advisory Panel (STAP). Staff from the STAP Secretariat in Washington, DC and Nairobi provided logistical support. We would also like to thank the WorldFish Center for hosting the workshop. Alphis Ponniah, formerly of the WorldFish Center, provided extraordinary help with planning and convening the workshop. Additionally, Devin Bartley, Fee-Chon Low and Ryan Hill provided input into workshop planning. We all benefited from the workshop facilitation by Brian Stenquist of Meeting Challenges Inc; he was instrumental in setting the tone of collaboration and deliberation among co-authors that continued throughout the writing of this book.

We are grateful to the following organizations whose various expressions of support made this book possible: the Minnesota Sea Grant College Program supported by the National Oceanic and Atmospheric Administration (NOAA), National Sea Grant Office (NSGO) of the US Department of Commerce, the Department of Fisheries, Wildlife and Conservation Biology of the University of Minnesota and a Pew Fellowship in Marine Conservation to A.R.K.; the CSIRO Division of Marine and Atmospheric Research's Australasian Invasive Animal Cooperative Research Centre and the Murray Darling Basin Commission; Shanghai Fisheries University; and the Edmonds Institute. We are deeply indebted to the China Center of the University of Minnesota for supporting critically important travel by three co-editors. Shanghai Fisheries University was a generous host of the editors for a week-long meeting in June 2006.

On behalf of all the co-authors, we are especially indebted to the Series Editors and peer reviewers of this book. Eric Hallerman and Peter Schei led the peer review process, and 19 anonymous peer reviewers reviewed one or more chapters. Special thanks go to Eric Hallerman for also providing extensive reviews of chapters. Kelly Paulson, Blake Ratner and Igor Solar took additional time to provide helpful suggestions for chapters other than the ones they co-authored. Stu Hann, a systems safety engineer, inspired the Safety First ideas discussed in parts of this book.

The authors thank their family and friends for their support and understanding when meeting book writing and editing deadlines. Finally, we are forever indebted to Wayne Barstad, Liana Francisty, and WanQi Cai for their unwavering moral support and care during long hours of editing work.

1 Introduction to Environmental Risk Assessment for Transgenic Fish

K.R. HAYES, A.R. KAPUSCINSKI, G. DANA, S. LI AND R.H. DEVLIN

Introduction

Fish are an important source of animal protein, particularly for people in developing countries (Fig. 1.1). As total catches of wild fish level off, with concurrent declines of many individual stocks, aquaculture production of finfish, molluscs, crustaceans and seaweeds has increased to help meet the protein demands of the developed and developing world's growing population (Fig. 1.2). To date, this global increase in aquaculture production has been supported by expanding facilities, selective breeding of relatively few suitable species and intensification of production through increases in quantity or quality of inputs (Delgado *et al.*, 2003). The last 20 years have also witnessed growing interest in the potential of transgenic methods to increase aquaculture production.

To date, the two primary aims of transgenic fish research have been to enhance aquaculture production and develop model species for basic research (Chapter 3, this volume).[1] Table 1.1 introduces the types of traits that are current or potential targets for transgenic modification, together with a range of hypothesized environmental effects.[2] Many such potential effects have been raised in the scientific and popular literature surrounding transgenic fish development. Seminal papers on classes of transgenic organisms (Tiedje *et al.*, 1989; Snow *et al.*, 2005), and transgenic fish in particular (Kapuscinski and Hallerman, 1990, 1991), identify a number of potential consequences of entry of transgenic fish into ecosystems and highlight ways in which to mediate such effects. These hypothetical effects and mediation mechanisms have been subsequently

[1] The world's first commercial transgenic fish, however, is a transgenic variety of zebrafish: the Glofish, R. The Glofish is produced in Florida fish farms and sold in US pet stores (Cortemeglia and Beitinger, 2006, available at: www.glofish.com).

[2] This overview is revisited in greater detail in subsequent chapters of this book, particularly in Chapters 3, 5 and 6.

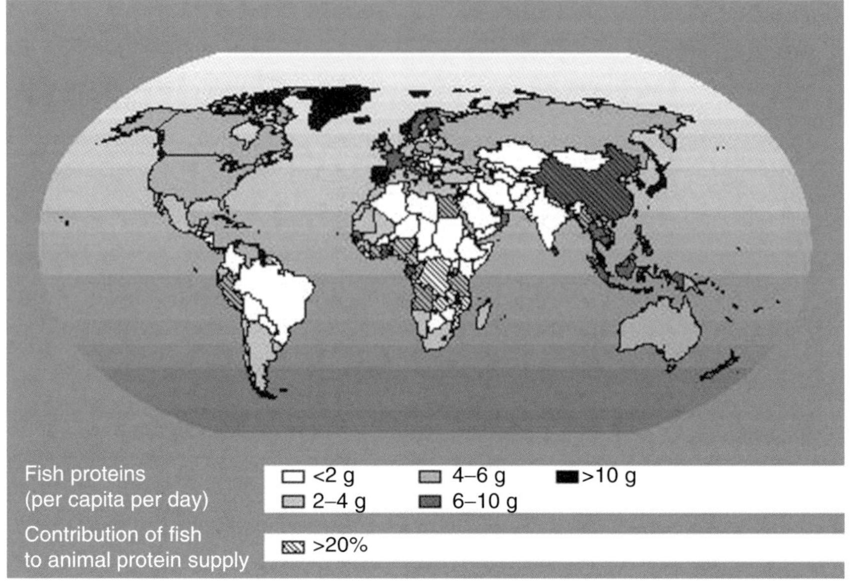

Fig. 1.1. Contribution of fish to animal protein supply in human food, average for 1999–2001. (Reprinted from FAO, 2004a. With permission.)

elaborated in risk assessment guidelines, the fate of fish transgenes modelled and containment strategies outlined, providing guidance to regulators grappling with decisions regarding transgenic fish (Devlin and Donaldson, 1992; ABRAC, 1995; Hallerman and Kapuscinski, 1995; Scientists' Working Group on Biosafety, 1998; Kapuscinski *et al.*, 1999; Muir and Howard, 1999, 2001, 2002; Maclean and Laight, 2000). Additional scientific papers have helped to

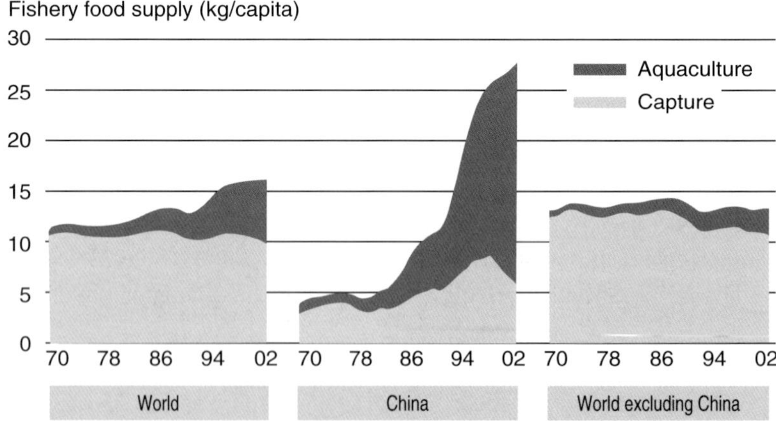

Fig. 1.2. Growth in global aquaculture production and its relative contribution to consumed food fish. (Reprinted from FAO, 2004a. With permission.)

Table 1.1. Examples of modified traits in current and future lines of transgenic fish and hypothetical environmental effects.

Class of trait	Target trait	Objectives for aquaculture applications	Hypothetical effects of transgenic fish entry into ecosystems
Somatic physiology	Growth	Shorten time to reach harvest size, thus shorten production cycle	Enhanced foraging activity and resource requirements that might heighten competition with wild fish
	Metabolism (e.g. nutritional capability)	Improve feed conversion efficiency, reduce dependence on essential nutrients and other feed ingredients (vitamins, pigmentation, high-quality lipids and protein sources)	Enhanced fitness and resilience to variable quality of food supplies that heighten competition with wild fish
	Disease resistance	Reduce mortality during aquaculture, enhance quality of the fish	Could alter natural mortality of populations, affecting population demographics
	Stress response	Reduce stress experienced by cultured fish	May make transgenic fish more vulnerable to predation, but could improve their recovery from acute or chronic stress situations
	Physiological tolerances	Increase range of environmental conditions for good growth and survival (e.g. cold water tolerance, salinity tolerance)	Could become invasive in a wider range of natural environments
	Novel protein expressed by transgene	Fish as biofactory to produce a pharmaceutical compound or other protein of commercial value	May make transgenic fish less vulnerable to predation, due to being less palatable or toxic to natural predators, but could reduce other fitness-related traits
Reproductive physiology	Gonad development	Sterilization to prevent gene flow to wild relatives	Single generation ecological effects from entry of sterile transgenic fish
	Sex determination	Produce monosex populations to enhance yields or to reduce gene flow to wild relatives	Single generation ecological effects when ecosystem lacks conspecifics; alter sex ratios in wild populations of conspecifics
	Maturation timing	Synchronize reproduction of brood stock	Alter ability to find and mate with sexually mature wild relatives
	Age at maturity	Shorten breeding cycles	Alter population growth rates, disrupt mating strategies of wild relatives
Morphology	Body shape	Improve dressed yield of harvested fish	Reduce swimming performance and fitness
	Body colour	Produce novel phenotypes for marketing	Alter hiding and camouflage ability; modify spawning success
	Cell and organ structure	Alter flesh quality	Variety of organismal effects with presently unknown ecological implications
Behaviour	Feeding behaviour	Reduce wastage of delivered feed	Alter prey utilization, antipredator behaviour and dominance hierarchies
	Reproductive behaviours	Block reproductive behaviour	Alter reproductive fitness and hence population fitness

inform risk assessment, but important gaps still exist in our understanding of the potential environmental effects of transgenic fish and shellfish (Kapuscinski, 2005). There remains a pressing need for systematic, science-based guidance on how to assess and manage potential environmental risks as transgenic lines of fish and shellfish approach commercialization.

The importance of risk assessment and management methods based on sound scientific evidence is reflected in a number of international agreements, most notably Article 15 and Annex III (both on risk assessment), Article 16 (on risk management) and Article 22 (on capacity building) of the Cartagena Protocol on Biosafety (CBD, 2000). The Protocol is designed to protect biological diversity from the potential risks posed by living modified organisms (LMOs)[3] and recognizes the need to increase scientific capacity in risk assessment and risk management methods. However, the Cartagena Protocol, along with a number of similar international and national agreements, provides only limited guidance on how to conduct scientifically sound risk assessments and, when needed, how to design and apply appropriate risk management (Hayes, 2002a; Kapuscinski, 2002).

This book is the third in a series specifically designed to support the aims of the Cartagena Protocol. The series aims to help developing countries strengthen their own scientific and technical capacity to address the environmental biosafety of genetically modified organisms (GMOs). Each book serves as a scientifically peer-reviewed tool to help different groups around the world conduct their own assessments and draw their own conclusions about the environmental biosafety of a particular GMO. The first two books address cases of genetically modified crops that are used commercially in at least one country and are under serious consideration in others (Hilbeck and Andow, 2004; Hilbeck *et al.*, 2006).

Objectives, Structure and Purpose of the Book

The purpose of this book is to strengthen the understanding of the science needed to inform environmental biosafety policy and regulation for transgenic animals in aquaculture, particularly in developing countries. It contains detailed guidance on how to develop and conduct a scientifically defensible environmental risk assessment and, when appropriate, implement effective risk management strategies, for transgenic fish. This book recognizes that information gaps may prevent analysts from conducting a completely quantitative risk assessment. It therefore presents qualitative and quantitative methodologies for assessing the ecological risks associated with transgenic fish on a case-by-case basis, with a high level of scientific quality.

Environmental risk assessments are conducted in many different ways around the world, and the methodologies presented in this book can be applied across this diversity of approaches. The book is structured around the following main topics:

[3] The Protocol on Biosafety uses the term 'living modified organism' (LMO), but this book uses the more common terms of 'genetically modified organism' (GMO) and 'transgenic organism'.

- Analytical-deliberative risk assessment framework that recognizes the need to conduct high-quality scientific analysis, while involving relevant stakeholders at appropriate points, and provides tools to enable this participation in a transparent and systematic fashion (Chapter 1, this volume);
- Problem Formulation and Options Assessment (PFOA), a process for science-guided multi-stakeholder involvement in environmental risk assessment (Chapter 2, this volume) in which stakeholders identify what societal need may be addressed by the transgenic fish, consider other technology options, deliberate on benefits and risks of the identified options and develop recommendations for decision makers;
- Overview of transgenic fish being developed for aquaculture, focusing on lines close to commercialization, and a review of relevant transgenic fish and shellfish research (Chapter 3, this volume);
- Presentation of scientific methodologies from appropriate disciplines (i.e. molecular biology, population genetics and ecology) for assessing the environmental effects of transgenic fish on wild relatives, other aquatic species and ecological processes in natural bodies of water (Chapters 4–6, this volume);
- Description of the types of uncertainty in environmental risk assessment and a range of qualitative and quantitative tools used to incorporate this uncertainty into risk assessment calculations (Chapter 7, this volume);
- Current and future methods for physical and biological confinement of transgenic fish (Chapter 8, this volume);
- How to design and implement an effective monitoring system for selected end points in an environmental risk assessment (Chapter 9, this volume);
- Summary of the book's key messages and a discussion of capacity-building needs for environmental risk assessment and management, particularly in developing countries (Chapter 10, this volume).

This book does not address food and human health safety of transgenic fish because these issues are outside the scope of the series. In-depth discussion of appropriate risk assessment methodologies for these issues would require a different interdisciplinary set of co-authors. Questions about the food safety of transgenic fish and their products are of great interest to consumers and private and public sector entities. Food safety assessment methods for transgenic fish were recently summarized by a group of international experts, who also noted that the unintended entry of certain transgenic fish into natural water bodies could raise food or human health safety questions (FAO/WHO, 2004). Furthermore, the book does not give guidance on drafting or evaluating biosafety policy; nor does it attempt to reach conclusions about the potential benefits or risks of any specific transgenic fish. Rather, it presents a synthesis of new and existing scientific methodologies that different parties can apply when conducting their own environmental risk assessments of transgenic fish.

The risk assessment framework and many of the methodologies presented here can be applied not only to transgenic fish but also to other ways of increasing aquaculture production. Selectively bred strains of fish, such as genetically improved farmed tilapia (GIFT), for example, are being developed through a wide variety of partnerships to benefit the poor, and they are being adopted and considered for aquaculture in many countries (Asian Development Bank, 2005). It is important to recognize that selectively bred strains raise many of the same environmental concerns as transgenic fish, such as the potential adverse effects of gene flow on wild relatives and fish communities in aquatic ecosystems.

Definitions and Types of Risk Assessment Methods

Risk can be defined as the likelihood of harm occurring as a result of some behaviour or action (including no action) (NRC, 1996). Harm refers to undesirable consequences to humans and the things that they value. Hazard can be defined as an act or phenomenon that, under certain circumstances, could lead to harm (The Royal Society, 1983) or, alternatively, as a substance's or activity's propensity to produce harm (Hayes, 1998). Hazard is often perceived solely as a function of a substance's intrinsic properties, but, as emphasized in the definition above, it is more usefully conceptualized as a function of both the intrinsic properties of a substance and circumstance.

Risk assessment is a process for determining the frequency and consequences of harmful events. The consequences of the harmful events in question are by definition adverse[4] and are expressed in terms of assessment end points, which are an explicit expression of the values that interested parties are trying to protect by undertaking the risk assessment process. Environmental risk assessment is distinguished from human health risk assessment by its ecological (e.g. extinction of endangered species) rather than human health (e.g. fatalities and injuries) end points. Risk analysts often distinguish between assessment end points (what they are trying to protect) and measurement end points (what they can actually measure), extrapolating from one to the other for the purposes of the risk assessment.

All risk assessments can be broadly divided into predictive or retrospective (Suter, 1993). Retrospective risk assessments seek to identify the causes and characteristics of harmful events that have already occurred. They seek patterns of response and are described in the literature with terms such as empirical, deductive or correlative assessments. Predictive risk assessments attempt to predict the likelihood and consequences of harmful events that have yet to occur. They seek to synthesize understanding and are described in the literature as theoretical, inductive or mechanistic assessments. Environmental risk assessments for transgenic fish fall into the predictive category because (as per the date of publication of this book) transgenic fish have yet to enter natural water bodies.

Risk assessments can be further classified by the statistical nature of their output as: qualitative, semi-quantitative or quantitative. Qualitative risk assessments produce nominal (e.g. species lists) or ordinal (e.g. high, medium or low) outputs based on expert and stakeholder opinion. Semi-quantitative risk assessments produce interval variables having a small number of discrete values (e.g. an invasion risk score such as 1–10, 10–50, >50), whereas quantitative risk assessments

[4] Hence, one considers the chance, not the risk, of winning the lottery.

produce continuous risk estimates that may or may not be grouped into categories. This classification, however, is not necessarily distinct – many assessments advocate a mixture of qualitative and semi-quantitative or quantitative approaches (Koehn, 2004). The strengths and weaknesses of each type of assessment, together with relevant examples, are discussed in more detail below.

Qualitative risk assessment

Qualitative risk assessments are flexible and can incorporate diverse sources and types of information. Current standards for qualitative risk assessment are relatively easy to implement and can incorporate the opinions of experts and stakeholders with diverse backgrounds and training (see, e.g. AS/NZS, 2004). Qualitative risk assessments express the likelihood and consequences of end points as high, medium or low, combining these estimates in a risk matrix that provides the overall risk estimate. Scientific literature contains a few examples of qualitative risk assessments applied to aquaculture escapees in general (Peeler and Thrush, 2004; Naylor *et al.*, 2005) or transgenic fish in particular (ABRAC, 1995; Scientists' Working Group on Biosafety, 1998; Maclean and Laight, 2000). All these examples make qualitative risk assertions, but they differ both in the categorization of likelihood and consequence (high, medium, low, etc.) and the rules for combining these measures to estimate risk. This variety of approaches is inevitable given the scope for interpretation of the various national and international guidelines and standards (OECD, 1986; ERMA, 1999; CBD, 2000).

The scientific credibility of a qualitative risk assessment is largely determined by the expertise of the group performing the assessment and the manner in which the group's opinions and predictions are elicited, combined and prioritized. All individuals exhibit a range of psychological behaviours that have a profound influence on qualitative risk estimates. Each person's judgement is affected by personal experience, level of understanding and control over the outcome of the proposal (e.g. to release transgenic fish), fear of the outcome and who ultimately bears the burden of risk. Furthermore, when individuals assess risks subjectively, they are often influenced by cognitive bias (overconfidence in one's ability to predict), framing effects (judgements of risk are sensitive to the prospect of personal gain or loss), anchoring (the tendency to be influenced by initial estimates) and insensitivity to sample size (Burgman, 2001, 2005).

The 'psychological frailties' described above can lead to unfounded certainty; both 'experts' and laypersons tend to be more confident about their predictions than they should be. Therefore, qualitative assessments may not err on the side of conservatism even when they purport to do so (see Ferson and Long, 1995, and examples therein). Furthermore, the same qualitative assessment, conducted by different groups, can reach opposite conclusions when presented with the same data, for no apparent reason (Hayes, 2003). These effects undermine the repeatability, transparency and scientific credibility of qualitative risk assessments.

There are a number of hazard analysis and risk assessment techniques that help qualitative risk assessments maintain scientific credibility by formalizing, for example, the hazard prioritization process (Box 1.1). These techniques

Box 1.1. Ways to help maintain the scientific credibility of qualitative environmental risk assessment.

Be representative and encourage shared understanding of scope
Identify all relevant experts and stakeholders and seek to include them in the assessment via the Problem Formulation and Options Assessment (PFOA) process (Chapter 2, this volume). Use the PFOA framework to ensure that the spatial and temporal scope of the assessment is understood by all. Clearly define all predictive terms (e.g. high, medium or low likelihood and consequence) relevant to the scope of the assessment.

Minimize complexity by appropriate end point selection
Establish a careful balance between reality, complexity and stakeholder concerns by choosing assessment end points that are clearly relevant to these concerns, but occur earlier (rather than later) in event chains that link exposure (e.g. release of transgenic fish) to effect (e.g. decline of a native fish population).

Avoid predictive bias and maintain transparency
Use structured elicitation and aggregation techniques to help avoid 'psychological frailties' such as insensitivity to sample size, overconfidence, judgemental bias and anchoring (Burgman, 2001, 2005). Use formal prioritization procedures, such as the analytical hierarchy process (Saaty, 2001), when prioritizing hazards or combining the predictions of different stakeholders. Keep a careful record of the process, methods and predictions of the assessment.

Identify all possible hazards
Use structured hazard identification techniques such as influence diagrams (Hart *et al.*, 2005), fault tree analysis (Haimes, 1998; Hayes, 2002b), failure modes and effects analysis (Palady, 1995; Hayes, 2002c), HAZOP (Royal Commission on Environmental Pollution, 1991; Kletz, 1999) or hierarchical holographic modelling (Haimes, 1998; Hayes *et al.*, 2004) to rigorously and systematically identify all possible hazards.

Monitor and test predictions
Predict measurable effects and monitor these with sufficient sensitivity (Chapter 9, this volume) to both test the risk assessment predictions and generate additional data needed to empirically characterize variability and help reduce incertitude. These monitoring activities are appropriate for approved actions (e.g. approving a transgenic fish for aquaculture) because they support iterative improvement of retrospective risk assessments of the approved action and prospective risk assessment of other similar proposals.

Qualitative modelling
Use qualitative modelling to ensure that conceptual models of environmental systems are stable, to test for internal consistency and optimal complexity and to identify critical interactions within the system (Dambacher *et al.*, 2003a,b).

Peer review
Seek an independent peer review of the risk assessment and its results.

encourage consistent, systematic evaluation, help expose assumptions and value judgements, and thereby maintain transparency in the assessment. However, these techniques are rarely used because of the additional time commitment that they require. Finally, qualitative risk assessments cannot effectively express uncertainty (Chapter 7, this volume) without 'numerical translation' through fuzzy set theory (Hobbs *et al.*, 2002; Ramsey and Veltman, 2005) or Bayesian approaches (Hayes, 1998; Ferson, 2005), i.e. without moving to semi-quantitative or quantitative risk assessment methods.

Semi-quantitative risk assessment

Semi-quantitative risk assessments convert qualitative predictions about likelihood and consequences of events into interval variables on a discrete range (e.g. potential ecological impact of a transgenic fish scored on a scale of 1–10). These assessments typically identify and score a set of risk factors, using the final aggregated score across all factors to estimate risk (see, e.g. Virtue *et al.*, 2001). Semi-quantitative risk assessments can use data from a variety of sources, but they are less flexible than qualitative approaches because they are constrained by the particular risk estimation method. Furthermore, there are no national or international standards for semi-quantitative risk estimation methods. Not surprisingly, examples relevant to transgenic fish (Kohler, 1992; Bomford and Glover, 2004; CABI, 2005; Copp *et al.*, 2005) exhibit a wide variety of risk factors, methods for scoring these factors and algorithms for aggregating these scores into a final risk estimate.

The scientific credibility of semi-quantitative score methods can only be ensured by retrospectively 'training' the risk assessment method (Box 1.2). Once trained, the risk assessment can be used to predict future high- or low-risk scenarios, so long as the environmental conditions or scope of future scenarios does not change significantly from those that generated the training data set. No training data sets currently exist for transgenic fish because, to date, none has been released into the environment. The credibility of semi-quantitative assessments can also be improved by ensuring that the risk calculation is completed independently by two or more experts.

Quantitative risk assessment

Quantitative risk assessment expresses the likelihood and consequences of undesired events on continuous scales relevant to the end point and scope of the assessment (e.g. Fig. 1.3). Retrospective assessments use a variety of statistical techniques to identify significant predictors of past events. The scientific literature is replete with numerous invasive species examples (Hayes and Barry, in press) but, to date, none of these techniques has been applied to environmental risk assessments of transgenic fish because none exists in the wild. Predictive assessments rely on these studies, or various other mathematical constructs (Crawford-Brown, 2001; Pastorok *et al.*, 2002), to predict likelihood

Box 1.2. Ways to help maintain the scientific credibility of semi-quantitative environmental risk assessment (in addition to Box 1.1).

Test the risk assessment against independent data sets
Where data permit, test the predictions of the risk assessment against as many known high- and low-risk situations as possible (e.g. instances of harmful and benign non-native fish introductions into the same environment being considered for transgenic fish). A common approach when designing retrospective risk assessments is to divide a data set into two halves. The first half is used to design the risk assessment, including amending the total risk factor scores to maximize the correct number of predictions and minimize the number of incorrect predictions. The second half is used to test the accuracy of the risk assessment (e.g. Pheloung *et al.*, 1999; Virtue *et al.*, 2001).

Obtain multiple responses
Capture diversity of expert opinion by ensuring that at least two, ideally more, experts complete the assessment simultaneously. Their parametric scores can be aggregated in a variety of ways (e.g. a simple average or by interval arithmetic) to maintain the minimum and maximum scores throughout the risk assessment process. This captures uncertainty and helps determine the range of plausible risk estimate.

Specify interval rather than point risk scores
Avoid a deterministic risk estimate by specifying interval, rather than point, estimates for the risk factor scores (e.g. Hayes *et al.*, 2005). Use interval arithmetic to aggregate the total scores in the manner dictated by the risk model. Examples of more complex approaches to uncertainty analysis for risk factor procedures are available in the literature (Hughes and Madden, 2003; Caley *et al.*, 2006). Again, this captures uncertainty and helps determine the range of plausible risk estimates.

Investigate model structure effects
Use sensitivity analysis to consider the effect of alternative model structures. Alternative model structures include alternative risk factors, alternative ways of calculating the final risk score or alternative ways of grouping the final risk scores into high, medium or low risk. Test whether risk estimates, or ideally decisions, would be substantially altered with alternative plausible model structures.

and consequence. Again, numerous examples relevant to the establishment (Bartell and Nair, 2003; Drake, 2004, 2005), spread (Drake and Bossenbroek, 2004; Neubert and Parker, 2004; Hastings *et al.*, 2005) and ecological impact (Paisley *et al.*, 1999; Leung *et al.*, 2002) of invasive fish are available, but there appear to be no equivalent examples for transgenic fish.

The scientific credibility of quantitative risk assessment is largely determined by:

- How practical the risk assessment models are (e.g. how well developed they are and how easy it is to estimate their parameters);
- How reliable the risk assessment models are (e.g. how realistic and relevant they are);
- How well the models are corroborated or verified (Rykiel, 1996; Bartell *et al.*, 2003).

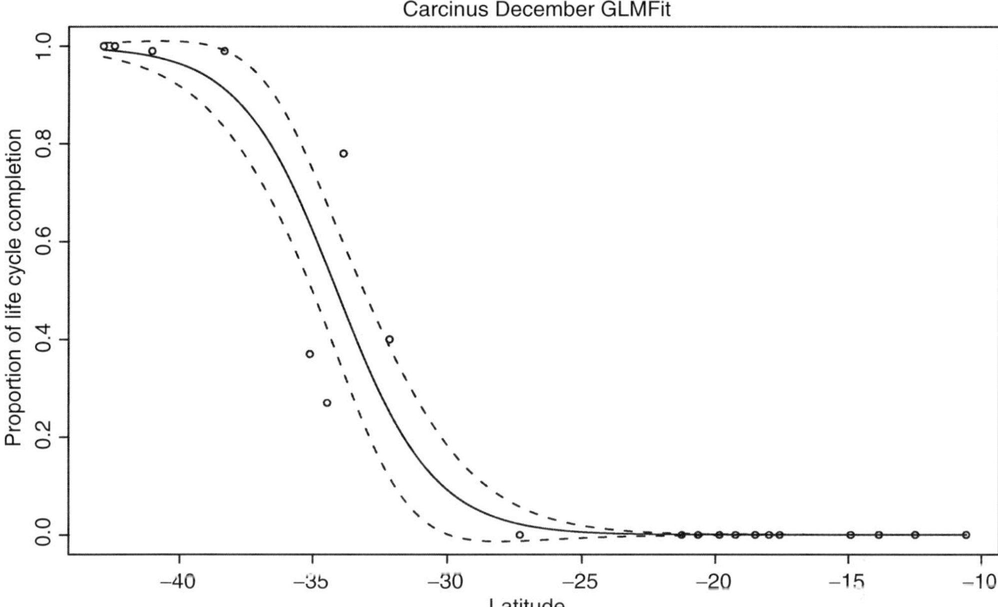

Fig. 1.3. Example of an output from a quantitative risk assessment. This example expresses the likelihood of an undesired event: survival of an invasive crab. The temperature tolerances of the European green crab (*Carcinus maenas* – a temperate water-invasive species) were compared to the time series of temperature in 21 locations in Australia to calculate the proportion of the crab's life cycle (the assessment end point) that could be completed at different latitudes in Australia in December (the open circles). A general linear model (GLM) was subsequently used to interpolate between these results (the solid line).The dotted lines are 95% confidence intervals that reflect the effects of variability in the temperature tolerance of *C. maenas* and the variability in sea temperatures at these latitudes in Australia. (From Hayes *et al.*, in preparation).

These criteria are linked to the complexity of the biological processes involved and the mathematical and statistical models used to represent them. These factors are ultimately dictated by the risk assessment end point(s). End points occurring at the end of long event chains (e.g. impacts on non-target species) necessitate more complex models than end points expressed at earlier points in an event chain (e.g. escape and survival of transgenic fish in the wild). Complex models generally require more data in order to estimate parameters and verify predictions, and they present greater challenges for uncertainty analysis. Analysts and stakeholders should therefore choose assessment end points and risk models carefully, balancing complexity, realism and the expectations of decision managers (Box 1.3). It is important to emphasize that quantitative risk assessment also is not immune from the 'psychological frailties' associated with qualitative assessments. For quantitative assessments, it is important that analysts do not bias the end point selection process towards, for example, overly simplistic end points that can be conveniently modelled. Analysts and stakeholders should aim for end points that are measurable and clearly relevant to the environmental values that they have agreed to protect.

Box 1.3. Ways to help maintain the scientific credibility of quantitative environmental risk assessment (in addition to Box 1.1).

Choose mathematical models carefully
Choose models that are well corroborated (i.e. well reviewed, widely used and well accepted by management agencies). Balance model realism and relevance against complexity and ease of use. Whenever possible, calibrate the model using site- or species-level data that are specific to the risk assessment problem at hand. Validate the model by comparing its predictions with independent laboratory and field observations (Chapters 6 and 9, this volume).

Identify and treat model uncertainty and variability
Investigate the effects of different types of uncertainty on the results of the risk assessment, particularly variability and model uncertainty (Chapter 7, this volume), and report the effects of these different types of uncertainty on the final risk estimate. Test whether risk estimates, risk assessment conclusions and, ideally, any risk management strategies would be substantially altered with alternative plausible model structures.

Framework and Process for Environmental Risk Assessment: Proposal for an Interactive Framework

Frameworks for environmental risk assessment differ across disciplines, not only in terms of their structure and methods, but also in the language used to describe the risk assessment process (Hayes, 1997). Most frameworks, however, have certain steps in common, such as hazard identification, exposure and effects assessment and risk communication. Frameworks for environmental risk assessment also differ between nations, particularly in terms of the party responsible for completing the assessment and the degree to which multiple stakeholders participate in different steps of the framework (see country-specific case studies relevant to transgenic fish in Chapter 2).

This chapter proposes a risk assessment and management framework (Fig. 1.4) that is highly interactive and participatory. It is designed to improve both the scientific validity of risk conclusions and the likelihood that stakeholders will trust the risk assessment and its conclusions. Our proposal suggests increased interaction between analysis and deliberation at various steps in the framework. It also places great emphasis on formal uncertainty analysis and on monitoring of the realized safety or risk of a project in order to inform risk management and future risk assessment. We recognize that widespread adoption of this proposed framework will require capacity building (see Chapter 10, this volume) and a commitment to this approach by relevant government authorities. It is important to note, however, that our proposed framework contains elements already used by different countries, and that methodologies presented in Chapters 2–9 can be applied within any existing risk assessment framework. We describe the steps of this framework below in order to stimulate debate, trial application, improvement and adaptation of our proposed approach to different regulatory and societal contexts.

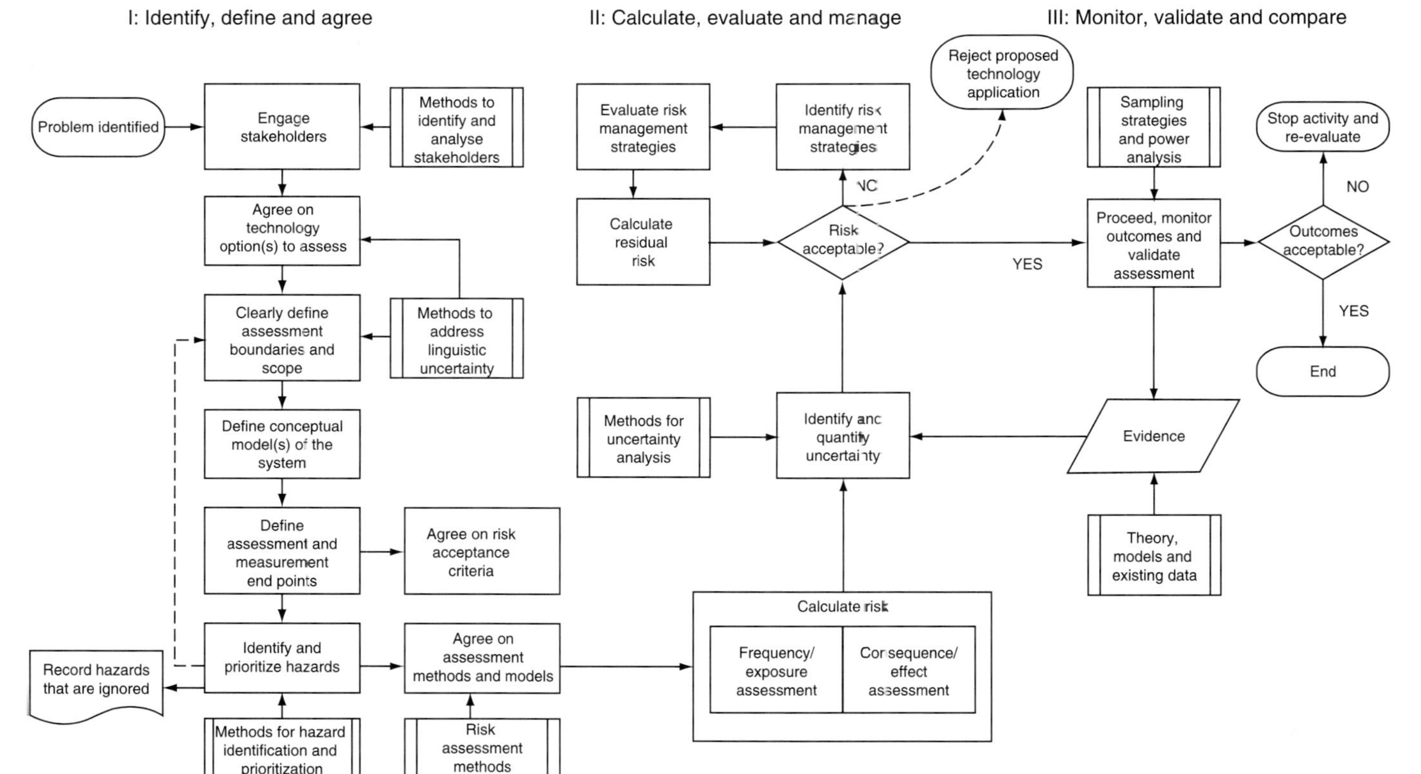

Fig. 1.4. Overview of the highly interactive environmental risk assessment framework highlighted in this chapter. The framework is divided into three stages. In the first stage, participants agree on an assessment option using the Problem Formulation and Options Assessment (PFOA) methodology, and then go on to define the scope of the assessment, agree on a conceptual model(s), identify assessment and measurement end points and culminate with an agreed-upon list of prioritized hazards. In the second stage, the risks and uncertainties associated with these hazards are assessed, risks are compared to predetermined acceptance criteria and, where necessary, risk management strategies are identified and evaluated. In the third stage, the predictions of the risk assessment are compared with reality, thereby generating additional data used to re-examine uncertainty in the risk assessment. Dashed lines indicate alternative actions the analyst could take.

Our proposed risk assessment framework can be broadly divided into three stages. The first stage (identify, define and agree, Fig. 1.5) aims to engage stakeholders, define the reasons stimulating the consideration of transgenic fish (Chapter 2, this volume), agree on the technology option to assess (e.g. one line of transgenic tilapia for production in two river basins), develop and analyse conceptual models of the system in question, identify and prioritize hazards and ultimately agree on which hazards to carry forward to the second stage of risk calculation. The methods and techniques employed in the first stage are qualitative, although they can draw upon relevant quantitative information (as discussed in the chapters identified in Fig. 1.5). If this first stage is performed carefully, it provides an effective, qualitative, 'first-tier' risk assessment within which stakeholders' values, conceptual models and hazard beliefs can be elicited and scientifically evaluated.

The second stage (calculate, evaluate and manage, Fig. 1.6) begins once a priority list of hazards and risk assessment methods have been agreed upon by stakeholders and scientists. This stage calculates the likelihood that the hazardous events identified in the first stage will actually occur, and their subsequent consequences, all expressed in terms of measurable impacts upon valued features of the environment (assessment end points) identified during the first stage. A thorough evaluation of the uncertainty associated with these calculations is critical to ensure an effective and 'honest' risk assessment (Burgman, 2005; see also Chapter 7, this volume). The second stage ends by comparing risk estimates with predetermined acceptance criteria (developed in the first stage) and, where necessary, identifying and evaluating risk management strategies.[5] Decision makers should determine whether to reject the proposed technology application if the risk assessment concludes that unacceptably high risks cannot be managed to acceptably low levels.

The third stage (monitor, validate and compare, Fig. 1.7) seeks to compare the predictions of the risk assessment with reality by monitoring the valued environmental features identified in the first stage. Importantly, this stage closes the regulatory loop, and provides additional empirical data for subsequent refinement of the risk assessment. Monitoring can provide new evidence for analysing and (wherever possible) reducing some of the uncertainty identified in the second stage. This allows an ongoing process of learning for risk assessors, stakeholders and policy decision makers, wherein results from the third stage of one risk assessment can improve the first and second stage of a retrospective assessment of the same case or predictive assessments of related cases. Real-world outcomes that are unacceptable and therefore not in agreement with the risk assessment's predictions indicate that the hazard analysis, risk calculation or risk management strategies are wrong and need to be revised in light of the new information. In either case, the risk assessment should be re-evaluated and decision makers should consider whether to stop the activity.

This highly interactive framework envisages stakeholder participation at key stages in the risk assessment, particularly the first stage. The framework

[5] This framework only considers methods for calculating risk and the effects of risk management strategies. The actual adoption of a risk management strategy and possible changes in response to monitoring are not explicitly addressed here.

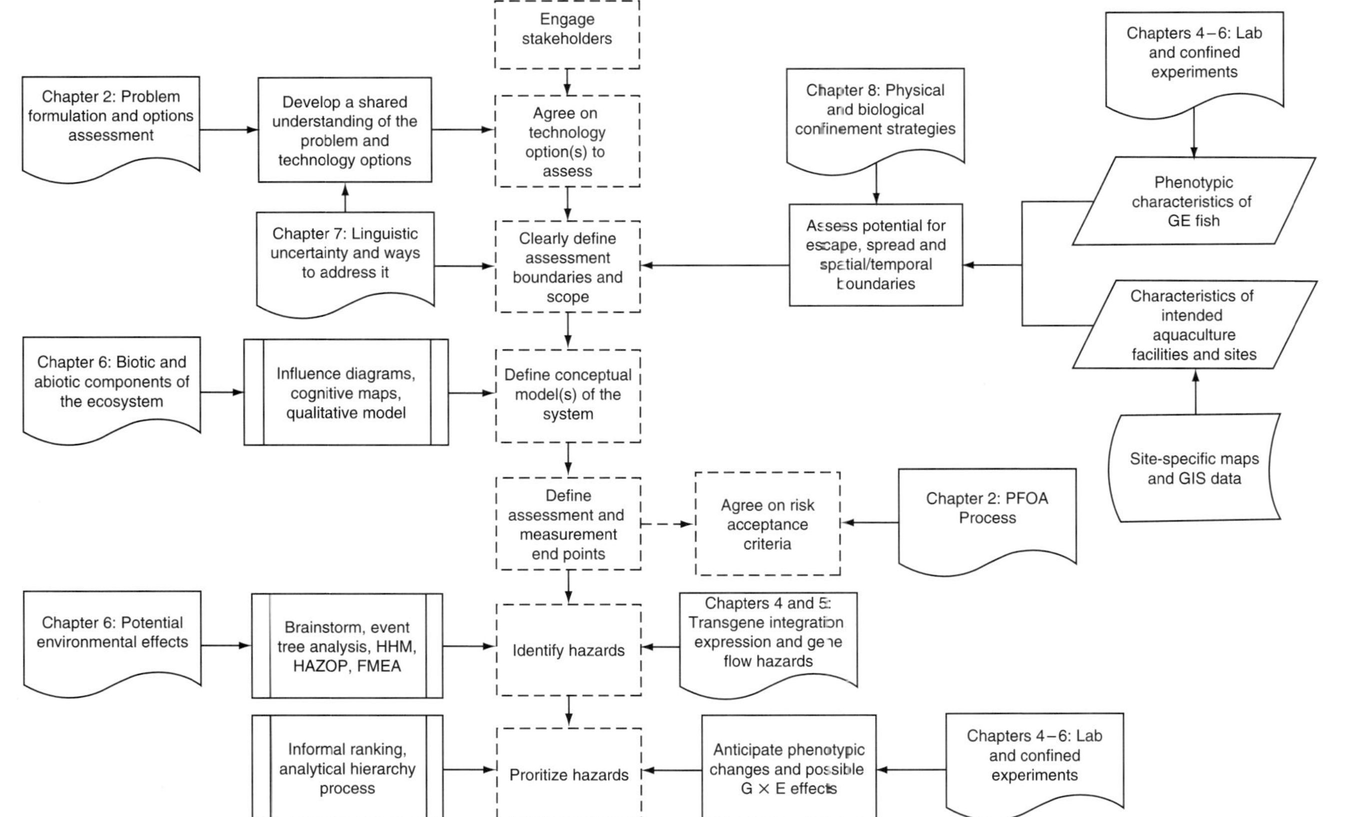

Fig. 1.5. A more detailed examination of the first stage (identify, define and agree) of the risk assessment framework outlined in Fig. 1.4, illustrating potential methods at each step of the assessment framework, together with the contribution of individual book chapters. Dashed lines around boxes denote steps in the risk assessment framework (Fig. 1.4).

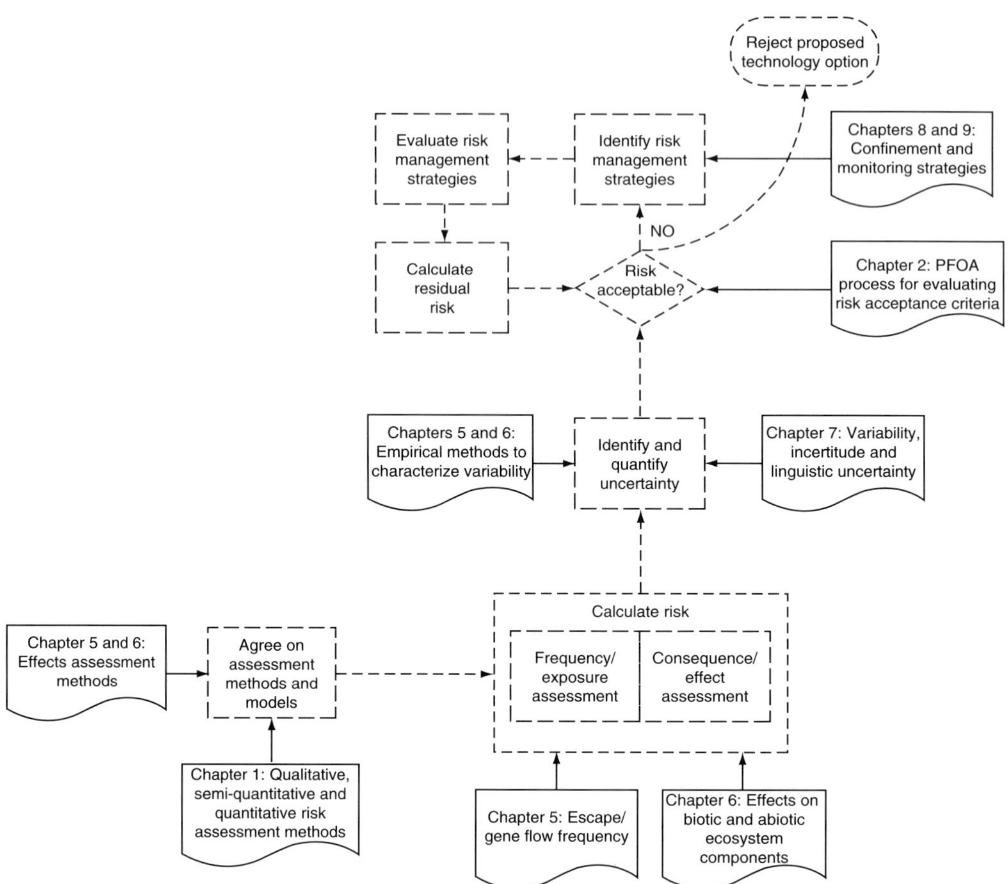

Fig. 1.6. A more detailed examination of the second stage (calculate, evaluate and manage) of the risk assessment framework outlined in Fig. 1.4, illustrating potential methods at each step of the assessment framework, together with the contribution of individual book chapters. Dashed lines around boxes denote steps in the risk assessment framework (Fig. 1.4).

emphasizes stakeholder participation as well as communication because risk assessment should be designed to protect inter alia stakeholder values and because stakeholders hold information that can be used to augment that provided by expert groups (scientists, analysts or consultants) who usually conduct a risk assessment (NRC, 1996). Furthermore, including stakeholders in the risk assessment process helps establish and maintain the legitimacy of the relevant agencies and their decisions on GMO regulation (NRC, 1996, 2002). We recognize, however, that the level of stakeholder participation in risk assessment will vary from case to case and from nation to nation. Individual nations will ultimately choose a level of stakeholder participation that is appropriate to their particular biotechnology governance structure. PFOA, described in Chapter 2, provides an appropriate venue to determine the level of stakeholder interaction in the risk assessment process.

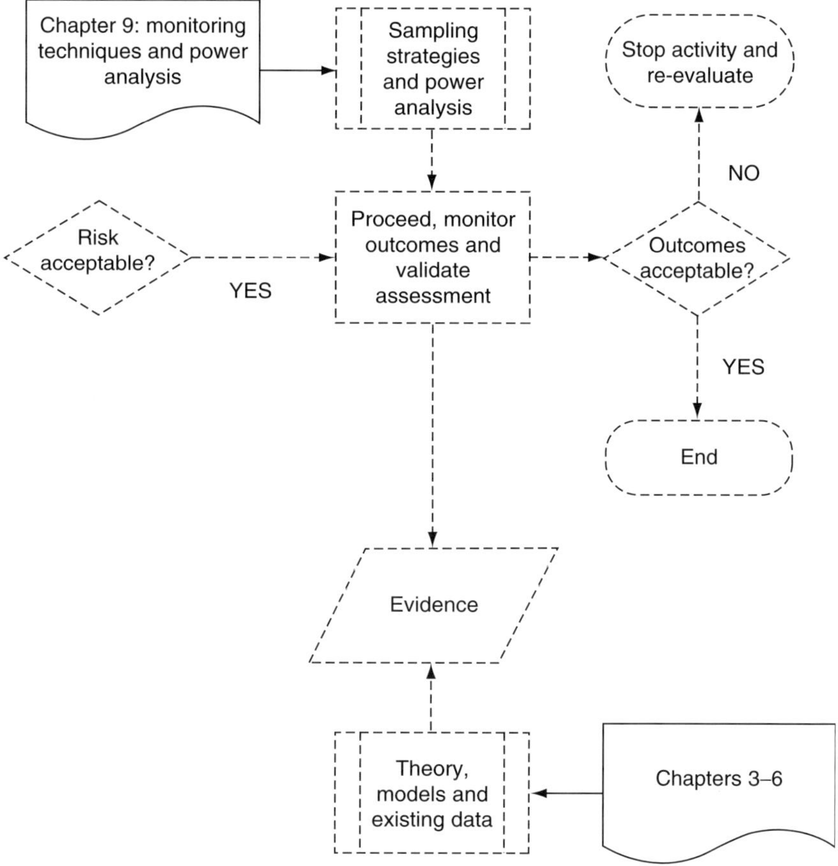

Fig. 1.7. A more detailed examination of the third stage (monitor, validate and compare) of the risk assessment framework outlined in Fig. 1.4, illustrating potential methods at each step of the assessment framework, together with the contribution of individual book chapters. Dashed lines around boxes denote steps in the risk assessment framework (Fig. 1.4).

The importance of stakeholder participation in environmental decision making is recognized in at least three international instruments governing GMOs, including the Rio Declaration on Environment and Development, the Aarhus Convention and the Cartagena Protocol on Biosafety (FAO, 2004b). This emphasis on stakeholder participation, however, is not reflected in many of the national and international risk assessment frameworks designed to address GMOs. Many existing frameworks highlight the importance of communication, primarily from decision makers to stakeholders, but none stresses an active participatory role (see, e.g. OECD, 1986; UNEP, 1995; DETR, 1999; ERMA, 1999; EU, 2001; OGTR, 2001).

In our proposed highly interactive framework, stakeholders would interact with various experts to help identify assessment options, define the boundaries of the assessment, develop conceptual models of the system in question and identify and prioritize hazards. Stakeholders may also provide evidence that contributes to uncertainty analysis, help identify risk management strategies and may also be active participants in monitoring strategies, comparing the predictions of the risk assessment with reality. Therefore, they would play an important role in many stages of the risk assessment process. The following sections discuss each of the risk assessment steps in more detail, in the hopes that this will stimulate efforts to incorporate interactive elements into existing frameworks for environmental risk assessment and management.

Identify, define and agree

Stakeholders must be identified, either by an authorizing body or a convening authority (see Chapter 2, this volume), before engaging them in the risk assessment process. Stakeholders can be identified using a variety of formal or informal methods such as stakeholder mapping (Glicken, 2000) and snowball sampling (in which stakeholders suggest other interested parties). Once identified, stakeholders initially engage in the PFOA stage where they, together with the group who will conduct the assessment, develop a shared understanding of the problem, help identify ways to address this problem and ultimately recommend a preferred option to be carried through the risk assessment process (Fig. 1.5). PFOA increases the chances of reaching agreement on these issues because it deliberately examines the underlying values and interests of different stakeholder groups. PFOA ultimately aims to introduce scientific information as a means to reach a common understanding between analysts and stakeholders and then to present decision makers with an agreed-upon set of recommendations (Chapter 2, this volume). This approach, however, requires a much greater level of information exchange, fostered by a much more dynamic relationship between stakeholders and analysts, than is traditionally practised in environmental risk assessment.

The case-specific characteristics of the preferred option (e.g. the use of transgenic fish or selectively bred fish to increase aquaculture production in region *X* for next *Y* years) set the stage for defining the geographical boundaries and temporal scope of the risk assessment. In the context of this book, the geographical boundaries of the assessment are determined by the environments into which transgenic fish may escape or be released. Chapters 4–6 provide information to help determine the phenotypic properties of transgenic fish affecting their potential to spread into aquatic environments, while Chapter 8 discusses physical and biological confinement options. Each of the chapters, together with site-specific maps and geographic information system (GIS) data, will help analysts and stakeholders determine the most appropriate geographical boundaries of the risk assessment.

Once the geographical boundaries of the assessment are determined, analysts and stakeholders should develop a shared understanding of the socio-environmental system into which the transgenic fish may be farmed or may escape, ultimately leading to a conceptual model of this system. The conceptual model should identify the biotic (e.g. native and non-native fish communities), abiotic (e.g. physio-chemical properties of water bodies) and socio-economic (e.g. current aquaculture operations) parts of the system, their interdependencies and potential responses following the introduction of a transgenic fish. Chapter 6 provides detailed guidance on identifying the relevant abiotic and biotic components of the system. Conceptual models may initially be developed as influence diagrams (Cox *et al.*, 2003) or cognitive maps (Ozesmi and Ozesmi, 2003). Qualitative analysis of conceptual models of the system (Dambacher *et al.*, 2003a,b) will help determine whether stakeholder and expert beliefs are consistent with existing sources of scientific information.

The conceptual model needs to strike a balance between simplicity and complexity of its components, making it manageable without stifling the rigour of the risk assessment, particularly during the hazard identification stage. This involves aggregating the components and processes of the environment and the characteristics of transgenic fish into sensible categories. Risk analysts may identify a long list of genetic, physiological and behavioural traits of transgenic fish that could potentially interact with another equally long list of biological, physical and chemical attributes of the receiving environment. Qualitative modelling can help identify optimal levels of aggregation, abstraction and simplification of ecosystem components and processes. For example, one task might be to identify whether or not multiple piscivorous species targeted by capture fisheries can be treated as one functional group without altering the model's representation of the overall dynamics of the system and its response to a stress such as the introduction of transgenic fish. Chapter 6 provides guidance on the key characteristics of transgenic fish and thereby helps analysts group these characteristics into sensible categories. Identifying the optimal level of simplification of both the stressor (i.e. transgenic fish) and the receiving environment is an important precursor to the hazard identification stage.

It is not necessary, and often inadvisable, to agree on a single conceptual model. If divergent opinions within stakeholder and expert groups result in two or more different conceptual models, and all are consistent with existing scientific information, then analysts and regulatory agencies should consider whether or not all models need to be carried through the risk assessment and analysed as model uncertainty (Chapter 7, this volume). This approach requires analysts to consult with stakeholders and also to employ assessment methods, such as qualitative modelling, that explicitly capture and propagate stakeholders' beliefs through the risk assessment process.

The conceptual model developed at this stage provides the context within which analysts and stakeholders can define the risk assessment end points. Assessment end points are often expressed in terms of impacts on species; for example, maintaining the population of commercially valuable or endangered species at a specified level within a given time frame. However, end points can be expressed at various levels of biological organization – from individual organisms to the entire ecosystem (e.g. Fig. 9.1) – and can include impacts on fundamental ecosystem processes and species that are not directly valuable to humans (Asian Development Bank, 1990; Suter, 1993). The first stage of the risk assessment framework also calls for stakeholders and analysts to define risk acceptance criteria as soon as the assessment end points are determined (Fig. 1.4). These criteria should be specified in terms of unacceptable changes in assessment or measurement end points. This can be a difficult and contentious step because assessment end points are value expressions, and a diverse group of stakeholders will neither hold the same values nor tolerate the same level of impact on them. The PFOA process (Chapter 2, this volume) is very important at this step because it provides an opportunity to examine stakeholder values and reach agreement on end points and acceptance criteria. Risk acceptance criteria should ideally be defined in quantitative terms in order to minimize linguistic uncertainty associated with terms such as 'low risk' (Chapter 7, this volume).

Identifying assessment end points that are both relevant to stakeholder concerns and that allow risks to be calculated with reasonable certainty is a significant challenge. End points relevant to stakeholder concerns usually occur at the end of complex chains of events. Consider, for example, the risks associated with introgression of transgenes into wild populations (Chapter 5, this volume). Stakeholders may, quite reasonably, suggest that the analyst determines whether a commercially valuable conspecific population will become extinct. This assessment end point, however, occurs at the end of a long chain of events starting with the escape or introduction of mature or immature transgenic fish (e.g. Figs 5.1 and 9.1). The complexity of all risk assessment problems increases as the number of steps between exposure (escape or release of transgenic fish) and harm (extinction of a conspecific population) increases, and sometimes (such as in the example above) these steps may take many decades to complete and may be influenced by other stressors. Risk assessments for transgenic fish are further confounded by pleiotropic effects, undermining the ability to accurately extrapolate laboratory observations to the field (Chapter 6, this volume). Risk analysts need to carefully balance the relevancy of assessment end points with their ability to calculate risks with reasonable bounds of certainty. Analysts are advised to work with stakeholders to, where possible, seek simpler, intermediate-level end points over specific periods of time (e.g. number of transgenic fish that escape from contained facilities per annum) that are still close enough to their concerns to allow them to set risk acceptance criteria, but not so complex as to preclude confident analysis. Analysts must be careful, however, not to oversimplify the problem or unduly bias the end point selection process.

Assessment end points that are embedded within a scientifically consistent conceptual model of the ecosystem provide an ideal platform for hazard identification and prioritization. Hazard identification requires the analyst to compare the genetic, physiological and behavioural traits of transgenic fish with the biological, physical and chemical attributes of the receiving environment and ask the question 'what can go wrong?'. Hazards are usually identified using informal brainstorming activities, which often culminate in a 'hazard checklist' (Hayes, 2002a). There are, however, a number of more rigorous, systematic hazard identification techniques (mentioned in Fig. 1.5, Box 1.1) that, while more time-consuming, are much less likely to overlook hazards. We recommend that analysts apply one or more of these techniques in order to fully utilize the combined knowledge base of stakeholders and experts, thereby identifying potentially hazardous situations associated with the production, use and release of transgenic fish. The temporal scope of the hazard analysis is determined by the risk assessment end points and, in this context, could include evolutionary and landscape changes if the time horizon of the assessment is very long.

The first stage of our proposed risk assessment culminates with a prioritized list of hazards. Systematic hazard identification techniques typically produce long lists of ways things might go wrong (see, e.g. Hayes *et al.*, 2004), and this list needs to be prioritized if the risk assessment is to proceed in an efficient manner. There are various methods for prioritizing hazards (Burgman, 2005), but it is important to note that the analyst can complete the first stage, and a large part of the second stage, of the risk assessment by prioritizing hazards using qualitative assertions about their likelihood and consequences. These assertions will provide the basis for a rigorous and defensible qualitative risk assessment so long as the first stage, including the hazard prioritization, is carefully completed and documented.

Calculate, evaluate and manage

The second stage of the risk assessment commences once there is an agreed-upon and prioritized list of potential hazards. The aim of the second stage is to estimate, qualitatively or quantitatively, the likelihood that the hazardous events will occur and what their consequences will be, expressed in units relevant to the assessment or measurement end point in question, over a time period agreed upon during the definition of the boundaries and scope of the assessment. Qualitative risk calculations typically culminate in a risk matrix based on qualitative assertions about the likelihood and consequences of potential hazards (e.g. AS/NZS, 2004), and, as noted earlier, these assertions can form part of the hazard prioritization stage. A quantitative risk calculation should culminate with a frequency distribution of undesired consequences (or equivalent cumulative probability distribution) for each assessment end point, showing the complete range of a potential consequence on the horizontal axis and the expected frequency or cumulative probability of this consequence on the vertical axis (e.g. Fig. 1.3).

Qualitative and quantitative risk calculations should reflect as accurately as possible the effects of uncertainty in the risk assessment process (Chapter 7, this volume).The main reason to adopt quantitative risk assessment methods is to include the most 'honest' appraisal possible of the uncertainty associated with risk calculations, and thereby assist regulatory authorities in making well-informed decisions. Qualitative risk estimates are disadvantageous in this context because they do not adequately distinguish linguistic uncertainty from variability and incertitude (see Chapter 7, this volume). We recognize, however, that in many instances fully quantitative risk estimates for transgenic fish may be difficult to achieve or defend because of inter alia pleiotropy, genotype-by-environment interactions and other reasons discussed in this book (e.g. see Chapters 5 and 6, this volume). Analysts and stakeholders should consider a range of qualitative, semi-quantitative and quantitative risk assessment methods (see Boxes 1.1–1.3) and available data (from literature, laboratory experiments and field trials) to agree on an assessment approach.

If risk estimates are deemed unacceptable relative to predefined acceptance criteria, analysts should work with stakeholders to identify and evaluate risk management strategies (e.g. containment methods described in Chapter 8, this volume) or determine whether to recommend that decision makers end the project. It is important that analysts evaluate risk management strategies by considering how easy they are to implement and how effective they might be given the uncertainty associated with the original risk estimate. The analyst should ideally seek risk management strategies that are robust to the uncertainty associated with the original risk estimates.

Monitor, validate and compare

The third and final stage of our proposed risk assessment framework aims to monitor outcomes in the natural environments surrounding aquaculture operations that may use transgenic fish. The main aim of the third stage is to validate or refute the predictions of the risk assessment and provide an (ideally) early warning if environmental changes are inconsistent with the assessment risk predictions or risk management objectives. Observations that are inconsistent with the magnitude (or direction) of predicted change in the assessment end points indicate a need to re-evaluate the risk assessment or risk management strategies. This can be a difficult step because of the costs involved with reiterating some or all of the risk assessment steps.

It is important to consider the statistical power of the monitoring techniques employed when designing a monitoring programme. The aim is to avoid being 'blind' to important changes in measurement end points that cause type II errors, i.e. falsely concluding that there has been no effect on measurement end points because of insufficient sample size (Hill and Sendashonga, 2003). Chapter 9 provides extensive guidance on monitoring techniques for key measurement end points discussed in this book and the statistical power of these techniques (Fig. 1.7).

The third stage of the risk assessment also gathers additional information that can be used to improve future risk assessments, either retrospective assessment of the same case or predictive assessment of similar cases. For example, characterizing the variability of critical traits of escaped transgenic fish in the receiving environment provides empirical information on an important source of uncertainty in the assessment: genotype-by-environment interactions (Chapters 5 and 6, this volume). These data can also be used to distinguish between alternative conceptual models if two or more plausible models were identified in the first stage of the risk assessment, thereby addressing model uncertainty (Chapter 7, this volume). In this manner, both risk assessment and risk management strategies become adaptive, and can therefore be modified as additional empirical information that is generated through the monitoring programme (Kapuscinski *et al.*, 1999).

Chapter Summary

Environmental risk assessment aims to identify the full range of potential impacts on valued components of the environment, and to estimate the probability and magnitude of these impacts, following (for the purpose of this book) the accidental or intentional introduction of transgenic fish into aquatic ecosystems. The risk assessment should aim to use all relevant sources of information, both qualitative and quantitative, to identify and assess the chain of events leading from entry of transgenic fish into natural water bodies to potentially harmful environmental change. A risk assessment can employ qualitative, semi-quantitative and quantitative methods, depending on the situation. Chapters 2–9 present current methodologies for environmental risk assessment and management of transgenic fish that can be readily implemented within the existing risk assessment frameworks of different countries.

This chapter goes further by proposing a highly interactive framework for environmental risk assessment that is designed to improve an assessment's scientific validity and trust by stakeholders. Widespread adoption of such a framework will require considerable capacity building, as discussed in Chapter 10, as well as commitment to the approach by policy makers. The basic tenets of the proposed interactive framework are:

- Emphasis on relevant stakeholder and expert participation within an analytical-deliberative process;
- Recognition of a variety of qualitative, semi-quantitative and fully quantitative risk assessment methods;
- Awareness of the problems of linguistic uncertainty and 'psychological frailties' associated with qualitative risk estimates;
- Encouragement for using rigorous hazard analysis methods and qualitative modelling to help maintain the scientific quality of qualitative risk assertions;
- Recognition of the importance of variability and incertitude in risk estimates and the need to design risk management strategies that are, as far as practical, responsive to this uncertainty;
- Statistically defensible monitoring strategies that avoid type II errors and provide additional empirical information to iteratively improve risk estimates.

This book presents information at several levels of biological hierarchy, from the transgenic construct itself to the aquatic ecosystem, which will support the development and implementation of scientifically sound environmental risk assessment for transgenic fish. It represents an important milestone in the development and synthesis of scientific methodologies for use by different groups worldwide to conduct their own environmental risk assessments and to design their own risk management programmes for transgenic fish.

References

ABRAC (Agricultural Biotechnology Research Advisory Committee) (1995) *Performance Standards for Safely Conducting Research with Genetically Modified Fish and Shellfish.* Parts I and II. Document Nos. 95-04 and 95-05. United States Department of Agriculture, Office of Agricultural Biotechnology, Washington, DC. Available at: www.isb.vt.edu/perfstands/psmain.cfm

Asian Development Bank (1990) *Dealing with Uncertainty in Environmental Impact Assessment.* ADB Environment Paper No.7, Honolulu, Hawaii, USA.

Asian Development Bank (2005) *An Impact Evaluation of the Development of Genetically Improved Farmed Tilapia, and Their Dissemination in Selected Countries.* Asian Development Bank, Mandaluyong City, Philippines. Available at: www.adb.org/Publications

AS/NZS 4360 (2004) *Risk Management.* Standards Australia International Ltd, Sydney, Australia and Standards New Zealand, Wellington, New Zealand.

Bartell, S.M. and Nair, S.K. (2003) Establishment risks for invasive species. *Risk Analysis* 24, 833–845.

Bartell, S.M., Pastorok, R.A., Akcakaya, H.R., Regan, H., Ferson, S. and Mackay, C. (2003) Realism and relevance of ecological models used in chemical risk assessment. *Human and Ecological Risk Assessment* 9, 907–938.

Bomford, M. and Glover, J. (2004) *Risk Assessment Model for the Import and Keeping of Exotic Freshwater and Estuarine Finfish.* Bureau of Rural Sciences, Canberra, Australia. Available at: http://www.feral.org.au/feral_documents/finfish_risk_assessment.pdf

Burgman, M.A. (2001) Flaws in subjective assessments of ecological risks and means for correcting them. *Australian Journal of Environmental Management* 8, 219–226.

Burgman, M.A. (2005) *Risks and Decisions for Conservation and Environmental Management.* Cambridge University Press, Cambridge.

CABI (2005) *UK Non-native Organism Risk Assessment Scheme User Manual.* CABI Bioscience (CABI), Centre for Environment, Fisheries and Aquaculture Science (CEFAS), Centre for Ecology and Hydrology (CEH), Central Science Laboratory (CSL), Imperial College London (IC) and the University of Greenwich (UoG). Available at: http://www.defra.gov.uk/wildlife-countryside/resprog/findings/non-native-risks/pdf/user-manual.pdf

Caley, P., Lonsdale, W.M. and Pheloung, P.C. (2006) Quantifying uncertainty in predictions of invasiveness. *Biological Invasions* 8, 277–286.

CBD (2000) Cartagena Protocol in Biosafety to the Convention on Biological Diversity: Text and Annexes. Secretariat of the Convention on Biological Diversity, Montreal. Available at: http://www.biodivorg/doc/legal/cartagena-protocol-en.pdf

Copp, G.H., Garthwaite, R. and Gozlan, R.E. (2005) Risk identification and assessment on non-native freshwater fishes: concepts and perspectives on protocols for the UK. *Science Series Technical Report 129*, Center for Fisheries and Aquaculture (Cefas), Lowestoft, UK.

Cortemeglia, C. and Beitinger, T.L. (2006) Projected US distributions of transgenic and wildtype zebra danios, *Danio rerio*, based on temperature tolerance data. *Journal of Thermal Biology* 31, 422–428.

Cox, P., Niewohner, J., Pidgeon, N., Gerrard, S., Fischhoff, B. and Riley, D. (2003) The use of mental models in chemical risk protection: developing a generic workplace methodology. *Risk Analysis* 23, 311–324.

Crawford-Brown, D.J. (2001) *Mathematical Methods of Environmental Risk Modelling.* Kluwer Academic, Boston, Massachusetts.

Dambacher, J.M., Hiram, W.L. and Rossignol, P.A. (2003a) Qualitative predictions in model ecosystems. *Ecological Modelling* 161, 79–93.

Dambacher, J.M., Luh, H.K., Li, H.W., and Rossignol, P.A. (2003b) Qualitative stability and ambiguity in model ecosystems. *American Naturalist* 161, 876–888.

Delgado, C.L., Wada, N., Rosegrant, M.W., Meijer, S. and Ahmed, M. (2003) *Fish to 2020: Supply and Demand in Global Markets*. World Fish Center Technical Report 62, IFPRI (International Food Policy) and World Fish Center, Penang, Malaysia. Available at: http://www.lib.noaa.gov/docaqua/noaa_matrix_program_reports/03_siwa_msangi.pdf

DETR (Department of the Environment, Transport and Regions) (1999) Guidance on principles of risk assessment and monitoring for the release of genetically modified organisms, DETR/ACRE guidance note 12. Department of the Environment, Transport and Regions, London.

Devlin, R.H. and Donaldson, E.M. (1992) Containment of genetically altered fish with emphasis on salmonids. In: Hew, C.L. and Fletcher, G.L. (eds) *Transgenic Fish*. World Scientific Press, Singapore, pp. 229–265.

Drake, J.M. (2004) Allee effects and the risk of biological invasion. *Risk Analysis* 24, 795–802.

Drake, J.M. (2005) Risk analysis for species introductions: forecasting population growth of Eurasian ruffe (*Gymnocephalus cernuus*). *Canadian Journal of Fisheries and Aquatic Sciences* 62, 1053–1059.

Drake, J.M. and Bossenbroek, J.M. (2004) The potential distribution of zebra mussels in the United States. *BioScience* 54, 931–941.

ERMA (Environmental Risk Management Authority) (1999) Identifying risks for applications under the Hazardous Substances and New Organisms Act 1996, ER-TG-01-1 9/99. ERMA, New Zealand. Available at: http://www.ermanz.govt.nz/resources/publications/pdfs/ER-TG-01-1.pdf

EU (European Union) (2001) Directive 2001/18/EC of the European Parliament and of the Council of 12 March 2001 on the deliberate release into the environment of genetically modified organisms and repealing Council Directive 90/220/EEC. *Official Journal of the European Communities* L106 (17.04.2001):1–38.

FAO (Food and Agriculture Organization of the United Nations) (2004a) The state of world fisheries and aquaculture. Part 1: *World Review of Fisheries and Aquaculture*. FAO, Rome. Available at: www.fao.org/documents/show_cdr.asp?url_file=/DOCREP/007/y5600e/y5600e00.htm

FAO (Food and Agriculture Organization of the United Nations) (2004b) *Public Participation in Decision-making Regarding GMOs in Developing Countries: How to Effectively Involve Rural People*. Electronic Forum on Biotechnology in Food and Agriculture: Conference 11. Available at: http://www.fao.org/Biotech/C12doc.htm

FAO/WHO (Food and Agriculture Organization of the United Nations/World Health Organization) (2004) Safety assessment of foods derived from genetically modified animals, including fish, a joint FAO/WHO expert consultation on food derived from biotechnology, Rome, Italy, 17–21 November 2003. Available at: http://www.who.int/foodsafety/biotech/meetings/ec_nov 2003/en/

Ferson, S. (2005) *Bayesian Methods in Risk Assessment*. Applied Biomathematics, Setauket, New York.

Ferson, S. and Long, T.F. (1995) Conservative uncertainty propagation in environmental risk assessment. In: Hughes, J.S., Biddinger, G.R. and Mones, E. (eds) *Environmental Toxicology and Risk Assessment*, Volume 3. American Society for Testing and Materials, Philadelphia, Pennsylvania, pp. 97–110.

Glicken, J. (2000) Getting stakeholder participation 'right': a discussion of the participatory processes and possible pitfalls. *Environmental Science and Policy* 3, 305–310.

Haimes, Y.Y. (1998) *Risk Modelling, Assessment and Management*. Wiley, New York.

Hallerman, E.M. and Kapuscinski, A.R. (1995) Incorporating risk assessment and risk management into public policies of genetically modified finfish and shellfish. *Aquaculture* 137, 9–17.

Hart, B., Burgman, M., Webb, A., Allison, G., Chapman, M., Duivenvoorden, L., Feehan, P., Grace, M., Lund, M., Pollino, C., Carey, J. and McCrea, A. (2005) *Ecological Risk Management Framework for the Irrigation Industry*. Report to the National Program for Sustainable Irrigation. Water Studies Centre, Monash University, Clayton, Australia.

Hastings, A., Cuddington, K., Davies, K.F., Dugaw, C.J., Elmendorf, S., Freestone, A., Harrison, S., Holland, M., Lambrinos, J., Malvadkar, B.A., Moore, K., Taylor, C. and Thomson, D. (2005) The spatial spread of invasions: new developments in theory and evidence. *Ecology Letters* 8, 91–101.

Hayes, K.R. (1997) *Ecological Risk Assessment Review*. CRIMP Technical Report 13, CSIRO Division of Marine Research, Hobart, Australia.

Hayes, K.R. (1998) *Bayesian Statistical Inference in Ecological Risk Assessment*. CRIMP Technical Report 17, CSIRO Division of Marine Research, Hobart, Australia.

Hayes, K.R. (2002a) *Best Practise and Current Practise in Ecological Risk Assessment for Genetically Modified Organisms*. Interim report for the Australian Government Department of Environment and Heritage. KRA Project 1: robust methodologies for GMO risk assessment. CSIRO Division of Marine Research, Hobart, Australia.

Hayes, K.R. (2002b) Identifying hazards in complex ecological systems. Part 1: fault tree analysis for biological invasions. *Biological Invasions* 4, 235–249.

Hayes, K.R. (2002c) Identifying hazards in complex ecological systems. Part 2: infections modes and effects analysis for biological invasions. *Biological Invasions* 4, 251–261.

Hayes, K.R. (2003) Biosecurity and the role of risk-assessment. In: Ruiz, G.M. and Carlton, J.T. (eds) *Bioinvasions: Pathways, Vectors, and Management Strategies*. Island Press, Washington, DC, pp. 382–414.

Hayes, K.R., Gregg Peter, C., Gupta, V.V.S.R., Jessop, R., Lonsdale, M., Sindel, B., Stanley, J. and Williams, C.K. (2004) Identifying hazards in complex ecological systems. Part 3: Hierarchical Holographic Model for herbicide tolerant oilseed rape. *Environmental Biosafety Research* 3, 1–20.

Hayes, K.R., Sliwa, C., Migus, S., McEnnulty, F. and Dunstan, P. (2005) *National Priority Pests:* Part II, *Ranking of Australian Marine Pests*. Final report for the Australian Government Department of Environment and Heritage, CSIRO Division of Marine Research, Hobart, Australia.

Hayes, K.R., Barry, S.C. and Lawrence, E. (in prep) *Ballast Water Risk Assessment: Calculating the Probability of Survival in the Recipient Port using Life-cycle and Time-series Models.*

Hayes, K.R. and Barry, S.C. (in press) Are there any consistent predictors of invasion success? *Biological Invasions.*

Hilbeck, A. and Andow, D.A. (2004) *Environmental Risk Assessment of Genetically Modified Organisms, Volume I. A Case Study of Bt Maize in Kenya*. CAB International, Wallingford, UK.

Hilbeck, A., Andow, D.A. and Fontes, E.M.G. (2006) *Environmental Risk Assessment of Genetically Modified Organisms, Volume 2. Methodologies for Assessing Bt Cotton in Brazil*. CAB International, Wallingford, UK.

Hill, R.A. and Sendashonga, C. (2003) General principles for risk assessment of living modified organisms: lessons from chemical risk assessment. *Environmental Biosafety Research* 2, 81–88.

Hobbs, B.F., Ludsin, S.A., Knight, R.L., Ryan, P.A., Biberhofer, J. and Ciborowski, J.J.H. (2002) Fuzzy cognitive mapping as a tool to define management objectives for complex ecosystems. *Ecological Applications* 12, 1548–1565.

Hughes, G. and Madden, L.V. (2003) Evaluating predictive models with application in regulatory policy for invasive weeds. *Agricultural Systems* 76, 755–774.

Kapuscinski, A.R. (2002) Controversies in designing useful ecological assessments of genetically engineered organisms. In: Letourneau, D.K. and Burrows, B.E. (eds) *Genetically Engineered Organisms: Assessing Environmental and Human Health Effects*. CRC Press, Boca Raton, Florida, pp. 385–415.

Kapuscinski, A.R. (2005) Current scientific understanding of the environmental biosafety of transgenic fish and shellfish. *Revue Scientifique et Technique de l' Office International des Epizooties* 24, 309–322.

Kapuscinski, A.R. and Hallerman, E.M. (1990) Transgenic fish and public policy: I. Anticipating environmental impacts of transgenic fish. *Fisheries* 15, 2–11.

Kapuscinski, A.R. and Hallerman, E.M. (1991) Implications of introduction of transgenic fish into natural ecosystems. *Canadian Journal of Fisheries and Aquatic Sciences* 48(Suppl. 1), 99–107.

Kapuscinski, A.R., Nega, T. and Hallerman, E.M. (1999) Adaptive biosafety assessment and management regimes for aquatic genetically modified organisms in the environment. In: Pullin, R.S.V. and Bartley, D. (eds) *Towards Policies for Conservation and Sustainable Use of Aquatic Genetic Resources*. ICLARM Conference Proceedings, International Center for Living Aquatic Resources Management, Makati City, The Philippines, pp. 225–251.

Kletz, T.A. (1999) *Hazop and Hazan: Identifying and Assessing Process Industry Hazards*. Taylor & Francis, Philadelphia, Pennsylvania.

Koehn, J. (2004) Carp (*Cyprinus carpio*) as a powerful invader in Australian waters. *Freshwater Biology* 49, 882–894.

Kohler, C.C. (1992) Environmental risk management of introduced aquatic organisms in aquaculture. In: Sindermann, C. and Hershberger, B. (eds) *Introductions and Transfers of Aquatic Species*. Selected papers from a Symposium held in Halifax, Nova Scotia, Canada, 1990. ICES Marine Science Symposium 194, 15–20.

Leung, B., Lodge, D.M., Finnoff, D., Shogren, J.F., Lewis, M.A. and Lamberti, G. (2002) Anounce of prevention or a pound of cure: bioeconomic risk analysis of invasive species. *Proceedings of the Royal Society of London, Series B* 269, 2407–2413.

Maclean, N. and Laight, R.J. (2000) Transgenic fish: an evaluation of benefits and risks. *Fish and Fisheries* 1, 146–172.

Muir, W.M. and Howard, R.D. (1999) Possible ecological risks of transgenic organism release when transgenes affect mating success: sexual selection and the Trojan gene hypothesis. *Proceedings of the National Academy of Sciences USA* 96, 13853–13856.

Muir, W.M. and Howard, R.D. (2001) Fitness components and ecological risk of transgenic release: a model using Japanese medaka (*Oryzias latipes*). *American Naturalist* 158, 1–16.

Muir, W.M. and Howard, R.D. (2002) Assessment of possible ecological risks and hazards of transgenic fish with implications for other sexually reproducing organisms. *Transgenic Research* 11, 101–114.

Naylor, R., Hindar, K., Fleming, I.A., Goldburg, R., William, S., Volpe, J., Whoriskey, F., Eagle, J., Kelso, D. and Mangel, M. (2005) Fugitive salmon: assessing the risks of escaped fish from net-pen aquaculture. *BioScience* 55, 427–437.

NRC (1996) *Understanding Risk: Informing Decisions in a Democratic Society*. National Academy Press, Washington, DC. Available at: http://www.nap.edu

NRC (2002) *Environmental Effects of Transgenic Plants: The Scope and Adequacy of Regulation*. National Academy Press, Washington, DC. Available at: http://www.nap.edu

Neubert, M.G. and Parker, I.M. (2004) Projecting rates of spread for invasive species. *Risk Analysis* 24, 817–831.

OECD (Organisation for Economic Cooperation and Development) (1986) *Recombinant DNA Safety Considerations*. Organisation for Economic Cooperation and Development, Paris.

OGTR (Office of the Gene Technology Regulator) (2001) *Risk Assessment Framework for License Applications to the Office of the Gene Technology Regulator*. Office of the Gene Technology Regulator, Canberra, Australia.

Ozesmi, U. and Ozesmi, S. (2003) A participatory approach to ecosystem conservation: fuzzy cognitive maps and stakeholder group analysis in Uluabat Lake, Turkey. *Environmental Management* 31, 518–531.

Paisley, L.G., Karlsen, E., Jarp, J. and Mo, T.A. (1999) A Monte Carlo simulation model for assessing the risk of introduction of *Gryodactylus salaris* to the Tana River, Norway. *Diseases of Aquatic Organisms* 37, 145–152.

Palady, P. (1995) *Failure Modes and Effects Analysis: Predicting and Preventing Problems Before They Occur*. PT Publications, West Palm Beach, Florida.

Pastorok, R.A., Bartell, S.M., Ferson, S. and Ginzburg, L.R. (2002) *Ecological Modelling in Risk Assessment: Chemical Effects on Populations, Ecosystems and Landscapes*. Lewis Publishers, Boca Raton, Florida.

Peeler, E.J. and Thrush, M.A. (2004) Qualitative analysis of the risk of introducing *Gryodactylus salaris* into the United Kingdom. *Diseases of Aquatic Organisms* 62, 103–113.

Pheloung, P.C., Williams, P.A. and Halloy, S.R. (1999) A weed risk assessment model for use as a biosecurity tool for evaluating plant introductions. *Journal of Environmental Management* 57, 239–251.

Ramsey, D. and Veltman, C. (2005) Predicting the effects of perturbations on ecological communities: what can qualitative models offer? *Journal of Animal Ecology* 74, 905–916.

Royal Commission on Environmental Pollution (1991) *GENHAZ: A System for the Critical Appraisal of Proposals to Release Genetically Modified Organisms into the Environment*. HMSO, London, UK.

Rykiel, E.J. (1996) Testing ecological models: the meaning of validation. *Ecological Modelling* 90, 229–244.

Saaty, T.L. (2001) *The Analytical Hierarchy Process: Decision Making with Dependence and Feedback*. McGraw-Hill, New York.

Scientists' Working Group on Biosafety (1998) *Manual for Assessing Ecological and Human Health Effects of Genetically Engineered Organisms*. Part One: *Introductory Materials and Supporting Text for Flowcharts*. Part Two: *Flowcharts and Worksheets*. The Edmonds Institute, Edmonds, Washington, DC. Available at: http://www.edmonds-institute.org/manual. html

Snow, A., Andow, D., Gepts, P., Hallerman, E., Powers, A., Tiedje, J. and Wolfenbarger, L. (2005) Ecological risks and benefits associated with the environmental release of GMOs. *Ecological Applications* 15, 377–404.

Suter, G.W. (1993) *Ecological Risk Assessment*. Lewis Publishers, Boca Raton, Florida.

The Royal Society (1983) *Risk Assessment: Report of a Royal Society Study Group*. The Royal Society, London.

Tiedje, J.M., Colwell, R.K., Grossman, Y.L., Hodson, R.E., Lenski, R.E., Mack, R.N., and Regal, P.J. (1989) The planned introduction of genetically engineered organisms: ecological considerations and recommendations. *Ecology* 70, 298–315.

UNEP (United Nations Environment Programme) (1995) International Technical Guidelines for Safety in Biotechnology. United Nations Environment Programme, Nairobi, Kenya. Available at: http://www.biosafetyprotocol.be/UNEPGuid/Contents.html

Virtue, J.G., Groves, R.H. and Panetta, F.D. (2001) Towards a national system to determine the national significance of weeds in Australia. In: Groves, E.H., Panetta, F.D. and Virtue, J.G. (eds) *Weed Risk Assessment*. CSIRO Publishing, Canberra, Australia, pp. 125–150.

2 Problem Formulation and Options Assessment: Science-guided Deliberation in Environmental Risk Assessment of Transgenic Fish

K.C. Nelson, Z. Basiao, A.M. Cooper, M. Dey,
D. Fonticiella, M. Lorenzo Hernandez, S. Kunawasen,
W. Leelapatra, S. Li, B.D. Ratner and M.I. Toledo

Introduction

Problem formulation and options assessment (PFOA)[1] is one cornerstone of an environmental risk assessment for transgenic organisms. It is public deliberation about a transgenic organism, and provides a rational, science-guided planning process by which multiple stakeholders can assess shared needs, evaluate risks related to a variety of future options and make recommendations to decision makers about policies that reduce societal risks and enhance the benefits provided by various options. It establishes the framework for interaction between stakeholders and scientists, guides deliberation among stakeholders and provides a linkage to biosafety governance. Public deliberation about transgenic organisms is often conflict-ridden; a PFOA process can help alleviate this tension through transparent, systematic and science-based discussion. A decision-making process can gain social legitimacy by incorporating a PFOA, and thereby society gains greater confidence in the decisions taken as a result of an environmental risk assessment.

To facilitate socially acceptable choices (NRC, 1996), countries must create a socially responsive risk assessment system. At its core, the PFOA discussion assesses whether a transgenic organism can address particular problems and, if so, under what conditions. PFOA focuses on the critical societal needs addressed by the transgenic organism and the risks associated with using it. Acting on societal needs and their associated risks requires informed reflection by a cross section of society's members. A deliberative process with multi-stakeholder

[1] The description of the PFOA methodology has been modified from that of Nelson *et al.* (2004) developed within the GMO ERA Project (2006); see Nelson and Banker (2007).

 Environmental Risk Assessment of Genetically Modified Organisms: Vol. 3. Methodologies for Transgenic Fish (eds A.R. Kapuscinski *et al.*)

participation allows members of society to participate in the evaluation of future alternatives and risks. A cross section of society – producers, consumer groups, industry, environmental representatives, policy makers, etc. – must have a medium to express their concerns and evaluate the future alternatives for addressing basic needs. Such a deliberative process will be increasingly important for resource-scarce nations, especially if public investment is involved, because a comparative reflection by a cross section of society may be beneficial to prioritize and target resources. The benefits and risks from the use of transgenic aquatic organisms may extend across national borders, and PFOA provides an opportunity for multi-country deliberations within a watershed or region. For example, transgenic fish aquaculture may address a societal need for one country, while escapes of these organisms may result in harm for other countries within the same water system.

Given the present uncertainties surrounding transgenic organisms and regular reports of new discoveries and developments, the system for conducting a deliberative discussion of risk issues should be flexible and be able to respond to a society's core values, concerns and needs. Countries accept risk with most major policy decisions, but work to reduce risk when possible. At the same time, the discussion is best served if it is guided by a scientifically based assessment and review (Gibbons, 1999; NRC, 2002; CBD, 2006). An environmental risk assessment should clearly delineate when a body of scientific knowledge, information and analysis can answer key questions, and when it cannot.

In most natural resource arenas, practitioners and scholars are implementing and evaluating diverse options for societal discussion about critical issues (O'Brien, 2000; Wondolleck and Yaffee, 2000). Specific contributions within the biosafety arena have detailed potential approaches to multi-stakeholder dialogues (McLean *et al.*, 2002; Glover *et al.*, 2003) and have evaluated their key attributes (Irwin, 2001) at international and national scales. We advise that societal discussion be a part of the process of setting priorities and strategies for aquaculture research and development, formulation of national biosafety frameworks (NBFs) and environmental risk assessment of biotechnologies such as transgenic fish. PFOA is applicable in all of these contexts, and it can be employed in an iterative fashion to incorporate feedback from changing societal values and advances in scientific knowledge.

Chapter Purpose

This chapter is intended primarily for policy makers, legislators, regulators and other public officials involved in the design or implementation of biosafety policy. It provides an overview of PFOA for scientists, risk assessment specialists, industry representatives and other interested public stakeholders who may be involved in deliberation over transgenic fish technology. The chapter presents a tested approach for structured societal discussion, but one that is flexible enough for individuals and agencies to adapt to the specific social, political and regulatory context of their countries. Because of the need for flexibility when making biosafety policy decisions, PFOA will be utilized in various ways across countries. For example, some nations may design a highly interactive process at multiple stages of the environmental risk assessment to enhance exchange between scientists and stakeholders for improved scientific research and stakeholder deliberation; others may design a more consultative process with PFOA at critical stages of the risk assessment, emphasizing stakeholder information sessions about scientific findings and consultation with decision makers about societal opinions.

Specifically, the chapter explains PFOA, how it contributes to environmental risk assessment and assesses its advantages and challenges for the subject of transgenic fish biosafety. The presence of already-existing NBFs means that countries need to develop a country-specific deliberative process that fits the particular structure and authority of their relevant decision-making bodies and implementing agencies. For many national political systems, legitimate authorities exist for incorporating PFOA into their legislative or regulatory context, but there is ongoing debate about whether and how to modify policies and regulation for transgenic organisms (Munson, 1993; Miller, 1994; Hallerman and Kapuscinski, 1995; Sagar *et al.*, 2000; NRC, 2002). In some legislative or regulatory situations, PFOA can be incorporated into existing public consultative processes mandated for any regulation or policy. It may also be added as an alternative process, and supported by civil society, to inform the debate held by traditional decision-making bodies.

Consideration of how PFOA could be incorporated into diverse governance situations is explored in four country-specific case studies: Chile, Cuba, Thailand and China. The chapter outlines each country's policy and regulatory context and how PFOA could be applied in each case. The chapter also illustrates the sorts of issues that PFOA might address in these contexts, although it does not draw conclusions for any specific case.

Finally, the chapter contains a preliminary reflection about questions to consider when designing a country-specific PFOA, paying special attention to the capacity-building areas needed to strengthen a country's ability to conduct PFOA (see also Chapter 10, this volume, for further discussion of capacity-building needs).

In keeping with the broader goals of this book, PFOA is reviewed as one essential component of the package of environmental risk assessment tools for transgenic fish. The PFOA structures a process to link together stakeholder deliberation, scientific analysis and regulatory decision making. It therefore provides a context in which the risk assessment methodologies detailed in other chapters can be applied. PFOA espouses an interactive process of informing science and using the results of this book's methodologies to inform deliberation. For example, during the PFOA for a specific transgenic fish, stakeholders should reach a common understanding of potential problems and compare the attributes of future options in an organized manner. Stakeholders should review existing science and discuss the merits of the proposed organism, starting with recommendations at the level of transgene design and integration (identified in Chapter 4, this volume). Through deliberation, stakeholders should consider the positive contributions of the technology, as well as their concerns about potential harms: examples might be gene flow to wild relatives and its consequences and other ecological effects of transgenic organisms (as discussed in Chapters 5 and 6, this volume). Finally, the PFOA participants should come to a shared understanding of system-level changes across temporal and spatial scales that may result from, or be necessary for, the technology to produce the greatest benefits with the least risk; information important for methodologies are discussed in Chapters 7 and 8. In an iterative risk assessment process, PFOA recommendations can provide insights into risk assessment, risk management planning, risk communication strategies and post-approval monitoring, should the technology be utilized.[2]

PFOA Methodology and Environmental Risk Assessment

The PFOA methodology is one component of a rigorous environmental risk assessment process within a regulatory system, applicable for any country establishing a biosafety framework. Practitioners and scholars have tested numerous stakeholder participation techniques that serve as a methodological foundation for risk-based resource management (Grimble and Wellard, 1997; Kessler and Van Dorp, 1998; Schmoldt and Peterson, 1998; Biggs and Matsaert, 1999; Loevinsohn *et al.*, 2002; Traynor *et al.*, 2002; GMO ERA Project, 2006). PFOA was developed through the Genetically Modified Organisms Environmental Risk Assessment Project (GMO ERA Project), through a series of workshops in Kenya, Brazil and Vietnam (GMO ERA Project, 2006). In each workshop, PFOA section participants evaluated the PFOA methodology for its attributes and opportunities, and they made recommendations about how it would fit their country context (Nelson *et al.*, 2004; Capalbo *et al.*, 2006; Nguyen van Uyen *et al.*, Wallingford, UK, 2007). With each workshop, the participants refined the PFOA methodology, building a shared understanding across participants and countries.

The PFOA methodology is characterized by an emphasis on science, deliberation, transparency and legitimacy; the foundation of PFOA is a science-guided inquiry. Questions are answered with data, impacts are assessed with valid indicators and uncertainty is explicitly acknowledged (see also Chapter 7, this volume). A science-guided PFOA must be a deliberative process (Forester, 1999) designed to foster social reflection and discussion about genetically modified organisms (GMOs). A sound deliberative process is transparent, equitable, legitimate and, when possible, data-driven (Susskind *et al.*, 2000). Transparency allows for open communication of information between all parties and easily accessible reporting of decisions to the public (Hemmati, 2002). An equitable PFOA process means that information from the broadest spectrum of society must be included, with all stakeholders having the possibility to contribute. Civil society must perceive that there are sufficient avenues for input and consideration of diverse viewpoints and concerns. The PFOA gains legitimacy in the public eye when transparency and equitable input are central to the process.

[2] See page 39 for the timing and staging of PFOA within a risk assessment, as well as Figs 1.4, 1.5 and 1.6.

Public legitimacy must be supported by traditional legitimacy through sanctioning of PFOA recommendations by a formal political body that receives the recommendations. The PFOA deliberative process can be incorporated into a regulatory authority or legislative authority, and if it is not, there should be a means by which results from the PFOA inform government decision making and action.

The PFOA methodology comprises specific brainstorming, discussion and analytical components (Table 2.1). During this book's international book-writing workshop, the authors conducted an exercise with PFOA to understand the types of information, deliberation and shared understanding that might occur in a complete PFOA consideration involving a transgenic fish line; they explored steps 1–8 (Table 2.1) to develop examples of information that a PFOA can generate (Boxes 2.1–2.5).

Table 2.1. Problem formulation and options assessment (PFOA): designed for transgenic fish environmental risk assessment. (Adapted from Nelson *et al.*, 2004; GMO ERA Project, 2006.)

Initiating proposal

A. Proposal to use GMO
Any PFOA for transgenic organisms will be initiated by the request or suggestion that a particular GMO would be a beneficial alternative to the way things are currently being done in a particular aquaculture system.

B. Decision by regulatory body
Is there merit to moving forward to evaluate the GMO as a possible option, or is the initiating proposal premature? Yes/No

PFOA

Questions to be answered by all stakeholders and shared in the deliberative process.

Step 1: Problem formulation

Formulation of problem	**Basic human needs**	**Interests**
An unmet need that requires change	Food, shelter, safety	A stakeholder group's values, goals and perspectives

A. Whose problem is it? Whose problem should it be?
1. What needs of the people are not being met by the present situation?
2. What aspects of the present situation must be changed to meet the needs?

Step 2: Prioritization and scale

A. Is this problem a core problem for the people identified? (*Prioritization of the problem*)
1. Do the people recognize the problem as important to their lives?
2. What are the potentially competing needs of these people?
3. How do the identified needs rank in importance to these other competing needs?

B. How extensive is the problem? (*Scale of the problem*)
1. How many people are affected?
2. In what part of the country are these people located?
3. How large an area is affected by the problem?
4. How severe is the problem (local intensity)?

Continued

Table 2.1. *Continued*

Step 3: Problem statement

Shared understanding of the unmet need and its relative importance for a particular group of people.

Step 4: Recommendation by regulatory body

Do we move forward to identify options and conduct an options assessment?

Option Identification and Assessment Chart

Step 5 Options	Step 6 Characteristics	Step 7 Changes	Step 8 Effect on the system	
			Internal	External
Future alternatives	For problem solving	Required/anticipated	(Social, environmental, economic)	
Option A Option B Option C Etc.				

Step 5: Option identification

Brainstorm possible options as future alternatives to solve the identified problem; transgenic organisms would be one option.
This step can be completed by the multi-stakeholder group for the initial identification of options. The multi-stakeholder group can do steps 6–8, or a technical committee can develop a report that covers steps 6–8, and the multi-stakeholder group can use the document to begin their evaluation of options and modify the assessment.

Step 6: Assessment of the options in relation to the problem

Assessing the capability of potential solutions to solve the problem:
1. What are the characteristics of the 'technology' option? (i.e. transgene, aquaculture system, etc.)
2. What is the range of production systems, and what is the geographic region in which the option is likely to be used or to affect?
3. What is the efficacy of the 'technology' on the target?
4. What are the costs of the 'technology' within the production system?
5. What barriers to use exist? For example, is there a distribution system for fish eggs, fry or fingerlings in place; can the potential solution be integrated into present production; can aquaculture practitioners afford the potential solution?
6. How might the use of the option change current practices, such as breeding systems or feeding regimes (including water quality impacts)? What useful practices are reinforced by the potential?
7. What information is needed to show that the changes are likely to occur? Baseline data associated with the diversity of present practices should be used if they are available.
8. How will anticipated changes in aquaculture practices affect the needs identified in steps 1 and 2?

Step 7: Changes required and anticipated for a specific option

1. What changes in *farm and aquaculture management* practices might contribute to the solution?
2. What changes in the *local community* might contribute to the solution?

Continued

Table 2.1. *Continued*

3. What changes in *government support* for farmers might contribute to the solution?
4. What changes in the *structure of aquaculture production* might contribute to the solution?
5. What other changes would likely be *needed to facilitate widespread use* of this option?

Step 8: Adverse effects

Potential adverse consequences from this option. Potential beneficial effects can be considered 'negative' adverse effects.
1. How might the potential solution affect the structure *of aquaculture or related infrastructure*?
2. How might the potential solution *reinforce poor practices or disrupt useful practices*?
3. What are the *potential adverse effects* of these changes internally and externally to the production system?
4. How will its use affect other nearby *production systems, such as capture fishing, other aquaculture or aquaculture–agriculture systems, and terrestrial farming* and non-aquaculture environments (can its use be restricted to a particular system or geographic region)?
5. Are any of these changes *difficult to reverse*, once they occur?

Step 9: Recommendation

The multiple stakeholder group should present their PFOA to the appropriate decision-making body.

In a real situation, the innovator of the transgenic fish line first submits to the appropriate oversight entity a proposal for research and development or production and use of a transgenic fish. The innovator is most often a private company, federal research institute or even an individual scientist. If the oversight personnel determine that the proposal is complete, they initiate the first phase of PFOA (Table 2.1, steps 1–4). The first step involves formulating the societal problem that the technology will address (Box 2.1). The problem is defined as an unmet need that requires change (Goldstein, 1993). During the problem formulation, the group assesses whether a problem truly exists based on the extent, severity and relative importance compared to other problems. For example, the proposed technology may address a disease problem that only affects fish in one river tributary or only reduces production by 5%, compared to a feeding problem that results in an 80% reduction in productivity. On the other hand, the problem could be extensive throughout most of the country's aquaculture systems, affecting large- and small-scale production systems, with substantial loss in productivity. Based on the problem statement developed by the participants, decision makers decide whether there is sufficient reason to continue with the options assessment phase of PFOA (Table 2.1, steps 5–8) and an environmental risk assessment (methodologies discussed in other chapters of this book) or whether the societal problem is not sufficiently severe to merit proceeding to an assessment of options to address it.

If a decision is made to proceed, the PFOA continues with a comparative approach to options assessment. The range of options for addressing the problem are considered future alternatives. Such alternatives can include: options in

Box 2.1. Examples of country-specific problem statements.

Thailand: There is no problem with the use of tilapia aquaculture to produce protein for human consumption.
China: There is a need for enhanced protein production and exports through the use of tilapia in aquaculture.
Chile: There is low diversification in fish production and high costs of production due to the low market prices for export fish. New technology would help the large-scale aquaculture industry diversify production.
Cuba: There is low production of fish for export, high costs of production due to feeding costs, and increasing internal demand for protein given the growing human population. Increased protein availability would help the entire island population.

use now (status quo) and continued into the future, options that exist but could be more widely implemented or new options, such as use of the GMOs. These options are compared for their attributes, potential ability to address the problem identified, changes required to implement the option and their potential adverse effects to inform the environmental risk assessment (Table 2.1, steps 5–8 and examples in Boxes 2.2–2.5). Working through the options assessment

Box 2.2. Examples of options to increase production of protein-rich tilapia.

1. Selectively bred tilapia: e.g. genetically improved farmed tilapia (GIFT)
2. Transgenic growth-enhanced tilapia
3. Improved management of stocking density and water quality

Box 2.3. Examples of attributes to compare across options.

1. Feeding behaviour
2. Growth response
3. Feed consumption
4. Growth rate: juvenile to adult
5. Time to reach commercial weight
6. Weight harvested per year
7. Protein synthesis
8. Age at maturity
9. Absolute fecundity
10. Sex ratio
11. Reproductive sterility
12. Sex-based growth differences
13. Costs
14. Colour and quality of fillets
15. Taste

Box 2.4. Brief examples of country-specific system changes identified during the options assessment.

In General
1. All options would require good water treatment and drainage systems that do not presently exist in some countries or regions.
2. Increased cost of transgenic tilapia may make their use inaccessible to lower income farmers
3. Changes in best practices would be necessary for a 'domesticated species' option to remain profitable

Cuba-specific
1. Transgenic growth-enhanced tilapia would require an effective containment system, which does not presently exist

Box 2.5. Brief examples of possible option-specific adverse effects identified during the options assessment.

For transgenic growth-enhanced tilapia
1. Loss of external markets and trade in European Union or Japan
2. Harm to feral tilapia populations
3. Loss of biodiversity and ecosystem services
4. Loss of aquaculture access to poorest farmers

For selectively bred tilapia
1. Harm to feral tilapia
2. Harm to locally adapted breeds – e.g. those with cold tolerance

would require a minimum of two meetings: one to work with existing information to inform data gathering and analyses done in the environmental risk assessment (which may include new studies as discussed in Chapters 4–6, this volume); and one after completion of data gathering and analyses, allowing the group to consider the new information in its final options assessment. In a highly iterative process, the PFOA group could meet at multiple points during the environmental risk assessment, in order to increase engagement with scientists. After a complete analysis, the PFOA group makes a recommendation (Table 2.1, step 9) to decision makers to:

1. Continue the full environmental risk assessment for proposed commercialization of the technology;
2. Continue research and development (which may often include risk assessment research);
3. Halt public investment and support for research and development of the technology.

PFOA's contribution to ecological risk assessment, management and communication

PFOA contributes to ecological risk assessment by discussing and assessing adverse effects and comparative options as future alternatives, as well as by creating links with risk management and communication. PFOA begins by identifying the societal problem that may be addressed by the new transgenic organism, and it lays the foundation for ecological risk assessment. The options assessment steps in PFOA (see Table 2.1) are particularly useful to organize group consideration of the best available science and to share in the evaluation of perceived and potential harms for risk assessment. The risk assessment contribution is primarily based on two crucial components incorporated in the PFOA during steps 6–8 (Table 2.1). The first contribution is the formulation of the problem created by the technology (NRC, 1983, 1996). For example, identifying the problem created by, or associated with, the technology could include asking the following questions: What are the system impacts of the technology? Who will be impacted? What priority do stakeholders place on the adverse effects? This is the identification of potential hazards or harm[3] by multiple stakeholders based on existing scientific information and diverse value systems. The ability of PFOA to accurately express how society evaluates the adverse effects is influenced by how representative the selected stakeholders are. This type of problem formulation, one that incorporates multiple stakeholders, values and sources of information, informs the entire environmental risk assessment (see Chapter 1, this volume, Fig. 1.4).

The second contribution of PFOA to risk assessment is the identification of potential options/technologies as future alternatives. The proposed action, in this case use of a transgenic fish, is never the only possible option. Risk assessment depends entirely on an appropriate specification of future alternatives (including taking no action), so that comparative risk can be assessed and the appropriate controls for risk assessment science can be defined and used (Andow and Hilbeck, 2004). Assuming that not everyone is equally well informed, part of this process will be iterative, with ongoing learning and capacity building. As a result, the option attributes, system characteristics and problem formulation will be revisited and modified.

PFOA contributes to risk management by identifying possible technology or system changes necessary for the options (i.e. future alternatives) to be successful, in this case minimizing hazard and harm. Finally, rather than conceptualizing risk communication as the final stage in a linear process, the PFOA frames risk communication at multiple points in the risk assessment. For example, in the staging of PFOA, the first meeting involves a transparent communication of the problem and initial risk assessment framing prior to the full scientific assessment. A second meeting provides transparent communication of the comparison of alternative technologies, their potential harms and information for

[3] Harm is defined as the social construction of, and shared agreement on, potential loss based on social values (Nelson and Banker, 2007). See Chapter 1 for definitions of risk and hazard.

risk management design, allowing discussion to transition from risk assessment to management. This meeting can also provide information on risk communication strategies, public education and future participation opportunities.

Timing and staging considerations for PFOA

An important decision in PFOA design is the consideration of when it occurs during an environmental risk assessment and how many meetings are necessary. In Brazil and Vietnam, workshop participants decided that two meetings were the minimum number for environmental risk assessment, and more would be necessary if the PFOA involved post-release monitoring (Capalbo *et al.*, 2006; Nguyen van Uyen *et al.*, Wallingford, UK, 2007). In these countries, the first meeting would occur after the initial petition for testing of a transgenic organism, but prior to laboratory testing and field testing, to formulate the problem addressed by the technology and identify options. The second meeting would occur to deliberate on the conclusions of the scientific risk assessment and to make recommendations to the regulatory body (examples of where PFOA could fit within existing country aquaculture-related biosafety regulatory frameworks are presented later in this chapter). Decisions about the number and placement of meetings during the environmental risk assessment have to balance the PFOA's contribution to the risk assessment with the efficiency of managing the process. Strategically placed meetings encourage an iterative process between the PFOA and other components of the risk assessment, which is a necessary characteristic of the methodology. Yet, too many meetings could become unmanageable, resulting in problems for maintaining continuity in participants over time and excessive costs in terms of stakeholder and decision maker time.

Regulatory decisions about biosafety will most likely be made on a national basis, except in instances where a country has ceded its regulatory authority to a regional economic integration body. In such cases, however, many of the elements needed to conduct a PFOA, including scientific information on the characteristics of a technology, alternatives and potential risks, may be adapted from elsewhere for use in deliberations at a national level. Another possibility is that countries within a region may choose to conduct an assessment jointly, providing a basis for separate decisions at the national level.

Advantages and Challenges of the PFOA Methodology

One of the main advantages of PFOA is that it provides a systematic way to integrate scientific evidence and other public interests to inform societal decision making. This distinguishes it from expert-driven review processes that rely exclusively on scientific expertise, as well as from open-ended public debate in which participants are often poorly informed about the state of the science. As a structured, deliberative process, PFOA allows stakeholders with different views to learn about the current state of scientific information, hear each other's ideas, identify points where they agree and disagree, and understand the rationale behind each other's perspectives. Unlike public debates where the positions of advocates and opponents of a technology are often simply reinforced, participants in a PFOA often adapt their views as their understanding of the issue deepens. This makes it less likely that any one perspective will dominate the deliberation.

Other advantages relate to the PFOA's emphasis on transparency and consideration of multiple options. By explicitly examining different options in addition to the technology being considered for adoption, the PFOA encourages participants to think in terms of the most appropriate way to meet an identified societal need. This enables stakeholders to participate in a process that determines whether a transgenic fish can safely address a particular problem. Both risks and benefits of future technology options are considered together. Because the process is transparent, it provides a way for interested parties to understand not only the conclusions or recommendations of a PFOA, but also the options that were considered, the supporting information and the steps taken to reach the recommendations. Because information used in PFOA needs to be presented in terms that are broadly understood, it is also relatively easy to link PFOA to a broader education effort to help the public understand the significance of risk assessment and biosafety.

Finally, PFOA offers potential advantages in terms of good governance – helping decision makers make wise decisions in the best interest of society and to make them efficiently. A good PFOA may support different ministries in agreeing on a common decision regarding a future technology. Although the process requires an upfront investment of time and effort, it ultimately may save time by allowing a regulatory body to make science-based decisions or avert extended public conflict. It may also lead to the identification of other priorities to guide technological innovation (e.g. focus research on the development of disease- or cold-resistance traits in selected fish strains, rather than on increased growth rates). An effective PFOA in one country can also serve as a resource for other countries in the region because both the information and the process are clearly documented for others to see.

The decision to adopt a PFOA process may present planning and capacity challenges in many countries. Introducing PFOA does not result in automatic buy-in from participants. For example, some participants may be resistant to change because it is unknown, threatens the benefits from current practice or appears to be too much trouble for uncertain benefit. It is unlikely that each participant will have the same level of institutional or organizational power, presenting situations of unequal power within meetings. Disruptive participants, who lack sincerity in their deliberation, may use the meetings to create extreme conflict and block action. Operating rules and tools for assessing and managing potentially disruptive behaviour will have to be developed so that everyone can have voice and influence. Representatives must express interest in the deliberative process in preliminary interviews and agree upon ground rules of civil behaviour, with the final recourse of expulsion if the group agrees that the representative is not deliberating in good faith.

Other concerns of regulators and stakeholders may include: Will the current regulatory body accept PFOA as part of its environmental risk assessment processes? If so, who should convene the PFOA so that it has appropriate legitimacy? How open should the PFOA be during the deliberations? What capacities are required to manage the process? Is there adequate capacity to ensure quality scientific and risk assessment information in the process? How much will it cost, and who will fund it? How long will the PFOA process take – and could conducting it mean missing the 'window of opportunity' for adopting a technology? These are all valid questions and each is addressed in later sections of this chapter.

Presentation of Case Study Regulatory Systems and PFOA

During this book's writing workshop, the authors, with their knowledge about diverse countries, evaluated the PFOA methodology and considered its possible contribution to their respective biosafety regulatory systems. They found a diversity of regulatory systems, with some countries being parties to the Cartagena Protocol on Biosafety (CPB) and some not. In several cases, countries had recently completed an extensive capacity-building process to establish an NBF. In other cases, national leaders were just beginning to think about the appropriate laws and regulatory systems for biosafety. How PFOA is used and where it is placed in relation to risk assessment will depend on a country's governance philosophy, evolving social contracts between the state and civil society, as well as current regulatory and legal systems. The following section is a summary of the reflections by the authors about their respective regulatory systems applicable to transgenic fish and an example of how PFOA could fit within each one. The country-specific cases include Chile, Cuba, Thailand and China (Figs 2.1–2.4). It is important to note that everyone agreed that in each country PFOA would be associated with environmental risk assessment; its proposed best placement may be different in each case because every country has a distinct regulatory and governance system.

Chile's biosafety framework

In June 2002, the National Commission for the Development of Biotechnology was created (Supreme Decree No. 164), and a year later, it was presented the National Policy for the Development of Biotechnology, setting objectives, guidelines and proposals for enterprise development, human resources formation, development of technological scientific capacities, regulatory and institutional frameworks, public participation and transparency. In August 2006, the General Law of Fishing and Aquaculture No. 18,892, was modified to regulate the import or culture of aquatic species that have been genetically modified, organisms that have been altered in non-natural ways through artificial crosses or natural recombination (Law No. 20,116). Under this legislation, the Undersecretary of Fisheries (SUBPESCA), as part of the Ministry of Economy, has the authority to approve the entrance of GMOs, based on an animal health study which includes an environmental impact assessment (Law No. 18,892, art. 12), evaluating whether the species are disease-free and do not represent

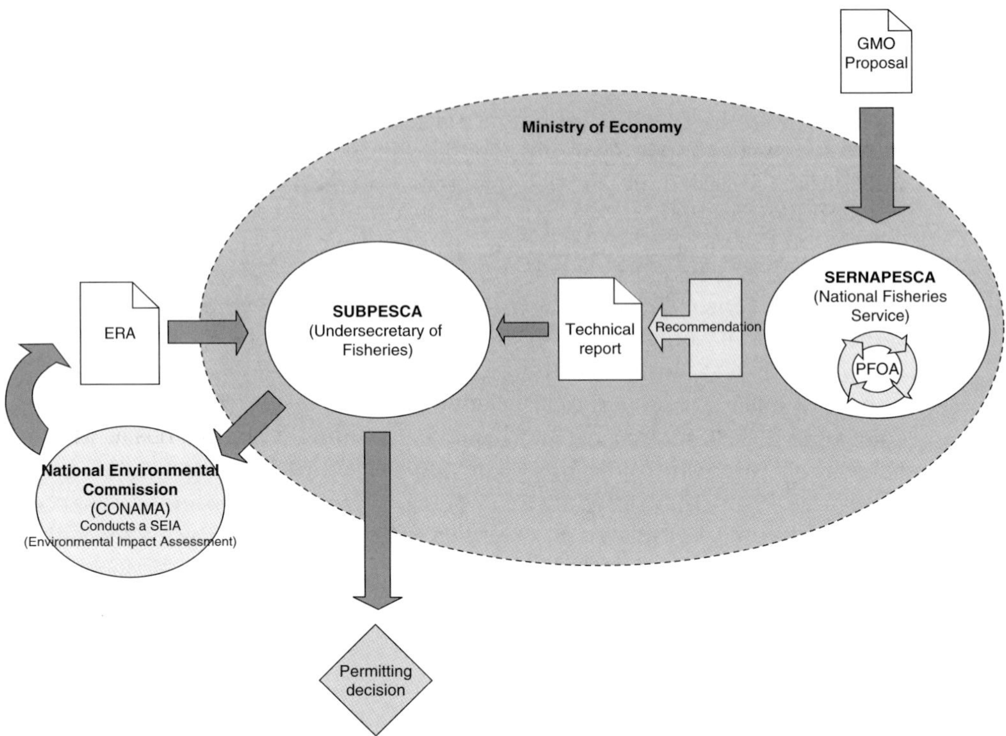

Fig. 2.1. Chile's Fisheries and Marine Regulatory System and possible location for Problem Formulation and Options Assessment (PFOA).

a risk of environmental impact. Future legislation will establish the procedures required to import, culture and market GMOs in order to avoid their propagation in the natural environment. The new regulation (Law No. 19,300) will also establish: (i) a roster of people involved in GMO activities; (ii) an accreditation system for the origin of the products; (iii) guarantees to ensure the repair of possible environmental damages; and (iv) penalties and fines regulations. In addition, a modification of Law No. 19,300 has been proposed. It would make obligatory the evaluation of the environmental impact of activities or projects that have the possibility of releasing GMOs or transgenic organisms into the environment, and identification of protected areas for 'clean' production, both organic and natural. This proposal is, as of April 2007, at the first constitutional proceeding level.

Based on the new text in Law (No. 20,116), the SUBPESCA, within the Ministry of Economy, is the institutional authority responsible for approval of GMO projects, on a case-by-case basis (Fig. 2.1). Although the SUBPESCA is the decision-making authority for aquaculture projects, the National Fisheries Service (SERNAPESCA, or Servicio Nacional de Pesca), also within the Ministry of Economy, is responsible for reviewing any proposal in their respective regional offices[4] (Fig. 2.1). In addition, when an aquaculture proposal is pre-approved by the SUBPESCA, it must be reviewed by the Environmental Impact Assessment System (SEIA, or Sistema de Evaluación de Impacto Ambiental) within the National Environmental Commission (CONAMA). As a component of the aquaculture project review, the PFOA methodology would provide an organized consideration of Chile's future options. At the level of the SERNAPESCA regional office, PFOA could provide recommendations to be included in the technical report for decision makers, assisting them in reaching a safe decision for any aquaculture proposal, regardless of whether transgenics are involved. Since the SERNAPESCA evaluates the risk of the proposed transgenic species, it should convene the PFOA at a regional level. Examples of stakeholders and agencies that could be consulted include CONAMA, SERNAPESCA, private industry, scientific associations, academics and marine authorities, depending on the specific case. In summary, PFOA has as its foundation the National Policy for the Development of Biotechnology and the Biosafety Law. The institutions involved in the review of an aquaculture proposal include SUBPESCA, SERNAPESCA (who would convene the PFOA) and CONAMA, as well as the Marine Undersecretary Commission for aquaculture projects in marine areas.

Cuba's biosafety framework

Cuba has two levels of environmental risk assessment. The first is a review by the biosafety staff of the Cuban agency proposing the biotechnology, such as a research institution. The second is the National Center for Biological Safety (NCBS) (Fig. 2.2). The Authorization Department within the NCBS evaluates the documentation prepared by the agency proposing use of a new GMO and sends their review to a group of experts from other ministries such as Health, Agriculture and Fisheries, as well as universities, among others. These experts are responsible for reviewing the risk assessment and evaluating the proposal. The risk assessment includes identifying all potential adverse effects, direct or indirect, immediate or long-term, based on known methodologies. The process also consists of the evaluation and characterization of risks, and it proposes options for risk management. In Cuba, risk evaluation norms specify that the assessment should be done in a transparent and scientific manner with the assistance of experts and the experience of international organizations associated with risk assessment. Risk assessment is done on a case-by-case basis because the nature and level of detail required can vary between transgenic organisms, depending on their previous use and the receiving environment.

The PFOA methodology could be incorporated into Cuba's current biosafety framework. It would be coordinated by the NCBS as one component of the environmental risk assessment process required by law. The expert group could review the responses to the PFOA questions as answered by stakeholders.

[4] The steps are contained in Undersecretary Resolution No. 290/1995 (more information is available at: www.supesca.cl and www.sernapesca.cl).

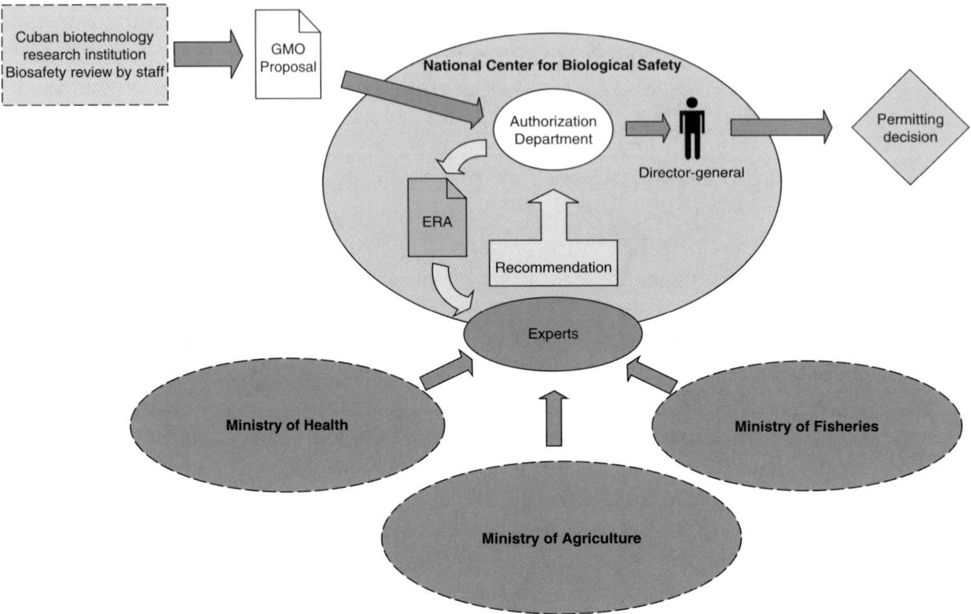

Fig. 2.2. Cuba's biosafety framework.

Based on the risk assessment results, PFOA findings and expert opinion, a final recommendation would be made to the Authorization Department and the Director of the NCBS for consideration and final decision. A Biological Safety License would be denied if: (i) adverse effects exist for human health, animal health or the environment in which a transgenic organism will be used; or (ii) the anticipated benefits do not justify the risks.

Thailand's biosafety framework

Under international agreements, Thailand is a party to the CPB. The Office of Natural Resources and Environmental Policy and Planning (ONEP), under the Ministry of Natural Resources and Environment (MONRE), manages all CPB-related matters, such as capacity-building projects for the NBF and the biosafety clearing-house (BCH). In 2001, Thailand's National Biosafety Framework (Thailand NBF) was developed by the Secretariat of the National Biosafety Committee (NBC), composed of representatives from government agencies such as the Department of Agriculture (DoA), Department of Fisheries (DoF), Department of Livestock Development (DoLD), Food and Drug Administration (FDA), Ministry of Commerce, the private sector and academia. A new NBC is to be appointed in 2007 within the new NBF, developed by ONEP and MONRE.

The first NBC (appointed in 1993) was housed within the Ministry of Science and Technology, and it was eliminated when CPB-related duties were moved to MONRE in 2002.

Guided by the previously mentioned policies, there are a range of institutions and processes that form the regulatory authority for GMOs in Thailand. At present, any research proposal for transgenic organisms that comes from a Thai institution is reviewed by the proposing institution's biosafety committee. If their Institutional Biosafety Committee (IBC) considers it appropriate, the proposal is endorsed. As of October 2006, Thailand had not allowed any transgenic aquatic organisms to be cultured for commercial purposes, as there is no existing procedure that allows the importation of transgenic aquatic organisms for release or commercial production. However, in the past few years, the DoF has received a few proposals to import transgenic fish for education and exhibition purposes. For these cases, the DoF used the existing regulatory system for alien species. When non-transgenic alien species are proposed for importation for aquarium use, they are considered as an invasive species or a disease and parasite carrier. Importers who want to import transgenic aquatic organisms would have to apply for permission from the relevant regulatory agency. The application, with information in accordance with Annex 1 of CPB, would be sent to the appropriate regulator's IBC for consideration.

The PFOA methodology would facilitate application and decision making for both importation and in-house research and development (Fig. 2.3). For importation, PFOA would help prioritize the environmental concerns to be evaluated after the proposal is submitted to the competent national authority. The science-based, environmental risk assessment process is conducted by an expert panel, and recommendations are submitted to the competent national authority for a final decision. For fisheries or aquatic animals, the relevant national authority is the DoF. At this stage, the competent national authority could also use a PFOA methodology to inform their decision-making process. For example, one or two meetings among stakeholders could be held before the final decision is made.

For research and development of transgenic organisms within the country, PFOA could be used for the entire application and approval process, starting from the design of transgenic fish and relevant safety evaluation to the final decision-making process. The research proposals would be submitted for approval to the IBC of the research institution. Once applications for commercial approval are made, they would go through the ERA process and be informed by the PFOA process, which would be conducted by the DoF and their IBC. Possible stakeholder representatives in Thailand include representatives from ministries, experts, academic institutions, fish farmer associations, conservation-oriented non-governmental organizations (NGOs) and industrial sectors, as well as the media and representatives from local administrative agencies. Special attention should be paid to timing, capacity building and information for the PFOA. The timing for the ERA should be specified in order to make the PFOA transparent. For example, there could be at least one stakeholder meeting within 3 months after receiving the application and prior to laboratory tests. After laboratory and field trials, at least two stakeholder meetings could be conducted within 1 year, i.e. before a final

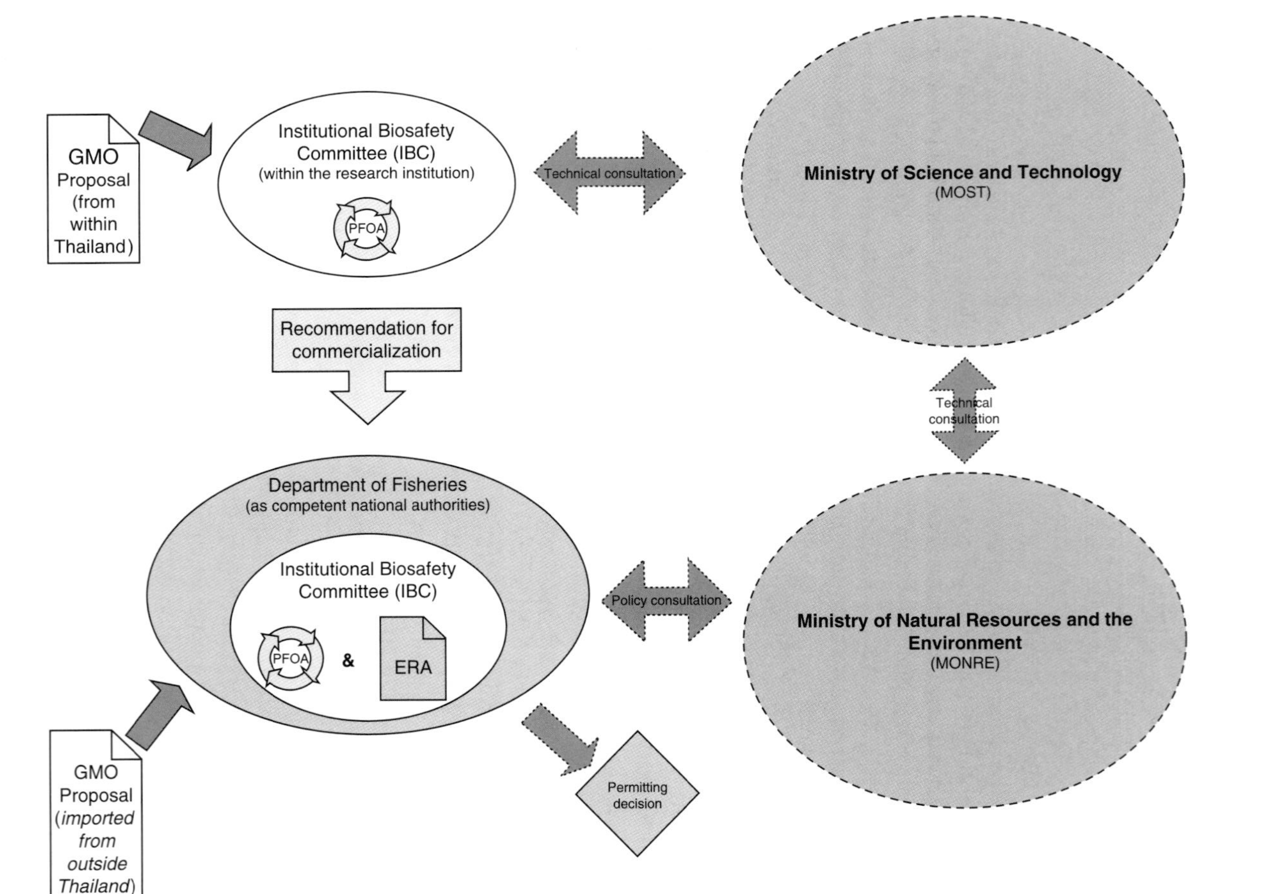

Fig. 2.3. Thailand's biosafety regulations and possible positions for PFOA in the case of a proposal from a research institute or an importer.

decision is made. To build capacity for biosafety and the PFOA, training is needed in the field of biosafety framework development. First, there should be training for both NBC and IBC members to have a common understanding of biosafety issues. Second, training is needed in the field of environmental risk assessment for the risk assessors, risk communicators and risk managers. Third, capacity building is necessary for biosafety research on environmental risk. Finally, it will be particularly important to establish access to the BCH within all competent national authorities and to strengthen all existing BCHs (UNEP, 2006).

China's biosafety framework

As a nation with advanced agriculture and aquaculture industries, China produces its own transgenic organisms. As of October 2006, China had field trials for transgenic fish under quarantine conditions, but there was no commercial production yet. The biosafety regulatory system for transgenic fish, from laboratory use to commercial production, is in place within the Ministry of Agriculture (MoA). To seek approval for a GMO, a national research institute or company must send a proposal for use of a transgenic organism to the National Committee on Biosafety in Agriculture (NCBA), located within the MoA. The NCBA is both the decision maker and risk assessor for transgenic crops, terrestrial animals and aquatic organisms. The Aquatic Organism Expert Group of the NCBA will: (i) evaluate all transgenic fish proposed for use in laboratories, field tests and commercial production in China; and (ii) perform the risk assessment/evaluation. Scientists and a few administrators comprise the Aquatic Organism Expert Group. During the risk assessment process, the Aquatic Organism Expert Group provides their recommendation to the NCBA for a vote on whether to approve the use of the transgenic organism for laboratory use, field testing or commercial production. No transgenic fish have been released for commercial production, so little is known by the authors about monitoring of aquatic transgenic organisms in China. Biosafety monitoring for release is urgently needed once current field testing is completed.

At this time, there is no process similar to PFOA within China's regulatory system for transgenic organisms. However, it would be useful, and there are two main ministries with which PFOA may interact (see Fig. 2.4): the MoA and the Ministry of Science and Technology (MOST). Within the MoA, there is an NCBA, and within this body there are the following groups: Crop Expert Group, Livestock Animal Expert Group and Aquatic Organism Expert Group. The authors see three possible options for convening a PFOA: (A) the Aquatic Organism Expert Group of NCBA, housed within the MoA; (B) within the National Fishery Technical Extension Center (NFTEC) and convened by the MoA; and (C) within the Chinese Academy of Fisheries Science (CAFS) and convened by the MoA.

Although not involved in the environmental risk assessment process, the MOST and the Committees of National Natural Science Foundation (NSFC) do direct and fund national-level research. The PFOA could impact the MoA, MOST and NSFC because the information generated by PFOA could affect

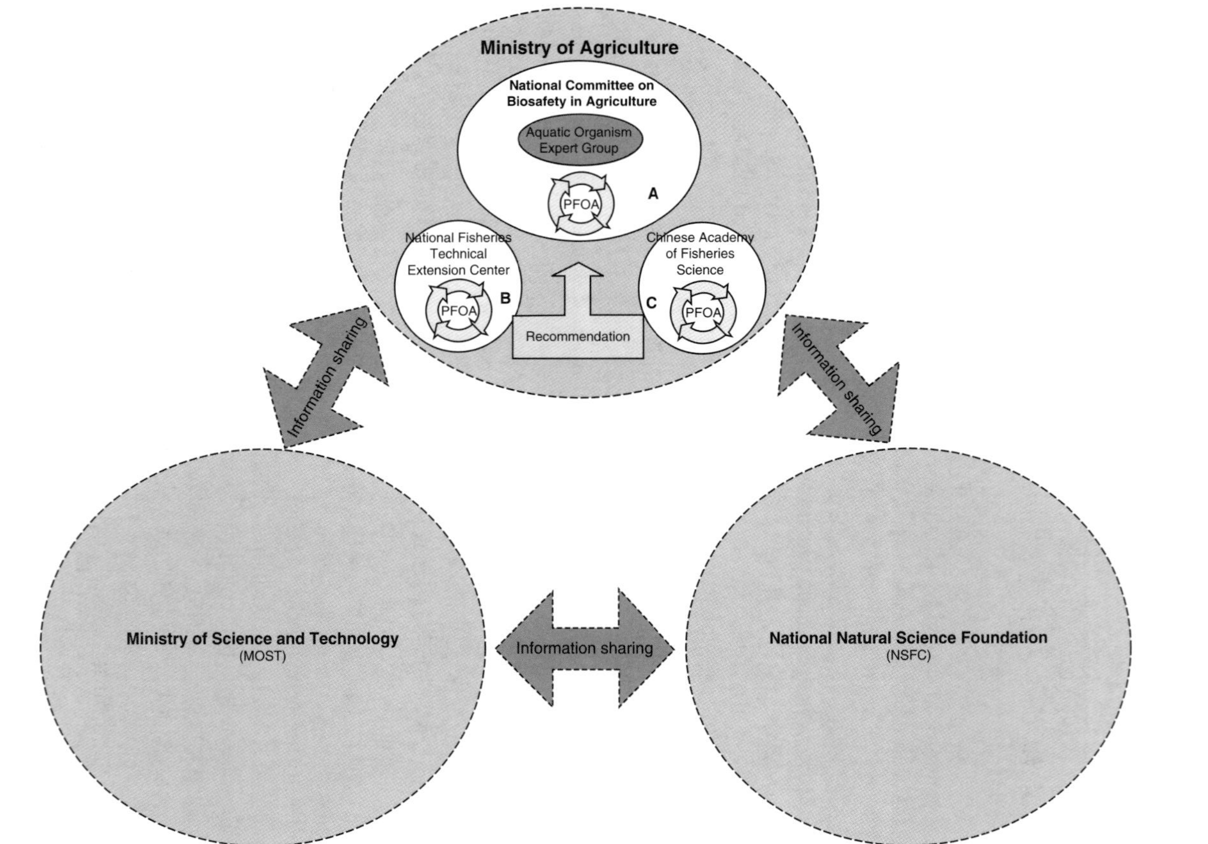

Fig. 2.4. Proposed role and placement for PFOA in China's national biosafety regulatory structure for aquatic organisms. There are three potential places PFOA could be placed: (A) the Aquatic Organism Expert Group of National Committee on Biosafety convened by the Ministry of Agriculture; (B) in the National Fishery Technical Extension Center and convened by Ministry of Agriculture; and (C) in the Chinese Academy of Fisheries Science and convened by the Ministry of Agriculture.

both research agendas and the NCBA's decision regarding the use of the technology. Although MOST is not involved with environmental risk assessment in China, it would be an excellent place to utilize the information and recommendations from PFOA and thus guide decisions about research spending. Therefore, PFOA could share information with the MOST and NSFC and provide advice and recommendations to the MoA, specifically at the level of the Aquatic GMO group of the NCBA.

Questions to Address when Designing a Country-specific PFOA

Application of the PFOA methodology to environmental risk assessment of transgenic fish is limited because the commercialization of transgenic fish for food production has only recently begun to be considered by governments. Yet, much can be learned from the applications of PFOA in other arenas (Nelson and Banker, 2007), from related experiences in environmental risk assessment and public deliberation and from the authors' collective expertise about the regulatory frameworks in a range of developing countries. Questions that teams are likely to confront when designing or implementing PFOA are presented below, along with a discussion of the ways to answer them.

Who will manage PFOA and who will pay for it?

Who will convene, manage and pay for PFOA is a critical issue, and it will depend on the individual country context. Any PFOA for transgenic organisms will be initiated by the request or suggestion that a particular transgenic fish would be a beneficial alternative to current practices in a particular aquaculture system. One option is for payment to be borne by the party with the greatest financial means or by the national government. The proposal may come from research institutions, relevant research universities, fish exporters/traders or from the developers and producers of a transgenic fish. Often, it is recommended that the party proposing the technology not be the institution responsible for convening and managing the process; this removes any appearance of bias. Techniques can be used to identify power differences and, at a minimum, monitor the deliberations so that agencies and parties with more power do not unduly influence outcomes (Susskind *et al.*, 1999). The basic principles of a PFOA process stipulate that it be convened by a legitimate authority (and therefore be transparent) to reduce political interference. For example, in Thailand, one possibility is that if the proposal comes from an importer, PFOA will be convened and managed by a government institution, but paid for by the importer and the producer of the transgenic fish. In an example from Cuba, the agency that would be the most appropriate for PFOA is the NCBS, while in Chile PFOA could be hosted by the SERNAPESCA within the Ministry of Economy. Finally, in an example from China – a nation that produces its own transgenic organisms – PFOA could be conducted by the NFTEC, CAFS or NCBA; all of these institutions are within the MoA.

Determining the PFOA stakeholder participants and observers

An essential element of the PFOA process is the involvement of a broad spectrum of stakeholders who are all allowed to contribute in the deliberative process. The identification and selection of the relevant stakeholders for the multi-stakeholder dialogue is particularly important for maintaining the public legitimacy of the proposed PFOA process. All interested and affected stakeholders, both powerful and marginalized, need to be included in the deliberation process. As shown in the PFOA workshops regarding transgenic crops in Kenya, Brazil and Vietnam (Nelson *et al.*, 2004; Capalbo *et al.*, 2006; Nguyen van Uyen *et al.*, Wallingford, UK, 2007), designers of the PFOA process need to carefully address:

- How to guarantee independence
- How to guarantee that conflicting interests will be represented
- How to identify relevant stakeholders
- How to assure legitimacy of selected representatives
- How to prepare participants with information

Determining who the stakeholders are and understanding their perspectives is the only way to develop an effective PFOA process. A good start to identifying relevant stakeholders to get involved in the multi-stakeholder dialogue on transgenic fish technology is to ask questions such as:

- Who can provide knowledge or information about transgenic fish?
- Who makes decisions on the use of transgenic fish?
- Who is most likely to be affected by transgenic fish?
- Who is troubled about transgenic fish?
- Who has a stake in the successful adoption of transgenic fish?
- Who can disseminate information regarding transgenic fish?

The key to achieving a balance between the different interests of the various stakeholders is to be fair in the method used to select stakeholder representatives. Since PFOA is a science-guided, transparent learning process involving multiple sectors and disciplines (Forester, 1999; Susskind *et al.*, 2000; Hemmati, 2002), a cross section of society should participate in, or observe, the process. As they identify potential participants, conveners should be vigilant against the easy path of 'appointing friends' or 'the loudest voices' as a quick way to organize PFOA.

Which stakeholders should participate and what role can they play?

Various stakeholders will have different roles to play in the deliberative process. Their diverse contributions should be made during an open discussion free from prejudgement. In this manner, they can contribute to a rational, science-guided planning process by which countries can assess whether production of a transgenic fish can address a priority societal problem (i.e. an unmet need that requires change). If so, they can help determine whether it can be safely produced by contributing to the environmental risk assessment of the transgenic organism and by providing a comparison of multiple options (e.g. growth hormone-enhanced tilapia, genetically improved farmed tilapia (GIFT) and domesticated tilapia). In doing so, they can help design policies and regulations to reduce risks and enhance the benefits provided by the options. A wide variety of relevant stakeholder representatives could join PFOA on transgenic fish, and they include the following:

Scientists or developers of transgenic fish
Individuals responsible for developing transgenic fish can contribute their scientific knowledge and skills to the PFOA discussion. Many of the questions raised by other stakeholders will require relevant knowledge and scientific information about transgenic fish. If the data are not yet available or are insufficient, knowledge about scientific methodologies and tools will be necessary for planning future research necessary to answer such questions.

Risk assessment specialists
These could include geneticists, ecologists and conservation biologists, who could raise important ecological issues or provide scientific information on the potential adverse effects of the transgenic fish on the environment and the ecosystem (see Chapters 5 and 6, this volume, for more detailed discussion).

Hatchery operators or seed producers
Stakeholders from this sector could raise issues about how transgenic eggs and juvenile fish would affect local seed production and availability in terms of market competition. They may raise concerns about what types of physical and biological confinement measures are available and their ease of implementation if they opt to produce transgenic fish (Chapter 8, this volume). They could also raise questions about the uncertainties involved with, and costs versus benefits of, producing transgenic fish.

Grow-out operators
Representatives from this sector could provide information on existing farming systems using non-transgenic fish. Their likely concerns or questions would revolve around how they would need to modify or change their current farming systems to comply with the required level of confinement for using the transgenic fish in aquaculture facilities (see Chapter 8, this volume). Complete physical confinement may be feasible in land-based facilities such as tanks and cages, but not in open systems (cages and pens) as commonly used in Thailand and China. They could also raise the question about how to get the best economic return from using transgenic fish given the additional costs (e.g. implementing more stringent confinement).

Small-scale farmers
The main concern of small-scale farmers will likely be whether it is economically feasible for them to adopt transgenic fish in extensive farming systems and whether

there will be added production costs, such as transgenic fish fry (which may cost more than currently used fry), fertilizers, artificial feeds and confinement facilities.

Importers, trade organizations and industry
Representatives from this sector will raise issues of market share and profitability. They may question how transgenic fish could affect their existing trade with developed countries, especially with regard to the acceptability of transgenic fish to consumers in their target markets. They may ask about which economic sector would benefit the most from transgenic fish production.

Environmental and consumer groups
These stakeholders likely will raise health and environmental issues regarding transgenic fish. Environmentalists may want to understand the interaction between the new technology, native species and ecosystems. Consumer groups may want to discuss potential impacts of transgenic fish on human health, household nutrition, labelling or food prices.

Indigenous or ethnic minorities
Such stakeholders may raise the question of whether transgenic fish could affect fish that are socially, culturally or economically important in people's lives or affect their communities in other ways.

Religious groups
These stakeholders may have ethical concerns about the creation and consumption of transgenic fish. PFOA would be a good venue for them to learn the science behind transgenic fish from involved scientists, and for scientists to better understand religious concerns.

Involvement of some combination of the various representatives mentioned above will be necessary to create a transparent, science-guided and equitable process for PFOA in any country. It is also important to note that not all participants may be stakeholders in the discussion; some may simply be observers of PFOA. These include the following groups: *teachers, public educators and extension agents* could play secondary roles in PFOA. They could observe the deliberation process and consider how to incorporate the findings of the PFOA and transmit their understanding of transgenic fish to their students or the general public. *Lawyers and general policy makers* could be either stakeholders or observers and provide the group with information on existing regulation and national policies that deal with transgenic fish. They could also provide information on existing economic and diplomatic relationships with other countries that would affect trade and biosafety governance related to transgenic fish. *Biosafety decision makers* may want to be involved in, or observe, PFOA (e.g. NBC representatives). In most countries with established NBCs, the NBC is either an approving or endorsing body for transgenic organisms and will be the most likely recipient of the PFOA recommendations. It will be helpful for them to learn about the process of PFOA by observing it in action, leading to a better understanding of the deliberation and reasoning informing the recommendations generated by PFOA. This understanding could facilitate their decision making about transgenic fish regulation and biosafety. In a country with a free press, a unique category of observers could include *media representatives*. Although they may have their own interests and constraints, the media serves as an important vehicle for public debate and communication. Information provided to the media should be science-based, appropriate, easy to understand and timely. In addition, it is important to consider how reporters may perceive, understand and interpret what they hear and see during PFOA in order to avoid sensationalizing issues.

Ensuring that stakeholders' views are considered and incorporated throughout the PFOA process

It is important to ensure that the views and concerns of interested and affected parties are incorporated into the PFOA process. Documented experiences of multi-stakeholder processes (Susskind *et al.*, 1999; Nelson *et al.*, 2004; Capalbo *et al.*, 2006) can be used as a basis for the deliberative process for transgenic fish, and this process should be guided by scientific data whenever possible. When communication of information is open and stakeholder participation is encouraged, multiple stakeholders may more easily express their views. PFOA uses a deliberative process involving discussion, reflection and persuasion to communicate and work towards the greatest possible consensus among multiple stakeholders. Consensus-building is becoming a popular democratic form of decision making (O'Brien, 2000; Wondolleck and Yaffee, 2000; Irwin, 2001; McLean *et al.*, 2002; Glover *et al.*, 2003). A chairperson or a team leader who is respected, credible and highly skilled in his or her field should lead the PFOA as the chair or convener. Stakeholders should view this person as fair, competent and not beholden to any interest group. The chair would introduce and endorse the facilitator, an individual skilled in facilitating problem solving and developing consensus among disparate interests. The facilitator must be trusted, as this person is responsible for guaranteeing transparency in the deliberation process. This facilitator could either be selected by the stakeholders from among themselves or from a different sector of society with a neutral interest regarding transgenic fish.

Managing the information, the PFOA process and the decision

Information management is essential for a transparent and informed deliberation during a PFOA. In many countries, there is limited scientific expertise available to provide information and support to conduct a PFOA. Although assistance is available to enable countries to comply with their obligations under the Cartagena Protocol, the cost of obtaining that expertise from outside may be unacceptably high. The technology under consideration or the environmental conditions under which the fish would be farmed may be very similar from one country to another, particularly within the same geographic region. For these reasons, conveners and managers of a PFOA should seek efficiencies of scale by drawing on the best available regional information and expertise. This would ensure that data collection or analyses are not replicated unnecessarily. Before the PFOA, a database of existing studies on transgenic fish should be organized and integrated. Existing international and national information, scientific studies and data can serve as a resource or reference. In many cases, countries may need to use regional information if country-specific studies have not been conducted. Over time, the PFOA will contribute to a 'community of practice' (Wegner and Snyder, 2000), establishing regional and international databases in which reports, experiences and processes are shared with all.

During the PFOA, a facilitator and a recorder should work in tandem with the stakeholders to create a visual record of the points while they are being discussed. This helps the group retain a visual memory of the key points as the discussion evolves and serves as a contribution to a written permanent record of the discussion summary and decisions. The entire group will contribute ideas and data in the assessment of transgenic fish, and such a record will help ensure that each participant's opinion has been considered. Real-time recording can be done via computers and projected on screens or written on large sheets of paper for everyone to see. Sophisticated computer equipment is not necessary to facilitate this process. In addition, the stakeholder group also may use drawings, illustrations or maps to help people recall what they have discussed. In order to maintain transparency and legitimacy, it is important to have an ongoing visual representation of what the group has discussed and agreed upon. These records ultimately need to be summarized and reviewed by all participants. Finally, a report is developed for the appropriate decision-making authority and signed by all the stakeholders in the PFOA process. If consensus is not achieved and the group develops a report based on a strong majority (75% or more of the participants agree), the group should make every effort to include a presentation of the diversity of opinions. Any minority opinion should be documented and explained.

Maintaining the legitimacy of the PFOA process

To conduct a successful PFOA, the responsible institution must have the capacity and the ability to maintain legitimacy with both the public and a formal political body. Science can contribute to the legitimacy of the process if it is used to inform decisions and shared openly with both the public and political bodies. Public legitimacy is largely dependent upon public opinion, and it is obtained differently in every culture (e.g. legitimacy is gained differently in China than it is in Thailand). However, the characteristics of the deliberative PFOA process should strengthen its legitimacy no matter which country. It is essential that the PFOA be transparent, equitable, representative and data-guided. Although each society defines these characteristics relative to their own norms and values, each characteristic's intrinsic value strengthens legitimacy regardless of who is conducting the PFOA. For example, each country must examine who their stakeholders are and how they should be incorporated into the process.

Public legitimacy through the inclusion of key stakeholders must be matched by political legitimacy through recognition of PFOA by the traditional decision-making authorities and governing bodies. This authority must also be trusted and credible in order to strengthen legitimacy. The regulatory authority must provide a means for the assessments made during the PFOA process to inform final decision making and agency action(s) regarding biosafety. Very often, this authority already exists within regulatory systems. However, designers of the PFOA will need to evaluate which authority will receive reports from the PFOA, as well as how many times and when the PFOA will meet during the environmental risk assessment process.

How can PFOA help with problems of uncertainty in risk assessment and societal discussion?

Uncertainty about a transgenic organism's effects on genetic diversity, other ecological impacts, changed genetic makeup, economic impacts, etc., is one of the greatest challenges for environmental risk assessment and biosafety (see also Chapter 7, this volume). The PFOA methodology can help identify and address societal uncertainty surrounding transgenic organisms. Societal uncertainties can stem from the lack of understanding or misinformation about the relevant science by members of society or about the relevant sectors of society by scientists. Uncertainty can be due to both scientific misinformation and misinformation about personal and social values. By presenting the best available scientific information in a timely fashion to all stakeholders, PFOA can help reduce some sources of uncertainty. Through group discussions, PFOA allows diverse stakeholders to reduce the misinformation and misinterpretation associated with conflict-ridden issues. It allows for clarification of their own values and provides an opportunity to listen to others, leading to an understanding of where stakeholders share common values and language and where they differ. It informs scientists about social systems and the construction of acceptable risk. It also allows scientists to hear the non-scientific concerns about the limits of scientific knowledge and thereby allows a richer description of potential incertitude in the risk assessment.

Developing capacities to conduct PFOA

The Cartagena Protocol mandates that in order to protect biodiversity in signatory countries, governments must have the capacity to develop country-specific biosafety laws and regulations. The PFOA methodology is one component of a rigorous process of environmental risk assessment that is necessary to support the creation of science-based and socially acceptable biosafety policies. Countries are encouraged to prioritize how they will begin the implementation of the PFOA methodology, as well as capacity-building needs. It is easy to imagine a PFOA that produces exhaustive, data-rich modelling reports based on multiple meetings by numerous stakeholders, but this may not be necessary, and it is certainly not where a country should start. Risk assessments of a new technology often evolve from being more qualitative to being more quantitative, as key gaps in information and methodologies are filled (see Chapter 1, this volume). Countries might start with using PFOA in a qualitative assessment and move later to using PFOA in more quantitative risk assessments. In the beginning, qualitative techniques can be used, based on the best available science. Techniques used during PFOA can become more sophisticated as new tools are developed for information sharing and for comparing options and as communication strategies (from web sites to Internet conferences) become more affordable.

Capacity building for PFOA can be relatively cost-effective. However, leaders must be dedicated to training regulators, scientists and stakeholders alike, to participating in a mutual experience as a learning cycle, as well as to using the PFOA products. Without preliminary training, participants may resort to more politically based, positional discussion. If leaders do not provide legitimacy for the outcomes of PFOA by using the reports to inform their biosafety decisions, the public will see PFOA as a pretence and not participate. In the beginning, workshops may be necessary to train participants in how to use the PFOA methods and how to find common agreement based on science, while allowing diversity in values. Facilitators can be trained to ensure reasonable progress and fair treatment of all participants. In time, country agencies will want to consider national databases to house information necessary to answer PFOA questions at a local scale, as well as linkages with regional and international databases.

An important step in capacity building is to assess existing knowledge and expertise for conducting risk assessment and management research in each country and corresponding responsible institutions. International teams of PFOA scientists or facilitators, from developing and developed countries, can help build capacity and training with neighbouring countries as they formulate their own PFOA methodology. This type of capacity building was demonstrated in the GMO environmental risk assessment workshops in Kenya, Brazil and Vietnam. One particular tool that can be used in PFOA capacity building is the *PFOA Handbook* (Nelson and Banker, 2007). This handbook is designed as a support tool for guiding individual countries through the construction and application of a customized PFOA. It introduces users to PFOA, guides them through techniques and resources that can assist them in formulating PFOA and makes suggestions for integrating PFOA into environmental risk assessment procedures for transgenic organisms. The Handbook's primary audience is government agencies and personnel responsible for conducting environmental risk assessments within a particular country, namely those considering or already using PFOA in their risk assessment procedures. Its secondary audience includes other previously mentioned stakeholders who may participate in the actual PFOA. In this capacity, it helps participants prepare for, and participate in, a PFOA more effectively.

Chapter Summary

Problem formulation and option assessment (PFOA) through multi-stakeholder deliberation strengthens the science of risk assessment and the integration of risk assessment findings into decision making and societal discussions about GMOs. PFOA begins with a discussion about the problem that will be addressed by proposed uses of transgenic fish in order to understand the importance of this problem from a societal perspective (Table 2.1). This places the risk assessment within a societal consideration of how technology will serve the broader goals of a nation and its people. To strengthen science informing a risk assessment, PFOA provides a question-based, iterative deliberation between stakeholders, scientists and regulators that allows an improved understanding of comparative future alternatives: options for addressing the societal problem, one of which would be the new transgenic fish technology. By comparing future alternatives, PFOA generates a shared understanding of adverse effects and system dynamics based on scientific data, as well as values and ethics. Finally, this broad deliberation within PFOA provides recommendations that bring science into the more complex societal discussion that informs decision makers.

The chapter highlights the authors' evaluation of PFOA and suggestions for how PFOA could fit within their country's respective biosafety frameworks. The authors saw both advantages and challenges in using the PFOA methodology. The principal advantage is the systematic approach to integrating science and public interests for informed decision making. It allows learning of different sources of knowledge and different points of view. By answering questions together, the deliberation avoids highly politicized debate based on predetermined positions. Another advantage is that it is transparent and accessible because it considers multiple options in terms that are broadly understood. It is possible to consider not only specific scientific findings, but also national concerns about biosafety. For example, participants can express values in the same deliberation where they discuss technological efficiency. Observers can understand the steps and discussion that precede the recommendations. As a result, PFOA contributes to good governance. The challenges for PFOA are that there may be resistance to opening up discussion to multiple stakeholders and concerns about how power imbalances could disrupt deliberation. In order for PFOA to be viewed as legitimate, regulators will have to accept PFOA as part of a risk assessment at a government level and have the capacity to conduct it along with all the other demands for scientific research. Agencies will have to give PFOA sufficient priority in risk assessment in order to allocate sufficient time and resources to the process.

Globally, there is a diversity of regulatory systems applicable to transgenic fish. The authors present four examples of how PFOA could fit, focusing on country-specific cases for Chile, Cuba, Thailand and China (Figs 2.1–2.4). It is important to note that although everyone agreed that PFOA would be associated with environmental risk assessment, its proposed best placement will be case-specific. In Chile, PFOA could be conducted at the level of the SERNAPESCA regional office and provide recommendations to be included in the technical report for the decision-making body (the SUBPESCA). In Cuba, PFOA could be incorporated into the current biosafety framework coordinated by the NCBS. In Thailand, the Department of Fisheries' Institutional Biosafety Committee could conduct PFOA when it received an application for import or commercialization of a transgenic fish. Finally, in China, there are three possibilities for convening PFOA: (A) the Aquatic Organism Expert Group of the National Committee on Biosafety housed within the MoA; (B) within the NAFTC and convened by the MoA; or (C) within the CAFS and convened by the MoA.

To support anyone implementing PFOA, the authors discussed preliminary questions that should be considered, ranging from initial design questions to how the PFOA recommendations can be incorporated into a traditional regulatory system. Specific questions include:

- Who will manage PFOA, and who will pay for it?
- How will we determine who should participate?
- Which stakeholders should participate, and what role can they play?
- How can we ensure that stakeholders' views are considered and incorporated?
- How will we manage information, process and final recommendation?
- How can we maintain the legitimacy of PFOA?
- How can PFOA help with problems of uncertainty in risk assessment and societal discussion?
- How can we develop capacities necessary to conduct PFOA?

PFOA is one essential component of the package of environmental risk assessment tools for transgenic fish. Countries around the world are considering how to respond to the challenges of promoting biosafety, and some have dedicated themselves to creating a comprehensive NBF. The authors encourage country regulators to incorporate a PFOA methodology within their environmental risk assessment process.

References

Andow, D.A. and Hilbeck, A. (2004) Science-based risk assessment for non-target effects of transgenic crops. *BioScience* 54, 637–649.

Biggs, S. and Matsaert, H. (1999) An actor-oriented approach for strengthening research and development capabilities in natural resource systems. *Public Administration and Development* 19, 231–262.

Capalbo, D.M.F., Simon, M.F., Nodari, R.O., Valle, S., Dos Santos, R.F., Coradin, L., De O. Duarte, J., Miranda, J.E., Dias, E.P.F., Le Quag Quyen, Underwood, E. and Nelson, K.C. (2006) Consideration of problem formulation and option assessment for Bt cotton in Brazil. In: Hilbeck, A., Andow, D.A. and Fontes, E.M.G. (eds) *Environmental Risk Assessment of Genetically Modified Organisms, Volume 2: Methodologies for Assessing Bt Cotton in Brazil*. CAB International, Wallingford, UK, pp. 67–92.

CBD (2006) Convention on biological diversity. Available at: www.biodiv.org

Forester, J. (1999) *The Deliberative Practitioner: Encouraging Participatory Planning Processes*. MIT Press, Cambridge, Massachusetts.

Gibbons, M. (1999) Science's new social contract with society. *Nature* 402, C81–C84.

Glover, D., Keeley, J., McGee, R., Newell, P., Da Costa, P., Ortega, A.R., Loureiro, M. and Lin, L.L. (2003) *Public Participation in National Biosafety Frameworks: A Report for UNEP-GEF and DFID*. Institute of Development Studies, Brighton, UK.

GMO ERA Project (2006) *International Project on GMO Environmental Risk Assessment Methodologies*. Available at: www.gmo-guidelines.info

Goldstein, I. (1993) *Training in Organizations*, 3rd edn. Brooks/Cole Publishing, Pacific Grove, California.

Grimble, R. and Wellard, K. (1997) Stakeholder methodologies in natural resource management: a review of principles, contexts, experiences and opportunities. *Agricultural Systems* 55, 173–193.

Hallerman, E.M. and Kapuscinski, A.R. (1995) Incorporating risk assessment and risk management into public policies on genetically modified finfish and shellfish. *Aquaculture* 137, 9–17.

Hemmati, M. (2002) *Multi-stakeholder Processes for Governance and Sustainability: Beyond the Deadlock and Conflict*. Earthscan, London.

Irwin, A. (2001) Constructing the scientific citizen: science and democracy in the biosciences. *Public Understanding of Science* 10, 1–18.

Kessler, J. and Van Dorp, M. (1998) Structural adjustment and the environment: the need for an analytical methodology. *Ecological Economics* 27, 267–281.

Loevinsohn, M., Berdegué, J. and Guijt, I. (2002) Deepening the basis of rural resource management: learning processes and decision support. *Agricultural Systems* 73, 3–22.

McLean, M.A., Frederick, R.J., Traynor, P.L., Cohen, J.I. and Komen, J. (2002) *A Conceptual Framework for Implementing Biosafety: Linking Policy, Capacity, and Regulation*. Briefing Paper # 47. International Service for National Agricultural Research, The Hague, The Netherlands. Available at: ftp://ftp.cgiar.org/isnar/publicat/bp-47.pdf

Miller, H.I. (1994) A need to reinvent biotechnology regulation at the EPA. *Science* 266, 1815–1818.

Munson, A. (1993) Genetically manipulated organisms: international policy-making and implications. *International Affairs* 69, 497–517.

NRC (National Research Council) (1983) *Risk Assessment in the Federal Government: Managing the Process*. National Academy Press, Washington, DC.

NRC (National Research Council) (1996) *Understanding Risk: Informing Decisions in a Democratic Society*. National Academy Press, Washington, DC.

NRC (National Research Council) (2002) *Environmental Effects of Transgenic Plants*. Committee on Environmental Impacts Associated with Commercialization of Transgenic Plants. Board on Agriculture and Natural Resources: Division on Earth and Life Studies: National Research Council. National Academy Press, Washington, DC.

Nelson, K.C. and Banker, M. (2007) *Problem Formulation and Options Assessment Handbook: A Guide to the PFOA Process and How to Integrate it into Environmental Risk Assessment (ERA) of Genetically Modified Organisms (GMOs)*. GMO-ERA Project (in press) Available at: www.gmo-guidelines.info

Nelson, K.C., Kibata, G., Muhammad, L., Okuro, J.O., Muyekho, F., Odindo, M., Ely, A. and Waquil, J.M. (2004) Problem formulation and options assessment (PFOA) for genetically modified organisms: the Kenya case study. In: Hilbeck, A. and Andow, D.A. (eds) *Environmental Risk Assessment of Genetically Modified Organisms, Volume 1: A Case Study of Bt Maize in Kenya*. CAB International, Wallingford, UK, pp. 57–82.

O'Brien, M. (2000) *Making Better Environmental Decisions: An Alternative to Risk Assessment*, 3rd edn. MIT Press, Cambridge, Massachusetts.

Sagar, A., Daemmrich, A. and Ashiya, M. (2000) The tragedy of the commoners: biotechnology and its publics. *Nature Biotechnology* 18, 2–4.

Schmoldt, D. and Peterson, D. (1998) Analytical group decision making in natural resources: methodology and application. *Forest Science* 46, 62–75.

Susskind, L., McKearnan, S. and Thomas-Larmer, J. (1999) *The Consensus Building Handbook: A Comprehensive Guide to Reaching Agreement*. Sage, Thousand Oaks, California.

Susskind, L., Levy, P.F. and Thomas-Larmer, J. (2000) *Negotiating Environmental Agreements: How to Avoid Escalating Confrontation, Needless Costs, and Unnecessary Litigation*. Island Press, Washington, DC.

Traynor, P.L., Fredrick, R.J. and Koch, M. (2002) *Biosafety and Risk Assessment in Agricultural Biotechnology: A Workbook for Technical Training*. The Agricultural Biotechnology Support Project. Institute of International Agriculture, Michigan State University, East Lansing, Michigan.

UNEP (2006) United Nations Environment Programme-Global Environment Facility Biosafety Projects. Available at: www.unep.ch/biosafety

Wegner, E. and Snyder, W. (2000) Communities of practice: the organizational frontier. *Harvard Business Review* January/February, 139–145.

Wondolleck, J.M. and Yaffee, S.L. (2000) *Making Collaboration Work: Lessons from Innovation in Natural Resource Management*. Island Press, Washington, DC.

3 Development of Transgenic Fish: Scientific Background

Y.K. NAM, N. MACLEAN, C. FU, T.J. PANDIAN AND M.R.R. EGUIA

Introduction

For thousands of years, people have sought to improve livestock breeds by selective breeding. Within the last century, with increasing knowledge of genetics, genes and genomes, a new idea germinated in the minds of animal breeders: might it be possible to modify and improve existing livestock one gene at a time? This could help reduce the likelihood of unforeseen negative genetic effects from selective breeding and focus efforts directly on obtaining phenotypic parameters of primary importance in agriculture or aquaculture. Indeed, it is now possible to produce new lines of genetically modified or transgenic animals with both unique geno- and phenotypes.

Transgenic animal technology began with the work of Gordon *et al.* (1980), who transformed mouse embryos by microinjecting purified DNA into the nuclei of fertilized eggs. Palmiter *et al.* (1982) further demonstrated this technique by introducing a DNA construct, consisting of a mouse metallothionein promoter sequence spliced to a rat growth hormone (GH) coding sequence, into the nuclei of newly fertilized mouse eggs. Some of the subsequent litter showed dramatic increase in growth compared to their non-transgenic siblings, and these fast-growing individuals contained the transgenic GH construct in their genomic DNA.

This pioneering work on transgenic mice, as well as several successes in genetic transformation of other animals, encouraged the application of the technology to fish. The earliest success in microinjecting cloned gene sequences into fish eggs was achieved simultaneously and independently by Maclean and Talwar (1984) with eggs of rainbow trout (*Oncorhynchus mykiss*), and by Zhu *et al.* (1985) with eggs of goldfish (*Carassius auratus*). These researchers soon demonstrated successful transgene integration, expression and transmission with rainbow trout and common carp (*Cyprinus carpio*), reported in a joint publication in 1987 (Maclean *et al.*, 1987). Since this early work, research has increased

Box 3.1. Historical landmarks of fish transgenesis for aquaculture.

1984 Maclean and Talwar (UK) report the first trial of gene transfer to fish with rainbow trout (*Oncorhynchus mykiss*) eggs.
1985 Zhu and his colleagues (China) achieve successful gene transfer into goldfish (*Carassius auratus*).
1987 Maclean, Penman and Zhu produce a joint publication demonstrating the first results on the transmission of a transgene in common carp (*Cyprinus carpio*) and rainbow trout.
1992 Du and his colleagues (Canada) publish the first demonstration of growth enhancement of transgenic Atlantic salmon (*Salmo salar*) using an 'all-fish' chimeric construct; Fletcher and his colleagues (Canada) report the transfer of an antifreeze protein gene to Atlantic salmon.
1994 Devlin and his colleagues (Canada) report the dramatic growth acceleration of coho salmon (*Oncorhynchus kisutch*) using an 'all-salmon' construct.
1996 Growth enhancement in transgenic tilapia (*Oreochromis niloticus*) is published by Martinez and her colleagues (Cuba).
1998 A UK research group led by Maclean reports fast-growing transgenic tilapia expressing a salmon GH gene.
2000 A 'clonal' version of an isogenically developed homozygous transgenic stock is first demonstrated in fish by Nam and his colleagues.
2001 A Korean group led by Kim reports the first fast-growing 'autotransgenic' diploid and triploid fish, in mud loach (*Misgurnus mizolepis*) containing an 'all-mud loach' construct.
2004 An ornamental transgenic zebrafish expressing fluorescence first appears in public markets, sold by a US company (Glofish, R.).

to involve many fish species and a wide range of phenotypes, including growth enhancement, disease resistance and cold tolerance. Most transgenic research using commercially important fish species has focused on growth improvement.

Early fish transgenesis research used only mammalian or viral sequences because the functions of genetic material from fish were poorly understood, and very few cloned fish genes were available for early work. However, beginning in the mid-1990s, information on fish gene sequences rapidly expanded due to the development of many sophisticated technologies in molecular biology and bioinformatics. Now it is possible to isolate and use fish genetic elements that are both homologous to a specific structural gene and a promoter region. Box 3.1 provides a brief historical background of landmark works with transgenic fish. Some of these achievements are considered in greater detail later in this chapter.

Overview of Important Steps in Fish Transgenesis

Figure 3.1 shows an overview of the steps involved in fish transgenesis to give readers some background information to understand the specifics of transgenesis discussed in this book. More technical details, especially regarding transgene

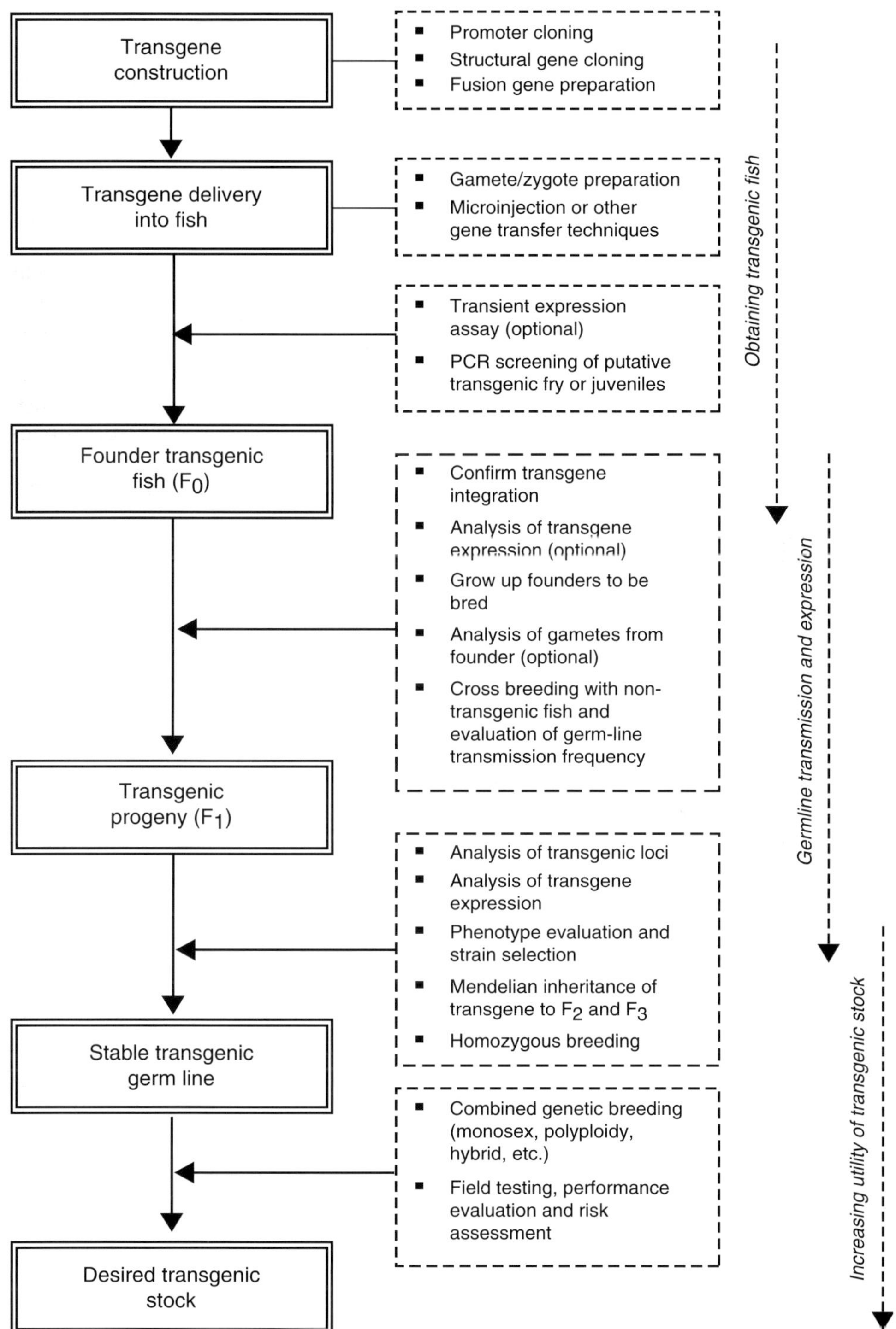

Fig. 3.1. A diagram to show important steps for transgenesis in fish.

construction, gene delivery and genomic integration, are discussed in Chapter 4. As shown in Fig. 3.1, the overall procedure for fish transgenesis is comprised of several important steps, beginning with transgene preparation and culminating with the establishment of a breeding programme to create a transgenic line. In this book, two terms, 'autotransgenic' and 'allotransgenic', are used to define all types of transgenic organisms. The term autotransgenic means that all the genetic elements in the transgene construct are from the same species, while allotransgenic denotes that some or all of the genetic material used in the construct is from a different species (Beardmore, 1997).

Designing the transgene construct

This step involves isolation of the coding and regulatory sequences of DNA to be inserted, followed by splicing together a structural gene and promoter sequences and preparation of multiple copies of the transgene construct with or without the vector backbone. Ideally, the promoter's ability to drive the expression of the structural gene is tested prior to full-scale transgenesis.

Introduction of the transgene into fish

Once the transgene construct is available, the first challenge in transgenesis is delivering the construct to fish zygotes or gametes. Usually, transgene delivery is done by microinjecting a DNA solution containing the transgene construct into one-celled embryos using fine microneedles. Several alternative methods, such as electroporation, can also be utilized to transform either sperm or eggs. Electroporation involves induction of transient pores in fish eggs by application of a short pulse of electricity, allowing the entry of exogenous DNA. In many aquaculture situations, the transgene is often inserted into a line previously subjected to selective breeding for one or more desirable traits (see related discussion in Chapter 5, this volume).

Identification of transgenic founder fish

Larvae from the injected eggs are then grown to a sufficient size for analysis. DNA samples from small amounts of tissue (e.g. blood or fin) can be obtained without adverse effects on fish viability. Transgene presence is usually analysed using polymerase chain reaction (PCR). Suspected transgenic fish (i.e. PCR-positive) are subjected to further analyses to determine genomic integration of the transgene. Transgene expression in the founder generation (F_0) can also be examined, preferably through the use of a non-lethal analytical method in order to keep as many founder individuals as possible alive.

Establishment of germ line expressing the transgene

After transgenic founders mature, their gametes are crossed with those from non-transgenic controls. The resultant genotypes are usually transgenic hemizygous. To isolate F_1 hemizygous transgenic fish, a portion of the resulting progeny from each cross undergoes PCR screening. Alternatively, transgene presence in founder gametes can be ascertained prior to making any crosses, identifying parents most likely to transmit the transgene to their offspring. The F_1 transgenics from multiple lines are identified and analysed in more detail to determine the integration and expression of the inserted transgene. Selected F_1 transgenic fish lines are further propagated to produce subsequent, true-breeding generations, and the persistence of transgenic traits at both the genomic and phenotypic level is examined at each generation.

Breeding of transgenic fish

Transgenic strains can be further manipulated to maximize their utility. Transgenic homozygous lines are needed for propagating a stable germ line, and they may be established by sister–brother mating, backcross mating or artificial parthenogenesis. Transgenic homozygous fish serve as broodfish because they transmit the transgene to their progeny at a high frequency, which facilitates the propagation of transgenic stock. Breeders may apply other genetic and reproductive control methods to transgenic fish lines, either separately or together, in order to combine a transgenic phenotype with other desirable traits achieved by traditional genetic breeding methods (e.g. the production of transgenic monosex stocks, transgenic hybrids or transgenic polyploids). However, the possibility of unwanted inbreeding depression or loss of genetic diversity should be carefully considered during these manipulations, including during the establishment of transgenic homozygous lines. Production characteristics of the established transgenic stock should be evaluated by field tests in a larger confined facility (see Chapter 8, this volume, for confinement methods). As indicated throughout this book, environmental risk assessment and management should be addressed prior to considering commercial use of a transgenic line in aquaculture.

It is important to note that, from the outset, transgenic fish research pursued either one of two major goals, although some experimental work encompassed both. The first goal was to improve understanding of gene regulation by studying promoter and enhancer activity and specificity; fish were used as model systems for basic work in molecular and developmental biology. The second goal was to use transgenic technology to develop a novel transgenic strain with desired traits, allowing for increased efficiency and capacity of farming practices in aquaculture. In line with the scope of this book, this chapter focuses on the second purpose of transgenic application (practical application of transgenic fish in aquaculture) rather than the first one (transgenic fish as model systems for basic science research).

Based on this focus, the objectives of this chapter are twofold. One is to provide background information on transgenic fish, focusing on transgenic fish strains that may be nearing commercial consideration, especially for aquaculture in developing countries. The second objective is to discuss what additional kinds of transgenic fish and shellfish might appear in the future based on projects currently in the early stages of research and development.

Status of Development

During the last two decades, more than 35 fish species have been subjected to transgenesis. Most transgenic research on commercially important fish species has targeted traits affecting growth, often through the transfer of GH gene constructs, because growth is one of the most important parameters in successful fish production. Growth enhancement can provide advantages for aquaculture by shortening production times, improving feed conversion efficiency (FCE) and controlling product availability. For this reason, most documented lines of transgenic fish closest to serious consideration for commercial use in aquaculture are GH-transgenics. This chapter highlights some important examples of transgenic fish development, but it does not aim to be a comprehensive review of all existing transgenic fish lines.

Some important examples include GH-transgenic common carp (*C. carpio*), tilapias (*Oreochromis* spp. and *Oreochromis niloticus*), Atlantic salmon (*Salmo salar*) and mud loach (*Misgurnus mizolepis*). Growth-enhanced common carp developed in China is already being held and tested in a confined pond system, and a selected transgenic line is nearing formal review for commercial approval. Transgenic tilapias developed in Cuba and the UK have been subjected to various theoretical and practical evaluations regarding their potential commercialization. At least one US/Canadian biotechnology company, AquaBounty Technologies™ (Waltham, Massachusetts), is pursuing commercial approval for growth-enhanced Atlantic salmon. Growth-enhanced mud loach, developed in South Korea, is the first fast-growing autotransgenic fish (utilizing genetic elements from their own species). These autotransgenic mud loaches have been subjected to further genetic manipulation, including sterilization and hybridization.

Pacific coho salmon (*O. kisutch*) has also undergone a great deal of growth-enhancement transgenesis. This strain contains an 'all-salmon' transgene made of a sockeye salmon (*Oncorhynchus nerka*) GH gene whose expression is promoted by a sockeye salmon metallothionein promoter. Genetic characteristics, production traits and physiological alterations of the transgenic coho salmon lines are described in a number of publications (Devlin *et al*., 2004, 2006), and this transgenic line has been used as a model system for public domain research, with particular emphasis on the ecological risks associated with growth-enhanced transgenic fish. This chapter does not describe the details of the GH-transgenic coho salmon lines, but it is important to note that information obtained from these salmon serves as a rich source of information for the phenotypic assessment of transgenic fish and their possible impacts on natural ecosystems (Chapters 5 and 6, this volume).

Carp

Common carp is a cosmopolitan species cultured in many countries and is one of the most important species in global aquaculture (FAO, 2003). China has played a major role in the development of fast-growing transgenic carps, and the most prominent progress on GH-transgenic common carp was achieved by transferring an 'all-fish' genomic construct (pCAgcGH) containing a grass carp (*Ctenopharyngodon idella*) growth hormone gene driven by a common carp β-actin promoter into common carp eggs (Wang *et al*., 2001). Fish developed from the microinjected eggs showed twofold growth enhancement as compared to the non-transgenic controls. The founder generation transmitted the transgene to their progeny, and enhanced growth was observed in subsequent generations. Information on these transgenic carps is available in Wu *et al*. (2003) and Fu *et al*. (2005).

Field trials were conducted using selected transgenic Chinese common carp lines in confined ponds to determine whether they have commercialization potential. A special pond-based facility was built to provide security and physical confinement. All of the water channels in and out of the ponds were carefully designed and strict measures were adopted to prevent the escape of transgenic fish. About 200 transgenic individuals showing significant growth enhancement were artificially spawned to produce F_1 transgenics. The F_1 transgenics from selected lines were reared in 1.67 ha ponds, and non-transgenics were reared as controls under the same conditions. Based on successive examinations of random samples of transgenic and non-transgenic carps, the transgenic individuals showed at least a 42% greater growth rate than non-transgenic fish. Most transgenic individuals reached marketable size within 5 months, while non-transgenic fish needed an additional 6–10 months to reach the same size. Furthermore, the gross feed conversion (total feed weight per unit of gained body weight) of the transgenics was 1.10, a significant improvement over food conversion of 1.35 in non-transgenics. Notably, the gonadosomatic index (GSI) of the transgenics was significantly lower than that of the non-transgenics, although there was no difference in the quality of gametes. Furthermore, sexual maturity of growth-enhanced carps was later than that of non-transgenic fish, indicating that this line of fast-growing transgenic carp does not have a younger age of sexual maturity (Fu *et al*., 2005).

Two main methods for producing sterile transgenic stock are being explored to minimize the potential risk of transgenic carp to local gene pools. One is the generation of triploid transgenics by crossing spontaneously occurring tetraploid carps, and the other is the blockage of proper translation of the gonadotropin-releasing hormone (GnRH) transcript using an antisense technique (see Chapter 8, this volume, for a detailed discussion of such biological confinement measures). The GH transgene construct was transferred to a spontaneously occurring tetraploid genotype, specifically, progeny from an interspecific cross between red crucian carp (*C. auratus*) and common carp. Transgenic tetraploids were then crossed with diploid carps to produce transgenic triploids (Zeng *et al*., 2000). Transgenic triploid carps were found to be functionally sterile, and they also had higher growth rates than non-transgenic carps.

Another transgenic sterilization method was tried in carps by introducing a construct comprised of the antisense GnRH cDNA, driven by a common carp β-actin promoter, into fast-growing transgenic carp. Theoretically, the expressed antisense GnRH transcripts prevent the proper translation of endogenous GnRH mRNA into functional polypeptides in the transgenic carp; their fertility can be recovered by administering exogenous GnRH. These transgenics are undergoing detailed evaluations with respect to both their degree of sterility and its reversibility (e.g. the recovery of fertility by administering exogenous GnRH) (Li, 2004). Other detailed examinations of physiological alterations, bioenergetic analysis, reproductive performance and food safety of transgenic carps have been carried out in physically confined pond systems (C. Fu, Fudan University, Shanghai, 2006, personal communication).

Tilapias

Tilapias are the second most important group of food fishes in the world (FAO, 2003). Different tilapia species are native to Africa and the Middle East (Lim and Webster, 2006), and a number of species are cultured in much of Asia, as well as parts of Europe and the Americas. They are robust and relatively free of diseases, tolerant of poor water quality and are capable of rapid growth, reaching sexual maturity in 6 or 7 months. Two research groups in Cuba and the UK have done considerable work on transgenic tilapias.

Cuban scientists have produced transgenic tilapia derived from an interspecific hybrid line, *O. hornorum* × *O. aureus*, using a tilapia growth hormone (tiGH) cDNA whose expression is driven by a cytomegalovirus (CMV) promoter. Patterns of ectopic expression of the tiGH construct were assessed in various transgenic tilapia tissues at both the transcript and protein level. Weak correlation between GH mRNA levels and growth performance was detected in such lines (Martinez *et al.*, 1996; Hernandez *et al.*, 1997).

In the Cuban research, transgenic male founder (F_0) fish bearing a single copy of the transgene were selected based on the previous molecular genetic analyses and performance tests. Stable Mendelian inheritance of the single transgene was observed up to the F_4 generation. The transgenic trait (growth enhancement) was persistent through these generations. The selected line exhibited a 60–80% growth improvement when judged at 9 months, depending upon culture conditions (Martinez *et al.*, 1996; Hernandez *et al.*, 1997; de la Fuente *et al.*, 1999). However, juvenile GH-transgenic tilapia from the same line did not show any significant growth stimulation during the first 5 weeks of the similar growth efficiency study, which may be the result of non-optimal growth conditions. The transgenics showed significantly better feed conversion than the non-transgenics. The transgenics exhibited roughly 3.6-fold less feed consumption than non-transgenic controls, with a lower feeding motivation relative to their non-transgenic counterparts (Martinez *et al.*, 2000).

The Cuban research group tested GH-transgenic tilapia's ability to compete for food with their non-transgenic siblings and feral (i.e. wild-type) counterparts (Guillén *et al.*, 1999). Transgenic and non-transgenic tilapia were reared in the laboratory for multiple generations, and the wild-type fish were collected from local ponds for these experiments. The group found that transgenic tilapia showed lower dominance status relative to non-transgenic tilapia. Wild-type tilapia had the highest dominance status compared to all other groups. These findings suggest that laboratory-reared transgenic fish may have inferior ability to compete for food resources. Assessing the ability of this line to persist in the wild requires more focused tests to assess gene flow from the transgenic line to non-transgenic tilapia (see Chapter 5, this volume) and to assess ecological interactions between the transgenic fish and other fish in the ecosystem (i.e. outside of the laboratory) (see Chapter 6, this volume). Cuban transgenic tilapia lines were also subjected to some food safety assessments, including nutritional evaluation of transgenics and consumption by non-human primates and human volunteers (Guillén *et al.*, 1999).

Transgenic tilapia have also shown better osmoregulatory ability than non-transgenic tilapia (Guillén *et al.*, 1999). This finding indicates that increased plasma levels of GH might provide an advantage for seawater adaptation in this transgenic line (see also Sakamoto *et al.*, 1997; Mancera and McCormick, 1998). Given that this line was founded from hybrid-origin parents, it is worth noting that the parental species can live and reproduce at moderate salinity (*O. aureus*) or high salinity (*O. hornorum*) (Talbot and Newell, 1957; Philippart and Ruwet, 1982; Stickney, 1986). Direct tests of survival and reproduction of this transgenic line under different salinities have not yet been reported.

Other transgenic tilapia research has been conducted using Nile tilapia (*O. niloticus*). This research used an ocean pout (*Macrozoarces americanus*) antifreeze promoter spliced to a chinook salmon (*O. tshawytscha*) GH coding sequence. One transgenic line carrying a single copy of the GH construct was ultimately selected based on laboratory evaluations of the transgene's integration, expression of growth traits and its germ-line transmission (Rahman *et al.*, 1998, 2000; Rahman and Maclean, 1999). Although this transgenic line expressed the heterologous GH at lower levels than other transgenic lines with multiple transgene copies, it demonstrated better overall growth performance. This transgenic line underwent field tests in Hungary, and in 7-month growth trials achieved two- to threefold growth enhancement (Figs 3.2 and 3.3). FCE was 20% greater in the transgenics, and a digestibility trial suggested that this line of transgenic tilapia utilized protein, dry matter and energy more efficiently than non-transgenic controls (Rahman *et al.*, 2001).

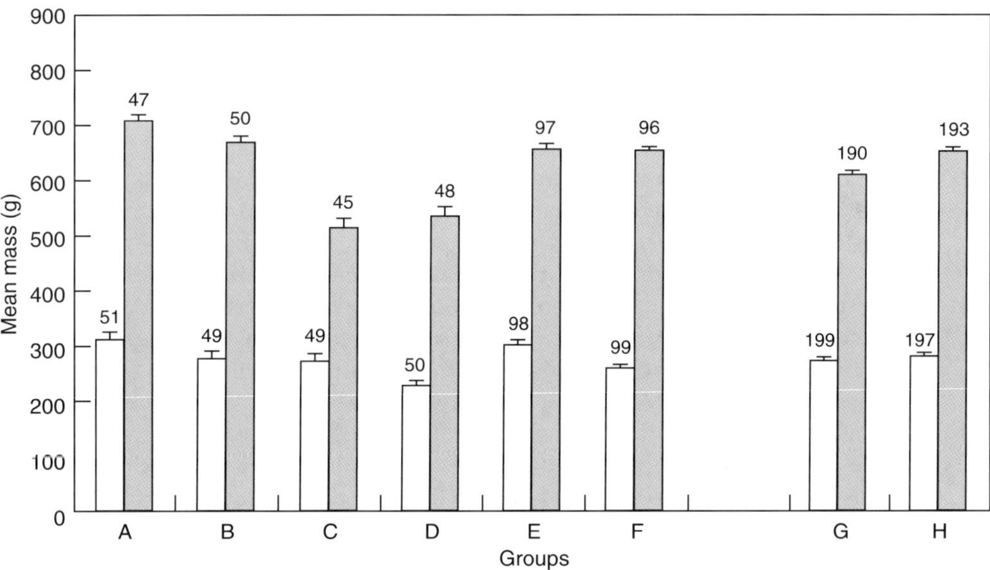

Fig. 3.2. Mean masses (±95% confidence level) of 30-week transgenic (filled grey bars) and non-transgenic full siblings (open bars) of second generation from the C86 tilapia line. A–D were reared as mixed batches in tanks 3, 4, 5 and 6, respectively; E, non-transgenic (tank 1) and transgenic (tank 2) which were reared separately; F, non-transgenic (tank 7) and transgenic (tank 8), which were reared separately; G–H, overall mean body mass of non-transgenic and transgenic tilapia, which were reared as mixed and separate batches, respectively. Numbers of individuals are given as numbers above bar. (Reprinted from Rahman *et al.*, 2001, with permission from Blackwell.)

Fig. 3.3. Two 30-week-old transgenic tilapias (*O. niloticus*; left) along with their non-transgenic siblings (right). Fish are from separate batches of line C86. The average mass of transgenics in this group was 653 g and that of non-transgenics 260 g. Scale = 15 cm. (Photograph from Rahman *et al.*, 2001, reprinted with permission from Blackwell.)

Additional research on Nile tilapia led to the production of entirely autotransgenic GH tilapia. These fish were produced using a tilapia β-actin promoter spliced to a tilapia GH gene sequence (all derived from *O. niloticus*). This work is ongoing, and long-term results are still awaited (Maclean *et al.*, 2002; Maclean, 2003).

Several other transgenic manipulations of Nile tilapia have been undertaken, in addition to growth enhancement. Producing sterile tilapia expressing an antisense gene against GnRH would have beneficial applications for biological confinement of transgenics. Sterile fish could then be 'rescued' for breeding purposes by injecting GnRH (see Chapter 8, this volume, for discussion of the antisense technique and biological confinement of transgenic fish). The depression of gonad development in transgenic tilapia expressing anti-GnRH was claimed by Maclean *et al.* (2002); achieving complete functionality and reversibility requires further research.

Atlantic salmon

Current global production of Atlantic salmon (*S. salar*) exceeds 1.0 million tonnes, and its production is ranked as the top aquaculture moneymaker (FAO, 2003). For this reason, a great deal of research has been devoted to improving the productivity of this species using a wide variety of techniques, including selective breeding, improved husbandry and reproductive control. From the late 1980s, transgenesis has been used to alter traits responsible for growth, leading to growth-enhanced transgenic salmon with shortened culture periods. The gene construct used to produce transgenic Atlantic salmon is the same one used in transgenic Nile tilapia, the opAFPGHc consisting of chinook salmon GH cDNA fused to an ocean pout antifreeze protein (AFP) promoter (Hew *et al.*, 1995). One private company (AquaBounty Technologies™) is pursuing commercial approval of the fast-growing transgenic salmon for aquaculture.

Growth-enhanced transgenic Atlantic salmon exhibit a two- to threefold rate of early growth relative to non-transgenic salmon. Several production traits of the transgenic line have also been examined, with particular emphases on feed utilization, body constituents, oxygen metabolism and food-securing behaviour. GH-transgenic salmon exhibited similar ability for feed utilization as non-transgenic salmon. There was also no difference between transgenic and non-transgenic fish in terms of digestibility coefficients for dry matter, crude protein and energy utilization (Cook *et al.*, 2000a). Transgenic salmon carcasses contained significantly higher moisture content, but lower absolute levels of body protein, dry matter, ash, lipid and energy, than those from non-transgenic fish. The higher observed moisture levels indicate a 10% improvement in gross FCE, despite similar capability in feed utilization, compared to non-transgenic control fish (Cook *et al.*, 2000a).

GH-transgenic Atlantic salmon showed more efficient oxygen metabolism over a pre-smolt body interval (8–55 g) than their non-transgenic counterparts; transgenic fish consumed less total oxygen (42% less) than non-transgenic salmon to reach the smolt stage. This reduction was due to their fast growth, although routine oxygen consumption rates (milligrams of oxygen per hour) were higher for transgenic fish (up to 1.7-fold) than for controls (Stevens *et al.*, 1998; Cook *et al.*, 2000b,c).

GH-transgenic Atlantic salmon exhibited significantly higher levels of appetite (spending more time for gathering food), greater daily feed consumption (2.1- to 2.6-fold) and larger intestinal folds (larger digestive surface area) as compared to similar-sized non-transgenic fish. It is believed that these alterations could subject GH-transgenic individuals to greater risks of predation in order to secure food (i.e. poor anti-predator behaviour) (Abrahams and Sutterlin, 1999; Stevens *et al.*, 1999). Chapter 6 discusses methodologies for testing foraging, anti-predator and other behaviours of transgenic fish as part of assessing their potential ecological effects.

Mud loach

Mud loach, a small freshwater fish species, is an important food fish in Korea. Besides its commercial importance, this species has many advantages as a model system for genomic and transgenic studies. It has a small adult body size (10 cm in total length), fast embryonic development (24 h at 25°C), short generation time (3–5 months), year-round spawning under controlled conditions and high fecundity (more than 10,000 eggs per female at one time). Mud loaches are also very hardy, with good tolerance to low oxygen, temperature fluctuations and diseases. Based on these characteristics, this species has been the subject of a number of genetic manipulations, including transgenesis and chromosome manipulation.

Mud loach growth traits can be manipulated readily through transgenesis. Fast-growing autotransgenic mud loach lines have been developed in South Korea. A transgenic construct containing the mud loach GH gene fused to a mud loach β-actin promoter was transferred into mud loach eggs by microinjection (Nam *et al.*, 2001b). Some lines of transgenic fish derived from the injected eggs showed dramatically accelerated growth, with a maximum of 35-fold faster growth than their non-transgenic siblings (Fig. 3.4). Many fast-growing transgenic individuals showed extraordinary gigantism beyond the normal body size found in wild stocks.

More than 30 transgenic lines were propagated in order to examine their growth performance. The growth of subsequent generations was dramatically accelerated, up to 35-fold. However, the levels of enhanced growth were variable among transgenic lines (e.g. Fig. 3.4), suggesting that characteristics of the insertion site and transgene structure play an important role in the expression and consequent physiological effect of the transgene. Based on the examination of growth performance and GH transgene expression in various lines, there appeared to be a weak correlation between transgene expression and actual weight gain, which may be due in part to abnormalities, such as slender body shape and gigantism, or acromegaly (abnormal deposition of cartilage

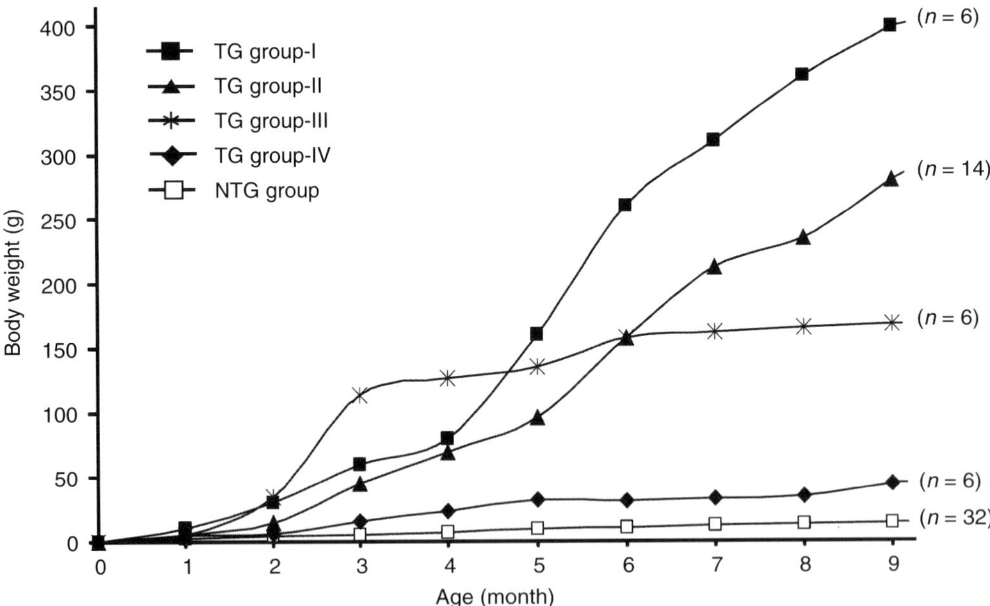

Fig. 3.4. Four distinct trends in body-weight growth from birth to 9 months in transgenic mud loach. The curves were generated by computing mean body weight of all lines with a similar growth pattern. The number of lines averaged for each growth curve is indicated in parentheses. The transgenic lines are TG groups I through IV, and the non-transgenic, communally grown control line is denoted as NTG. (Adapted from Nam *et al.*, 2001b.)

and bone), arising from overexpression of the transgene (Nam *et al.*, 2002a). Hence, maximum growth probably arises from the moderate levels of transgene expression rather than the highest levels of expression.

Transgenic mud loach reached marketable size (10 g) only 30–50 days after fertilization, compared to at least 6 months in non-transgenic mud loach. Improved FCE (up to 1.9-fold) was also observed, indicating that fast-growing transgenic mud loaches are metabolically more efficient than non-transgenics (Nam *et al.*, 2001b). Physiological alterations (i.e. abnormalities in body structure and some biochemical traits) varied between the transgenic lines and may have impacts on their long-term viability. Further evaluation of selected transgenic lines, with particular emphasis on nutritional characteristics, revealed that growth-enhanced transgenic mud loach may show higher protein content in their muscles than do non-transgenic fish (Nam *et al.*, 2003).

Several more genetic manipulations were carried out on fast-growing transgenic mud loaches. They include:

- Development of fast-growing stable homozygous transgenic lines obtained by induced interspecific androgenesis in order to produce stable isogenic lines (all individuals have the same genotype) capable of transmitting transgenes to their progenies at 100% frequency (Nam *et al.*, 2000, 2002a);
- Generation of transgenic polyploid stocks as a means for producing sterile transgenic fish necessary for biological confinement (Fig. 3.5; Nam *et al.*, 2001a);
- Generation of triploid transgenic hybrid loaches exhibiting complete sterility, with some desirable hybrid traits, including body colour (Nam *et al.*, 2004).

Fig. 3.5. Autotransgenic diploid (top pair) and triploid (middle three) mud loaches along with their non-transgenic sibling (bottom). (Photograph from Nam *et al.*, 2001a, reprinted with permission from World Aquaculture Society.)

Growth-enhanced transgenic mud loaches need to undergo field tests in larger confined facilities to confirm their enhanced performance under aquaculture production conditions. At the time of this writing, field tests on transgenic fish using facilities simulating the natural environment were not allowed in South Korea due to the concern about their escape (see Chapter 8, this volume, for discussion of approaches for confinement of large experimental systems to facilitate these kinds of field tests).

Several pilot projects on transgenic traits other than growth continue with mud loach. These include engineering disease resistance using anti-bacterial peptides or immunomodulator protein genes and metabolic engineering targeting the synthesis of essential micronutrients. Long-term goals for mud loach transgenesis are to manipulate multifunctional transgenic lines expressing more than two kinds of transgenes simultaneously in a controllable manner. If successful, such projects will enhance the utility of transgenic fish not only for theoretical purposes, but also for commercial interests in aquaculture and other bioindustries.

Key information about the transgenic lines closest to commercialization (carps, tilapias, Atlantic salmon and loach) is summarized in Table 3.1.

Other Future Applications of Transgenic Fish

In addition to the GH-transgenic lines described in the previous section, a number of other fish species have undergone transgenic manipulation of their growth traits. They are, however, in the earlier stages of development and farther from consideration for commercial use. More importantly, there are many other transgenic phenotypes (i.e. non-growth traits) actively undergoing transgenesis research. Possible candidate traits include disease resistance, sterilization for biological confinement, nutritional improvement and cold tolerance or freeze resistance. More recently, several groundbreaking studies have highlighted the potential for transgenic fish to function as biofactories (i.e. expressing transgenes producing pharmaceutical proteins) or environmental monitoring tools (i.e. as early indicators of aquatic pollution). Examples are described in more detail below.

Growth enhancement in other transgenic fish lines

Transgenic research has been done on an Indian major carp (rohu, *Labeo rohita*). Growth-enhanced transgenic rohu were produced in India via electroporated sperm-mediated transfer of a construct containing a grass carp β-actin promoter fused to endogenous rohu GH cDNA (Venugopal *et al.*, 2002). After confirming the transgene expression by reverse transcriptase PCR analysis, the rohu's growth was monitored for up to 60 weeks. Transgenic rohu grew approximately five times faster than their control siblings. The transgenic and control fish were fed 1.6% and 2.6% of their body weight, respectively (calculated as dry weight of feed to live fish weight), which resulted in FCE values of

Table 3.1. Summary of the traits of the transgenic lines close to commercial consideration.

	Key information							
	Gene construct							
Species	Regulator	Structural gene	Growth stimulation	Other physiological alterations	Performance tests in confined labs or ponds	Other comments	Country	Key references
Common carp (*Cyprinus carpio*)	Common carp β-actin promoter	Grass carp GH gene	Up to twofold	FCE improved (1.2-fold); GSI reduced and sexual maturity retarded	China – lab tanks, confined ponds	Triploid transgenics using spontaneously occurring 4N hybrid fish possible; nutritionally evaluated	China	Fu *et al.* (2005)
Hybrid tilapia (*Oreochromis hornorum* x *O. aureus*)	Cytomegalo-virus (CMV) promoter	Tilapia GH cDNA	Up to 1.8-fold	FCE improved (2.9-fold); increased osmoregulation ability detected	Cuba – lab tanks, confined ponds	Lower dominance status; nutritionally evaluated	Cuba	Martinez *et al.* (2000)
Nile tilapia (*Oreochromis niloticus*)	Ocean pout antifreeze protein promoter	Chinook salmon GH cDNA	Up to threefold	FCE improved (1.2-fold)	UK – lab aquaria Hungary – lab tanks	Triploid transgenics tested; nutritionally evaluated	UK	Maclean *et al.* (2002); Rahman *et al.* (2001)

Continued

Table 3.1. *Continued*

	Key information							
	Gene construct							
Species	Regulator	Structural gene	Growth stimulation	Other physiological alterations	Performance tests in confined labs or ponds	Other comments	Country	Key references
Atlantic salmon (*Salmo salar*)	Ocean pout antifreeze protein promoter	Chinook salmon GH cDNA	Up to threefold	FCE improved (1.1-fold); OCR increased (up to 1.7-fold); larger digestive surface area in transgenics	Canada – lab tanks	Potential change in predator avoidance; nutritionally evaluated	Canada	Cook *et al.* (2000a)
Mud loach (*Misgurnus mizolepis*)	Mud loach β-actin promoter	Mud loach GH gene	Up to 35-fold	FCE improved (1.9-fold); gigantism	Not yet	Autotransgenic; triploid and triploid hybrid transgenics tested	Korea	Nam *et al.* (2001a,b)

FCE = feed conversion efficiency; OCR = oxygen consumption rate; GSI = gonadosomatic index.

159% and 73%, respectively (Pandian and Venugopal, 2005). This suggested that transgenic rohu consumed relatively less food but converted it more efficiently.

Other work with transgenic common carp, *C. carpio*, has been done by microinjecting embryos with a rainbow trout GH cDNA coding sequence, driven by a Rous sarcoma virus (RSV) promoter. Founder fish transmitted the transgene to subsequent generations at varying frequencies due to mosiacism. Some transgenic progeny showed growth enhancement up to 50% compared to the non-transgenic fish (Zhang *et al.*, 1990; Chen *et al.*, 1993; Dunham, 2004).

Many other salmonids have been engineered for enhanced growth performance. The gene construct used to produce growth-enhanced Atlantic salmon (opAFPGHc) also caused rapid growth when introduced into Pacific salmon. These include coho salmon (*O. kisutch*), chinook salmon (*O. tshawytscha*), rainbow trout (*O. mykiss*) and cut-throat trout (*O. clarki*) (Devlin *et al.*, 1995). Other growth enhancement successes have been reported in Arctic charr (*Salvelinus alpinus*). A salmon GH gene (*O. nerka* GH1), under the control of CMV, piscine metallothionein B (*O. nerka* MT) or histone 3 (*O. nerka* H3) promoters, resulted in accelerated growth (up to 14-fold) in transgenic charrs (Pitkanen *et al.*, 1999b). The nutritional and metabolic functions of a selected line of fast-growing transgenic charr (generated using the CMV promoter) have also been evaluated (Krasnov *et al.*, 1999a).

Ayu (*Plecoglossus altivelis*) have also undergone growth enhancement experiments, using a rainbow trout GH gene, fused to a carp β-actin promoter transferred via sperm-mediated electroporation. At 5 months, transgenic ayu showed a twofold increase in body weight and a 1.3-fold increase in body length relative to their non-transgenic counterparts (Cheng *et al.*, 2002). Several other species belonging to widely divergent taxa are also undergoing research for GH transgenesis, although these transgenic lines are in preliminary stages with varying degrees of success (for reviews, see Zbikowska, 2003; Dunham, 2004; Kapuscinski, 2005).

Transgenic manipulation of non-growth traits in fish

Disease resistance

Disease resistance is one of the most significant aquaculture-related traits that transgenesis could improve. However, it is not yet clear which genes should be manipulated to improve resistance to viral, fungal or bacterial diseases. A pioneering attempt to improve disease resistance in fish was demonstrated by Hew *et al.* (1995), involving Atlantic salmon expressing a rainbow trout-derived lysozyme gene. A second attempt involved the production of transgenic lines of channel catfish (*Ictarulus punctatus*) expressing a silkmoth-derived cecropin gene (Dunham *et al.*, 2002). Cecropins are broad-spectrum antimicrobial proteins expressed by some insects. Additionally, Zhong *et al.* (2002) reported that transgenic grass carp showed increased resistance to grass carp haemorrhage virus infections when expressing a human lactoferrin gene. All of these attempts are in preliminary stages of research.

Cold tolerance and freeze resistance

Cold tolerance is distinct from freeze resistance and refers to the ability of fish such as tilapia or carp to survive, and even thrive, at temperatures ranging from 2°C to 10°C. Tolerance of low temperatures is clearly important during the winter months, and serious losses of some fish species occur during this season. For example, the culture of various carp species is limited in the north of the Yangtze River in China due to their poor survival rate in winter. Modifying the antifreeze protein might improve cold tolerance in fish by altering cell membrane permeability (Wu *et al.*, 1998), and gene transfer attempts have been undertaken in goldfish (Wang *et al.*, 1995). Engineering fish to be more cold-tolerant may also be possible by overexpressing delta-9 desaturase, a protein known to be cold-inducible in carp and to alter cell membrane permeability (Tiku *et al.*, 1996).

Atlantic salmon also have been engineered to express antifreeze protein genes (Fletcher *et al.*, 1992; Hew *et al.*, 1995, 1999). Although the resulting antifreeze synthesis was insufficient to endow the salmon with freeze resistance, it remains a potential development of interest, particularly for fish culture in sea cages in North Atlantic, Arctic or Antarctic waters.

Metabolism

Some fish reared in aquaculture, such as salmonids, are carnivorous and have to be fed with fish-derived products that are high in fat content and animal protein. This is an energy-intensive and ecologically harmful practice. For example, overfishing of capelin and sand eels from the North Sea to provide food for intensive salmon culture has already caused considerable ecological harm to sea bird populations (Holmes, 1996). A possible solution may be to modify the metabolic traits of cultured fish to enable them to use plant-derived carbohydrates. A Finnish research group reported that engineering salmonids with human glucose transporter type 1 (hGluT1) and rat hexokinase type II (rHKII) cDNA, driven by a CMV promoter, might enable them to use plant-derived feedstuffs more readily (Pitkanen *et al.*, 1999a; Krasnov *et al.*, 1999b).

Sterility

Sterility offers one possible means of biological confinement for transgenic fish and is an important risk management option for field trials or aquaculture production (Chapter 8, this volume). There are non-transgenic means of rendering fish sterile; the best known and most widely practised method is triploidy. Triploid induction, especially when combined with production of monosex fish, can yield a high frequency of functionally sterile individuals in many species, often with little or no loss in fish quality (Purdom, 1993; Benfey, 1999). It may also be possible to produce sterile fish through transgenesis. This would be of greatest practical benefit if the sterility were reversible, making it possible to restore fertility in broodstock. The most effective way to accomplish reversible sterility in fish is to target genes coding for sex hormones such as gonadotropin hormone (GnH) or GnRH (Maclean *et al.*, 2002). Because there are currently no means to 'knockout' (or delete) specific genes in fish, researchers have attempted to 'knockdown' (or reduce) gene expression by targeting the mRNA product of a gene. Of the three methods currently available to accomplish gene knockdown – RNA interference, ribozyme and antisense – only the latter has so far been shown to work in fish. Antisense against one of the GnRH genes induced sterility in both rainbow trout (Uzbekova *et al.*, 2000) and tilapia (Maclean *et al.*, 2002). In neither case has the technology shown widespread application, but transgenic reversible sterility remains a possibility for biological confinement of transgenic fish. More details on biological confinement can be found in Chapter 8.

Salinity tolerance

Salinity tolerance is another potential target for transgenic manipulation of fish. Freshwater is often limited or polluted, and fish production in either seawater or brackish coastal estuaries is often an attractive option. However, many fishes are unable to physiologically adapt to saltwater. This trait has not yet been a main target of transgenesis, although improved salinity tolerance has been reported as a side effect (pleiotropy) of the expression of GH transgenes. This is evidenced by the observation that GH-transgenic salmonids underwent smoltification (i.e. adaptation to saltwater) earlier than non-transgenic fish (Cook *et al.*, 2000b; Devlin *et al.*, 2000). Similar results were also found with a line of GH transgenic tilapia (Guillén *et al.*, 1999).

Biofactory systems

Fish have a number of attractive features as biofactory systems. They are cheap and fast-growing, with short generation times, and their transgenesis is easier to achieve than in birds or mammals. Practical developments in this area involve a line of tilapia engineered to produce human clotting factor VII. This line contains a transgene construct with a tilapia vitellogenin promoter driving a human clotting factor VII coding sequence (cDNA) (Hwang *et al.*, 2004). Vitellogenin is synthesized in the liver and is oestrogen inducible, especially in male fish. Factor VII is an internal clotting factor widely used after liver transplants and when treating injuries like ruptured spleens and internal gunshot wounds. Lines of transgenic tilapia are reared to adulthood, and the human protein is recovered and purified from the fish blood, and it does clot human red blood cells. Pohajdak *et al.* (2004) have also engineered tilapia to produce human insulin both in their serum and endocrine pancreases. The potential for fish eggs to serve as biofactories, and produce novel molecules, has been investigated by recovering glycosylated recombinant goldfish luteinizing hormone from transgenic trout (*O. mykiss*) embryos (Morita *et al.*, 2004).

Environmental monitoring

Transgenic fish can be engineered as extremely sensitive monitors of pollutants. One strategy is to insert genes coding for green fluorescent protein (GFP), whose expression is driven by promoters sensitive to specific pollutants. Blechinger *et al.* (2002) conducted such research with transgenic zebrafish (*Danio rerio*), who expressed GFP in the presence of low levels of heavy metals. Additional studies have succeeded in making transgenic medaka (*O. latipes*) express GFP in the presence of oestrogen and other endocrine-disrupting pollutants (Zeng *et al.*, 2005). There have also been several reports of research utilizing mutagenic bacteriophage-based vectors to detect *in vivo* mutations caused by aquatic pollutants in transgenic medaka (Winn, 2001).

Transgenic trials with molluscs and crustaceans

Crustaceans and molluscs comprise approximately one quarter of total aquaculture production, with both groups largely composed of high-priced commodity species (FAO, 2003). Therefore, increasing mollusc and prawn production is a good source of revenue for some nations. Increased production of high value species can be achieved through conventional means like improved husbandry methods and selective breeding, and transgenesis offers another option. Like fin fishes, molluscs and crustaceans are highly fecund and, for certain species, eggs and sperm can be obtained easily for *in vitro* fertilization, making them ideal for transgenesis. Transgenic research on molluscs and crustaceans could improve their growth rates, food conversion efficiency, survival, disease resistance and other economically important traits, such as pearl production in oysters. However, molluscs and crustaceans are generally less ideal for transgenic work than fin fishes. Problems such as very small female gametes (especially in molluscs), the difficulty of controlling reproductive cycles of crustaceans and the lack of suitable DNA sequence information constrain the application of transgenic techniques to these invertebrates (Harper *et al.*, 2003).

Powers *et al.* (1995, 1997) conducted early work on techniques for gene transfer and production of transgenic shellfish using red abalone (*Haliotis rufescens*). A transgene construct of either a β-galactosidase or salmon GH gene, driven by an abalone β-actin promoter, was transferred to fertilized abalone eggs through electroporation. Within the same year, Gendreau *et al.* (1995) developed a method for the transfer of the luciferase gene in brine shrimp (*Artemia franciscana*) and reported transient gene expression. In 1996, Lu *et al.* were the first to successfully effect gene transfer in a bivalve by introducing pantropic retroviral vectors into the dwarf surf clam (*Mulinia lateralis*). Several methods for gene transfer, integration, expression and transmission in transgenic aquatic invertebrates have been developed since then, with varying levels of success. Table 3.2 shows a list of the molluscs and crustaceans that have undergone transgenic research thus far. In spite of these numerous developments, the application of gene transfer technology to and production of transgenic molluscs and crustaceans still lags behind that of finfishes. This may be due in large part to relatively poor understanding about shellfish genes (and genomes), both in terms of gene structure and regulation.

Key Research and Capacity Needs for Further Development of Transgenic Fish

Despite the substantial progress of transgenic technology in fish during the last 20 years, there are still a number of issues that need to be resolved in order to increase the efficiency and accuracy of the technology (Box 3.2). This chapter

Table 3.2. Mollusc and crustacean species subjected to transgenic manipulation.

Species	Gene construct		Gene delivery method	Proposed objective	Country	Key references
	Regulator/promoter	Structural gene				
Red abalone (*Haliotis rufescens*)	*Drosophila* β-actin promoter	β-Galactosidase reporter gene	Electroporation	Methodology development	USA	Powers *et al.* (1995)
	Abalone actin promoter	Coho salmon GH cDNA	Electroporation	Growth promotion	USA	Powers *et al.* (1997); patent # US5675061
Japanese abalone (*Haliotis diversicolor suportexta*)	Ocean pout AFP promoter	Chinook salmon GH cDNA	Electroporation	Growth promotion	Taiwan	Tsai *et al.* (1997); Tsai (2000)
Pearl oyster (*Pinctada maxima*)	Rous sarcoma virus long terminal repeat (RSV-LTR)	Human insulin cDNA	Electroporation	Growth promotion and pearl production	USA	Paynter and Thibodeau (1996); patent # WO9615662
Oyster (*Crassostrea gigas*)	*Drosophila* heat shock protein 70; cytomegalovirus (CMV) promoter	Luciferase reporter gene	Particle bombardment	Methodology development	France	Cadoret *et al.* (1997)
Pearl oyster (*Pinctada maxima*)	Common carp β-actin promoter	Grass carp GH gene	Electroporation	Growth promotion	China	Hu *et al.* (2000)
Pearl oyster (*Pinctada fucata martensii*)	Mollusc prism protein and mantle protein promoters	Green fluorescent protein (GFP) fusion gene	Direct microinjection into gonad (adenovirus vector used)	Pearl production	Japan	Keizaburo *et al.* (2005); patent # US2005198697
Brine shrimp (*Artemia franciscana*)	*Drosophila* heat shock protein 70 promoter	Luciferase reporter gene	Biolistic bombardment	Methodology development	France	Gendreau *et al.* (1995)

Continued

Table 3.2. *Continued*

Species	Gene construct		Gene delivery method	Proposed objective	Country	Key references
	Regulator/promoter	Structural gene				
Tiger shrimp (*Penaeus monodon*)	Cytomegalovirus (CMV) promoter	β-Galactosidase reporter gene	Direct gene transfer into skeletal muscle	Methodology development	Brunei	Sulaiman (1999)
Tiger shrimp (*Penaeus monodon*)	Cytomegalovirus (CMV) promoter	Bacterial alkaline phosphatase gene	Electroporation	Methodology development	Taiwan	Tseng *et al.* (2000)
Kuruma prawn (*Penaeus japonicus*)	Information not available	Mutant tRNA gene (not essential for this experiment)	Microinjection; bombardment; electroporation in embryos	Methodology development	USA	Preston *et al.* (2000)
Giant freshwater prawn (*Macrobrachium rosenbergii*)	Cytomegalovirus (CMV) promoter	Green fluorescent protein (GFP) gene	Spermatophore-microinjection	Methodology development	Taiwan	Li and Tsai (2000)
Pacific white shrimp (*Litopenaeus vannamei*)	Shrimp β-actin promoter	β-Galactosidase reporter gene; enhanced green fluorescent protein (EGFP) reporter gene; Taura syndrome virus coat protein (sense or antisense)	Microinjection; electroporation	Disease resistance	USA	Sun *et al.* (2003); patent # WO03048325

Shrimp (*Litopenaeus schmitti*)	Cytomegalovirus (CMV) promoter	β-Galactosidase reporter gene	Electroporation (nuclear localizing sequence of SV40 T antigen involved)	Methodology development	Cuba	Arenal *et al.* (2004)
Dwarf surfclam (*Mulinia lateralis*)	Moloney murine leukemia virus long terminal repeat (MMLV-LTR); Rous sarcoma virus long terminal repeat (RSV-LTR)	Hepatitis B surface antigen; neomycin-resistant gene; β-galactosidase	Electroporation (pantropic pseudotyped retroviral vector used)	Methodology development	USA	Lu *et al.* (1996)
Crayfish (*Procambarus clarkii*)	Moloney murine leukemia virus long terminal repeat (MMLV-LTR); Rous sarcoma virus long terminal repeat (RSV-LTR)	Hepatitis B surface antigen; neomycin-resistant gene; β-galactosidase	Directly transforming immature gonads with replication-defective pantropic retroviral vectors	Methodology development	USA	Sarmasik *et al.* (2001)

only highlights these issues, as a more detailed discussion of transgene construction, integration and expression is presented in Chapter 4.

More efficient gene delivery methods

Microinjection is the primary method for introducing transgenes into fish eggs. However, many post-mortem studies of transgenic fish created through microinjection show that the frequency of transgene integration into host genomes is quite low. This shows that microinjection usually results in the random and late integration of multiple copies of the transgene into chromosomes. This causes the founder fish to have a high degree of mosaicism and unpredictable transgene expression and inheritance patterns. Unfortunately, no other practical gene delivery methods work well enough to replace microinjection in most fish species. However, a few gene delivery methods, including electroporation, may have potential in some species (Muller *et al.*, 1992; Powers *et al.*, 1992).

Co-injection of meganuclease

Improvements in gene transfer efficiency have been reported using the *I-SceI* meganuclease technique in model fish species (medaka, *Oryzias latipes*) (Thermes *et al.*, 2002). Meganucleases are sequence-specific endonucleases with large recognition sites, which can stimulate homologous gene targeting. Thermes

Box 3.2. Current technical limitations and drawbacks of fish transgenesis.

Transgene delivery and genomic integration

1. Rate of genomic integration of transgenes into host chromosomes is still quite low.
2. Eggs from some fish species are not suitable for microinjection, requiring alternative methods.
3. Transgene transfer via microinjection results in random integration of multiple copies of transgene.
4. Low frequency of germ-line transmission to progeny because most transgenic founders exhibit severe mosaicism caused by delayed transgene integration.

Transgene expression and transmission

1. Transgene expression is not under strict control due to lack of inducible or switchable vectors.
2. Levels of transgene expression are not uniform among individuals even within a single line.
3. Transgenic fish may suffer from pleiotropic transgene expression, which causes unintended phenotypic side effects.
4. Expression of transgene is frequently silenced in germ-line cells.
5. Inheritance of transgenes in subsequent generations is not always stable; meiotic drive or rearrangement of transgene can occur.

et al. (2002) flanked the transgenes with two *I-SceI* meganuclease recognition sites and co-injected together with the *I-SceI* meganuclease. They observed highly efficient transgenesis, demonstrated by uniform promoter-dependent expression and improved frequencies of both transgene integration and germ-line transmission, in the *I-SceI* meganuclease-injected group. There have been similar findings with other model fish species, including zebrafish (*D. rerio*) (Grabher *et al.*, 2004), and with a tetrapod vertebrate model, *Xenopus* (Ogino *et al.*, 2006; Pan *et al.*, 2006). The inclusion of *I-SceI* meganuclease during microinjection might offer a new way to overcome the main weakness of traditional microinjection: low efficiency of genomic integration and random incorporation of multiple transgene copies into host chromosomes. However, further research on the effectiveness of *I-SceI*-based transgenesis should be conducted with other fish species.

Retroviral, transposon or somatic nuclear transfer-based delivery

Several research groups have also demonstrated the potential for retroviral constructs or transposon-based vectors to increase transgene integration efficiency from microinjection (for reviews, see Hackett and Alvarez, 2000; Amsterdam and Becker, 2005; Tafalla *et al.*, 2006). However, as discussed in greater detail in Chapter 4, use of viral or transposon sequences may increase the risks associated with transgenesis. Recently, a somatic cell nuclear transfer technique (termed the 'Dolly technique') was demonstrated as a possible tool to produce transgenic fish, just as in mammals. Cultured zebrafish fish cells were manipulated *in vitro*, and nuclei were isolated from the transformed cells and transferred to enucleated eggs from donor zebrafish (Lee *et al.*, 2002). Unlike in the mammalian system, nuclear transfer in fish does not require reimplantation *in utero*, meaning that generation of transgenic fish via nuclear transfer could be much easier than in mammals. Furthermore, because most fish species have high fecundity, the availability of donor eggs does not pose the same limiting factor as in mammals.

Cell-based gene delivery

Another possible strategy for cell-based gene delivery involves embryonic stem cells (ESCs) or primordial germ cells (PGCs). A German research group tried to establish medaka embryonic stem cells for possible genetic engineering of ESC to generate transgenic fish (Hong *et al.*, 2000). More recently, a US research group showed progress on ESC-based transgenesis, including the culture of ESC, production of germ-line chimeras and targeted homologous recombination in ESCs (Fan and Collodi, 2002; Fan *et al.*, 2004, 2006). PGC-based techniques have been studied for fish transgenesis and other bioengineering applications (Yoshizaki *et al.*, 2000, 2003). These two techniques are valuable for bridging *in vitro* and *in vivo* manipulations for fish transgenesis; totipotent ESCs or germline competent PGCs can be manipulated *in vitro* and then used to form germline chimeras to generate individual fish with unique geno- and phenotypes. Further research is needed to increase the rate of success when producing viable transgenic fish using cell-based delivery techniques.

More powerful ways to induce or control expression of transgenes

The development of transgene constructs with sequences that provide stricter control of transgene expression (i.e. tissue-specific or inducible/switchable expression) has been a long-term goal in fish transgenesis. Ideally, sequences providing better control would replace viral promoters and regulators (e.g. CMV, SV40 or RSV-LTR) that are still being used to generate transgenic fish and shellfish. Although several piscine regulators, including the β-actin and antifreeze protein promoters, have been used in many transgenic fish lines, other promoter or regulatory elements of fish origin have limited availability despite their importance for providing more and better options for regulating transgene expression (Hackett and Alvarez, 2000). Several genetic elements or regulatory sequences have been proposed for developing tissue-specific or inducible transgenic constructs, but many of these ideas are based on theoretical or hypothetical possibilities (or design) rather than on empirical evidence from actual transgenic fish lines. Potential promoters for tissue-specific or inducible expression of transgenes in fish are listed in Box 3.3.

A genetic switch for regulating transgene expression is usually comprised of a specific inducer acting on a transcription factor that binds to the target promoter. The ideal inducible system should enable transgene expression to be switched on/off in a rapid and reversible manner. Any inducer compounds should also be safe for consumption or be eliminated or degraded during a pre-market withdrawal period. However, in many cases, the use of such inducible systems has been limited to cell culture with little success in actual transgenic animals (Yamamoto *et al.*, 2001; Rocha *et al.*, 2004). Inducer system limitations are often due to leaky expression, cellular toxicity, unstable transcripts and insensitivity to the inducer.

Box 3.3. Potential promoters for tissue-specific or inducible expression of transgenes in fish.

1. Vitellogenin promoter for possible liver-specific expression induced by oestrogen-like compounds in male transgenic fish (Maclean *et al.*, 2002).
2. Zona pellucida glycoprotein for oocyte-specific expression etc.
3. Vasa-like promoter for germ cell-specific expression (Yoshizaki *et al.*, 2000).
4. Warm temperature acclimation protein promoter allowing liver-specific expression, potentially controlled by thermal treatments (Kikuchi *et al.*, 1995).
5. Gulonolactone oxidase promoter for kidney-specific expression (Nam *et al.*, 2002b).
6. Skeletal α-actin or muscle creatine kinase promoter for muscle-specific expression (Ju *et al.*, 1999; Udvadia and Linney, 2003).
7. Cytochrome P-450 promoter with potential induction by aryl hydrocarbon (Rocha *et al.*, 2004).
8. Heat shock promoter with possible induction by heat treatment (Udvadia and Linney, 2003).
9. Tetracycline responsive element inducible by tetracycline or its analogue (Rocha *et al.*, 2004).
10. Oestrogen receptor promoter or oestrogen responsive element with potential induction by oestrogen-like compound (Udvadia and Linney, 2003).

More efficient and realistic transgene regulation in fish can be achieved only by further research on structural genes and regulatory elements. Collection of more baseline information on gene expression from fish tissues or organs is key to this research. Recent advances in several sophisticated tools for molecular genetics and bioinformatics provide guidance on how to select candidate genetic elements and improve regulation systems. Bulk isolation of expressed genes from different fish tissues by expressed sequence tag (EST) survey may be one good way to build knowledge on the transcriptional regulation of fish genes (Douglas *et al.*, 1999). Non-normalized transcription profiling (EST profiles) allows the approximate estimation of gene expression, which helps predict promoter strength for a target gene. Analysis of EST profiles of different tissues may be a good way to identify gene transcripts showing potential tissue-specific expression.

Microarray analyses, using either cDNA clones or oligonucleotides, can also show the transcription profile of a given genome, allowing rapid grouping of genes that show differential expression when exposed to a specific stimulus (Williams *et al.*, 2003; Peeters and Van der Spek, 2005). Regulation profiles of differentially expressed genes (DEGs), under different physiological conditions, could provide baseline information to inform the design of a transgenic inducible system. Another powerful technique is the suppression subtractive hybridization (SSH) method, which enables researchers to isolate the genes that are uniquely or differentially expressed under different cellular conditions (Alonso and Leong, 2002; Brown *et al.*, 2004; Tsoi *et al.*, 2004; Pinto *et al.*, 2005).

However, more work is required before these techniques can be applied in transgene construct design. Such work may include: (i) further mechanistic testing of selected genes for their expression under various biological and physical conditions; (ii) additional molecular manipulation of genes and promoters to improve control of regulatory strength and specificity; and (iii) the optimization of *in vivo* treatment conditions, especially regarding appropriate dosages of chemical inducers, treatment duration; and other physiological conditions of fish. Several of these advances are ongoing, especially in model fish species, but still need additional time and resources before the underlying complexities of gene regulation in fish are understood.

Chapter Summary

Transgenic manipulation of fish genomes offers a novel and efficient way to improve upon the successes and results of traditional breeding for aquaculture production. Many species have already undergone, or are currently undergoing, transgenic research for a wide variety of traits. Some of them, such as growth-enhanced carps, tilapias, salmons and mud loaches, are nearing consideration for commercial aquaculture use, with particular interest for developing country production. In addition to these four species, a wide variety of finfish and shellfish species and traits have also been targeted for research, with transgenic lines in early stages of research and development. Novel transgenic shellfish lines may be the next transgenic aquaculture species to approach commercial consideration. However, it is important to note that, despite the substantial progress of transgenic technology in fish, many unresolved issues, including their potential environmental risks, need further study and assessment. Consideration of methodologies for assessing and managing such environmental risks is presented in other chapters, providing ways to make fish transgenesis more predictable and controllable.

References

Abrahams, M.V. and Sutterlin, A. (1999) The foraging and anti-predator behaviour of growth-enhanced transgenic Atlantic salmon. *Animal Behavior* 58, 933–942.

Alonso, M. and Leong, J.-A. (2002) Suppressive subtraction libraries to identify interferon-inducible genes in fish. *Marine Biotechnology* 4, 74–80.

Amsterdam, A. and Becker, T.S. (2005) Transgenes as screening tools to probe and manipulate the zebrafish genome. *Developmental Dynamics* 234, 255–268.

Arenal, A., Pimentel, R., Garcia, C., Pimentel, E. and Alestrom, P. (2004) The SV40 T antigen nuclear localization sequence enhances nuclear import of vector DNA in embryos of a crustacean (*Litopenaeus schmitti*). *Gene* 337, 71–77.

Beardmore, J.M. (1997) Transgenics: autotransgenics and allotransgenics. *Transgenic Research* 6, 107–108.

Benfey, T.J. (1999) The physiology and behavior of triploid fishes. *Reviews in Fisheries Science* 7, 39–67.

Blechinger, S.R., Warren, J.T. Jr, Kuwada, J.Y. and Krone, P.H. (2002) Developmental toxicology of cadmium in living embryos of a stable transgenic zebrafish line. *Environmental Health Perspectives* 110, 1041–1046.

Brown, M., Davies, I.M., Moffat, C.F., Robinson, C., Redshaw, J. and Craft, J.A. (2004) Identification of transcriptional effects of ethynyl oestradiol in male plaice (*Pleuronectes platessa*) by suppression subtractive hybridization and a nylon macroarray. *Marine Environmental Research* 58, 559–563.

Cadoret, J.P., Boulo, V., Gendreau, S. and Mialhe, E. (1997) Promoters from *Drosophila* heat shock protein and cytomegalovirus drive transient expression of luciferase introduced by particle bombardment into embryos of the oyster *Crassostrea gigas*. *Journal of Biotechnology* 56, 183–189.

Chen, T.T., Kight, K., Lin, C.M., Powers, D.A., Hayat, M., Chatakondi, N., Ramboux, A.C., Duncan, P.L. and Dunham, R.A. (1993) Expression and inheritance of RSVLTR-rtGH1 complementary DNA in the transgenic common carp, *Cyprinus carpio*. *Molecular Marine Biology and Biotechnology* 2, 88–95.

Cheng, C.-A., Lu, K.-L., Lau, E.-L., Yang, T.-Y., Lee, C.-Y., Wu, J.-L. and Chang, C.-Y. (2002) Growth promotion in ayu (*Plecoglossus altivelis*) by gene transfer of the rainbow trout growth hormone gene. *Zoological Studies* 41, 303–310.

Cook, J.T., McNiven, M.A., Richardson, G.F. and Sutterlin, A.M. (2000a) Growth rate, body composition and feed digestibility/conversion of growth-enhanced transgenic Atlantic salmon (*Salmo salar*). *Aquaculture* 188, 15–32.

Cook, J.T., McNiven, M.A. and Sutterlin, A.M. (2000b) Metabolic rate of pre-smolt growth-enhanced transgenic Atlantic salmon (*Salmo salar*). *Aquaculture* 188, 33–45.

Cook, J.T., Sutterlin, A.M. and McNiven, M.A. (2000c) Effect of food deprivation on oxygen consumption and body composition of growth-enhanced transgenic Atlantic salmon (*Salmo salar*). *Aquaculture* 188, 47–63.

de la Fuente, J., Guillen, I., Martinez, R. and Estrada, M.P. (1999) Growth regulation and enhancement in tilapia: basic research findings and their applications. *Genetic Analysis: Biomolecular Engineering* 15, 85–90.

Devlin, R.H., Yesaki, T.Y., Biagi, C.A., Donaldson, E.M., Swanson, P. and Chan, W.-K. (1994) Extraordinary salmon growth. *Nature* 371, 209–210.
Devlin, R.H., Yesaki, Y.T., Donaldson, E.M., Du, S.J. and Hew, C.L. (1995) Production of germline transgenic Pacific salmonids with dramatically increased growth performance. *Canadian Journal of Fisheries and Aquatic Sciences* 52, 1376–1384.
Devlin, R.H., Swanson, P., Clarke, W.C., Plisetskaya, E., Dickhoff, W., Moriyama, S., Yesaki, T.Y. and Hew, C.L. (2000) Seawater adaptability and hormone levels in growth-enhanced transgenic coho salmon, *Oncorhynchus kisutch. Aquaculture* 191, 367–385.
Devlin, R.H., Biagi, C.A. and Yesaki, T.Y. (2004) Growth, viability and genetic characteristics of GH transgenic coho salmon strains. *Aquaculture* 236, 607–632.
Devlin, R.H., Sundstrom, L.F. and Muir, W.M. (2006) Interface of biotechnology and ecology for environmental risk assessments of transgenic fish. *Trends in Biotechnology* 24, 89–97.
Douglas, S.E., Gallant, J.W., Bullerwell, C.E., Wolff, C., Munholland, J. and Reith, M.E. (1999) Winter flounder expressed sequence tags: establishment of an EST database and identification of novel fish genes. *Marine Biotechnology* 1, 458–464.
Dunham, R.A. (2004) *Aquaculture and Fisheries Biotechnology: Genetic Approaches*. CAB International, Wallingford, UK.
Dunham, R.A., Warr, G., Nichols, A., Duncan, P.L., Argue, B., Middleton, D. and Liu, Z. (2002) Enhanced bacterial disease resistance of transgenic channel catfish, *Ictalurus punctatus*, possessing cecropin genes. *Marine Biotechnology* 4, 338–344.
Du, S.J., Gong, Z., Fletcher, G.L., Shears, M.A., King, M.J., Idler, D.R. and Hew, C.L. (1992) Growth enhancement in transgenic Atlantic salmon by the use of an 'all-fish' chimeric growth hormone gene construct. *Biotechnology* 10, 179–187.
Fan, L. and Collodi, P. (2002) Progress towards cell-mediated gene transfer in zebrafish. *Briefings of Functional Genomics and Proteomics* 1, 131–138.
Fan, L., Crodian, J. and Collodi, P. (2004) Production of zebrafish germline chimeras by using cultured embryonic stem (ES) cells. *Methods in Cell Biology* 77, 113–119.
Fan, L., Moon, J., Crodian, J. and Collodi, P. (2006) Homologous recombination in zebrafish ES cells. *Transgenic Research* 15, 21–30.
Fletcher, G.L., Davies, P.L. and Hew, C.L. (1992) Genetic engineering of freeze-resistant Atlantic salmon. In: Hew, C.L. and Fletcher, G.L. (eds) *Transgenic Fish*. World Scientific, London, pp. 190–208.
FAO (Food and Agriculture Organization of the United Nations) (2003) Fisheries Global Information System (FIGIS). FI Programme Websites. FAO, Rome, Italy. Available at: http://www.fao.org/figis/servlet/static?dom = org&xml = FIGIS_org.xml
Fu, C., Hu, W., Wang, Y. and Zhu, Z. (2005) Developments in transgenic fish in the People's Republic of China. *Revue Scientifique et Technique de l' Office International des Epizooties* 24, 299–307.
Gendreau, S., Lardans, V., Cadoret, J.P. and Miahle, E. (1995) Transient expression of luciferase reporter gene after biolistic introduction into *Artemia fransciscana* (Crustacea) embryos. *Aquaculture* 133, 199–205.
Gordon, J.W., Scangos, G.A., Plotkin, D.J., Barbosa, J.A. and Ruddle, R.H. (1980) Genetic transformation of mouse embryos by microinjection of purified DNA. *Proceedings of the National Academy of Sciences USA* 77, 380–384.
Grabher, C., Joly, J.S. and Wittbrodt, J. (2004) Highly efficient zebrafish transgenesis mediated by the meganuclease I-SceI. *Methods in Cell Biology* 77, 381–401.
Guillén, I.I., Berlanga, J., Valenzuela, C.M., Morales, A., Toledo, J., Estrada, M.P., Puentes, P., Hayes, O. and de la Fuente, J. (1999) Safety evaluation of transgenic tilapia with accelerated growth. *Marine Biotechnology* 1, 2–14.

Hackett, P.B. and Alvarez, M.C. (2000) The molecular genetics of transgenic fish. In: Fingerman, M. and Nagabhushanam, R. (eds) *Recent Advances in Marine Biotechnology, Volume 4*. Science Publishers, Enfield, New Hampshire, pp. 77–145.

Harper, G.S., Brownlee, A., Hall, T.E., Seymour, R., Lyons, R. and Ledwith, P. (2003) Global progress toward transgenic food animals: a survey of publicly available information. In: *Global Research and Development Targeting Transgenic Food Animals*. CSIRO Livestock Industries, Queensland, Australia.

Hernandez, O., Guillen, I., Estrada, M.P., Carbrera, E., Pimentel, R., Pina, J.C., Abad, Z., Sanchez, V., Hidalgo, Y., Martinez, R., Lleonart, R. and de la Fuente, J. (1997) Characterization of transgenic tilapia lines with different ectopic expression of tilapia growth hormone. *Molecular Marine Biology and Biotechnology* 6, 275–364.

Hew, C.L., Fletcher, G.L. and Davies, P.L. (1995) Transgenic salmon: tailoring the genome for food production. *Journal of Fish Biology* 47, 1–19.

Hew, C., Poon, R., Xiong, F., Gauthier, S., Shears, M., King, M., Davies, P. and Fletcher, G. (1999) Liver-specific and seasonal expression of transgenic Atlantic salmon harboring the winter flounder antifreeze protein gene. *Transgenic Research* 8, 405–419.

Holmes, R. (1996) Blue revolutionaries. *New Scientist* 152, 32–36.

Hong, Y., Chen, S. and Schartl, M. (2000) Embryonic stem cells in fish: current status and perspectives. *Fish Physiology and Biochemistry* 22, 165–170.

Hu, W., Yu, D.H., Wang, Y.P., Wu, K.C. and Zhu, Z.Y. (2000) Electroporation of sperm to introduce foreign DNA into the genome of *Pinctada maxima* (Jameson). *Sheng Wu Gong Cheng Xue Bao* 16, 165–168.

Hwang, G., Müller, F., Rahman, M.A., Williams, D.W., Murdock, P.J., Pasi, K.J., Goldspink, G., Farahmand, H. and Maclean, N. (2004) Fish as bioreactors: transgene expression of human coagulation factor VII in fish embryos. *Marine Biotechnology* 6, 485–492.

Ju, B., Xu, Y., He, J., Liao, J., Yan, T., Hew, C.L., Lam, T.J. and Gong, Z. (1999) Faithful expression of green fluorescent protein (GFP) in transgenic zebrafish embryos under control of zebrafish gene promoters. *Developmental Genetics* 25, 158–167.

Kapuscinski, A.R. (2005) Current scientific understanding of the environmental biosafety of transgenic fish and shellfish. *Revue Scientifique et Technique de l' Office International des Epizooties* 24, 309–322.

Keizaburo, M., Johji, M. and Nozomu, I. (2005) Transgenic mollusk and method for producing the same. US Patent (publication # 2005198697). Available at European Patent Office website: http://ep.espacenet.com/

Kikuchi, K., Yamashita, M., Watabe, S. and Aida, K. (1995) The warm temperature acclimation-related 65-kDa Protein, *Wap65*, in goldfish and its gene expression. *Journal of Biological Chemistry* 270, 17087–17092.

Krasnov, A., Agren, J.J., Pitkanen, T.I. and Molsa, H. (1999a) Transfer of growth hormone (GH) transgenes into Arctic charr (*Salvelinus alpinus* L.). II. Nutrient partitioning in rapidly growing fish. *Genetic Analysis: Biomolecular Engineering* 15, 99–105.

Krasnov, A., Pitkanen, T.I. and Molsa, H. (1999b) Gene transfer for targeted modification of salmonid fish metabolism. *Genetic Analysis* 15, 115–119.

Lee, K.Y., Huang, H., Yang, Z. and Lin, S. (2002) Cloned zebrafish by nuclear transfer from long-term-cultured cells. *Nature Biotechnology* 20, 795–799.

Li, S. (2004) Isolation of gonadotropin-releasing hormone from common carp (*Cyprinus carpio* L.) and primary study of the controlled reversible sterile transgenic fish. PhD dissertation. Institute of Hydrobiology, Chinese Academy of Sciences, Wuhan, China.

Li, S.S. and Tsai, H.J. (2000) Transfer of foreign gene to giant freshwater prawn (*Macrobrachium rosenbergii*) by spermatophore microinjection. *Molecular Reproduction and Development* 56, 149–154.

Lim, C. and Webster, C.D. (eds) (2006) *Tilapia: Biology, Culture, and Nutrition*. Food Products Press, Binghamton, New York.

Lu, J.K., Chen, T.T., Allen, S.K., Matsubara, T. and Burns, J.C. (1996) Production of transgenic dwarf surfclams, *Mulinia lateralis*, with pantropic retroviral vectors. *Proceedings of the National Academy of Sciences USA* 93, 3482–3486.

Maclean, N. (2003) Genetically modified fish and their effects on food quality and human health and nutrition. *Trends in Food Science and Technology* 14, 242–252.

Maclean, N. and Talwar, S. (1984) Injection of cloned genes into rainbow trout eggs. *Journal of Embryology and Experimental Morphology* 82, 187.

Maclean, N., Penman, D. and Zhu, Z. (1987) Introduction of novel genes into fish. *Biotechnology* 5, 257–261.

Maclean, N., Rahman, M.A., Sohm, F., Hwang, G., Iyengar, A., Ayad, H., Smith, A. and Farahmand, H. (2002) Transgenic tilapia and the tilapia genome. *Gene* 295, 265–277.

Mancera, J.M. and McCormick, S.D. (1998) Osmoregulatory actions of the GH/IGF axis in non-salmonid teleosts. *Comparative Biochemistry and Physiology*, Part B: *Biochemistry and Molecular Biology* 121, 43–48.

Martinez, R., Estrada, M.P., Berlanga, J., Guillen, J., Hernandez, O., Cabrera, E., Pimentel, R., Morales, R., Herrera, F., Morales, A., Pina, J.C., Abad, Z., Sanchez, V., Melamed, P., Lleonart, R. and de la Fuente, J. (1996) Growth enhancement in transgenic tilapia by ectopic expression of tilapia growth hormone. *Molecular Marine Biology and Biotechnology* 5, 62–70.

Martinez, R., Juncal, J., Zaldivar, C., Arenal, A., Guillen, I., Morera, V., Carrillo, O., Estrada, M., Morales, A. and Estrada, M.P. (2000) Growth efficiency in transgenic tilapia (*Oreochromis* sp.) carrying a single copy of an homologous cDNA growth hormone. *Biochemical and Biophysical Research Communications* 267, 466–472.

Morita, T., Yoshizaki, G., Kobayashi, M., Watabe, S. and Takeuchi, T. (2004) Fish eggs as bioreactors: the production of bioactive luteinizing hormone in transgenic trout embryos. *Transgenic Research* 13, 551–557.

Muller, F., Ivics, Z., Erdelyi, F., Papp, T., Vardi, L., Horvath, L. and Maclean, N. (1992) Introducing foreign genes into fish eggs by electroporated sperm as a carrier. *Molecular Marine Biology and Biotechnology* 1, 276–281.

Nam, Y.K., Cho, Y.S. and Kim, D.S. (2000) Isogenic transgenic homozygous fish induced by artificial parthenogenesis. *Transgenic Research* 9, 463–469.

Nam, Y.K., Cho, H.J., Cho, Y.S., Noh, J.K., Kim, C.G. and Kim, D.S. (2001a) Accelerated growth, gigantism and likely sterility in autotransgenic triploid mud loach *Misgurnus mizolepis*. *Journal of the World Aquaculture Society* 32, 353–363.

Nam, Y.K., Noh, J.K., Cho, Y.S., Cho, H.J., Cho, K.N., Kim, C.G. and Kim, D.S. (2001b) Dramatically accelerated growth and extraordinary gigantism of transgenic mud loach *Misgurnus mizolepis*. *Transgenic Research* 10, 353–362.

Nam, Y.K., Cho, Y.S., Cho, H.J. and Kim, D.S. (2002a) Accelerated growth performance and stable germ-line transmission in androgenetically derived homozygous transgenic mud loach, *Misgurnus mizolepis*. *Aquaculture* 209, 257–270.

Nam, Y.K., Cho, Y.S., Doulgas, S.E., Gallant, J.W., Reith, M.E. and Kim, D.S. (2002b) Isolation and transient expression of a cDNA encoding l-gulono-γ-lactone oxidase, a key enzyme for l-ascorbic acid biosynthesis, from the tiger shark *Scyliorhinus torazame*. *Aquaculture* 209, 271–284.

Nam, Y.K., Park, I.S. and Kim, D.S. (2004) Triploid hybridization of fast-growing transgenic mud loach *Misgurnus mizolepis* male to cyprinid loach *Misgurnus anguillicaudatus* female: the first performance study on growth and reproduction of transgenic polyploid hybrid fish. *Aquaculture* 231, 559–572.

Ogino, H., McConnell, W.B. and Grainger, R.M. (2006) Highly efficient transgenesis in *Xenopus* tropicalis using I-SceI meganuclease. *Mechanisms of Development* 123, 103–113.

Palmiter, R.D., Brinster, R.L., Hammer, R.E., Trumbauer, M.E., Rosenfeld, M.G., Birnberg, N.C. and Evans, R.M. (1982) Dramatic growth of mice that develop from eggs micro-injected with metallothionein-growth hormone fusion gene. *Nature* 300, 611–615.

Pan, F.C., Chen, Y., Loeber, J., Henningfeld, K. and Pieler, T. (2006) I-SceI meganuclease-mediated transgenesis in *Xenopus. Developmental Dynamics* 235, 247–252.

Pandian, T.J. and Venugopal, T. (2005) Contribution to transgenesis in Indian major carp *Labeo rohita*. In: Pandian, T.J., Strussmann, C.A. and Marian, M.P. (eds) *Fish Genetics and Aquaculture Biotechnology*. Oxford/IBH Publishing, New Delhi, India, pp. 1–20.

Paynter, K.T. Jr. and Thibodeau, F.R. (1996) Enhancing growth and pearl production in mollusks. *World Intellectual Property Organization* (publication # WO9615662). Available at European Patent Office website: http://ep.espacenet.com/

Peeters, J.K. and Van der Spek, P.J. (2005) Growing applications and advancements in microarray technology and analysis tools. *Cell Biochemistry and Biophysics* 43, 149–166.

Philippart, J.C. and Ruwet, J.C. (1982) Ecology and distribution of tilapias. In: Pullin, R.S.V. and Lowe-McConnell, R.H. (eds) *The Biology and Culture of Tilapias* (*ICLARM Conference Proceedings* 7). International Center for Living Aquatic Resources Management, Manila, Philippines, pp. 15–59.

Pinto, P.I.S., Teodosio, H.R., Galay-Burgos, M., Power, D.M., Sweeney, G.E. and Canario, A.V.M. (2005) Identification of estrogen-responsive genes in the testis of sea bream (*Sparus auratus*) using suppression subtractive hybridization. *Molecular Reproduction and Development* 73, 318–329.

Pitkanen, T.I., Krasnov, A., Reinisalo, M. and Molsa, H. (1999a) Transfer and expression of glucose transporter and hexokinase genes in salmonid fish. *Aquaculture* 173, 319–333.

Pitkanen, T.I., Krasnov, A., Teerijoki, H. and Molsa, H. (1999b) Transfer of growth hormone (GH) transgenes into Arctic charr (*Salvelinus alpinus* L.). I. Growth response to various GH constructs. *Genetic Analysis: Biomolecular Engineering* 15, 91–98.

Pohajdak, B., Mansour, M., Hrytsenko, O., Conlon, M., Dymond, C. and Wright Jr, J.R. (2004) Production of transgenic tilapia with Brockmann bodies secreting [desThrB30] human insulin. *Transgenic Research* 13, 313–323.

Powers, D.A., Hereford, L., Cole, T., Chen, T.T., Lin, C.M., Kight, K., Creech, K. and Dunham, R. (1992) Electroporation: a method for transferring genes into the gametes of zebrafish (*Brachydanio rerio*), channel catfish (*Ictalurus punctatus*) and common carp (*Cyprinus carpio*). *Molecular Marine Biology and Biotechnology* 1, 301–308.

Powers, D.A., Kirby, V.L., Cole, T. and Hereford, L. (1995) Electroporation as an effective means of introducing DNA into abalone (*Haliotis rufescens*) embryos. *Molecular Marine Biology and Biotechnology* 4, 369–375.

Powers, D.A., Hereford, L.M. and Gomez-Chiarri, M. (1997) Isolation and characterization of an actin gene from abalone. US Patent (publication # US5675061). Available at European Patent Office website: http://ep.espacenet.com/

Preston, N.P., Baule, V.J., Leopold, R., Henderling, J., Atkinson, P.W. and Whyard, S. (2000) Delivery of DNA to early embryos of the Kuruma prawn, *Penaeus japonicus*. *Aquaculture* 181, 225–234.

Purdom, C.E. (1993) *Genetics and Fish Breeding*. Chapman & Hall, London.

Rahman, M.A. and Maclean, N. (1999) Growth performance of transgenic tilapia containing an exogenous piscine growth hormone gene. *Aquaculture* 173, 333–346.

Rahman, M.A., Mak, R., Ayad, H., Smith, A. and Maclean, N. (1998) Expression of a novel growth hormone gene results in growth enhancement in transgenic tilapia. *Transgenic Research* 7, 357–369.

Rahman, M.A., Hwang, G., Razak, S.A., Sohm, F. and Maclean, N. (2000) Copy number dependent transgene expression in hemizygous and homozygous fish growth hormone gene. *Transgenic Research* 9, 417–427.

Rahman, M.A., Ronyai, A., Engidaw, B.Z., Jauncey, K., Hwang, G.-L., Smith, A., Roderick, E., Penman, D., Varadi, L. and Maclean, N. (2001) Growth and nutritional trials on transgenic Nile tilapia containing an exogenous fish growth hormone gene. *Journal of Fish Biology* 59, 62–78.

Rocha, A., Ruiz, S., Estepa, A. and Coll, J.M. (2004) Application of inducible and targeted gene strategies to produce transgenic fish: a review. *Marine Biotechnology* 6, 118–127.

Sakamoto, T., Shepherd, B.S., Madsen, S.S., Nishioka, R.S., Siharath, K., Richman, N.H., Bern, H.A. and Grau, E.G. (1997) Osmoregulatory actions of growth hormone and prolactin in an advanced teleost. *General and Comparative Endocrinology* 106, 95–101.

Sarmasik, A., Jang, I.-K., Chun, C.Z., Lu, J.K. and Chen, T.T. (2001) Transgenic live-bearing fish and crustaceans produced by transforming immature gonads with replication defective pantropic retroviral vectors. *Marine Biotechnology* 3, 470–477.

Stevens, E.D., Sutterlin, A. and Cook, T. (1998) Respiratory metabolism and swimming performance in growth hormone transgenic Atlantic salmon. *Canadian Journal of Fisheries and Aquatic Sciences* 55, 2028–2035.

Stevens, E.D., Wagner, G.N. and Sutterlin, A. (1999) Gut morphology in growth hormone transgenic Atlantic salmon. *Journal of Fish Biology* 55, 517–526.

Stickney, R.R. (1986) Tilapia tolerance of saline waters: a review. *The Progressive Fish-Culturist* 48, 161–167.

Sulaiman, Z.H. (1999) Direct gene transfer into skeletal muscle of seabass (*Lates calcarifer*) and black tiger prawns (*Penaeus monodon*). *Science Asia* 25, 73–75.

Sun, S.S.M., Sun, P.S. and Arakaki, K.L. (2003) Nucleotide sequences of shrimp beta-actin and actin promoters and their use in genetic transformation technology. *World Intellectual Property Organization* (publication # WO03048325). Available at European Patent Office website: http://ep.espacenet.com/

Tafalla, C., Estepa, A. and Coll, J.M. (2006) Fish transposons and their potential use in aquaculture. *Journal of Biotechnology* 123, 397–412.

Talbot, F. and Newell, B. (1957) A preliminary note on the breeding and growth of tilapia. *East African Agriculture Journal* 22, 118–121.

Thermes, V., Grabher, C., Ristoratore, F., Bourrat, F., Choulika, A., Wittbrodt, J. and Joly, J.S. (2002) I-SceI meganuclease mediates highly efficient transgenesis in fish. *Mechanisms of Development* 118, 91–98.

Tiku, P.E., Gracey, A.Y., Macartney, A.L., Benyon, R.J. and Cossins, A.R. (1996) Cold induced expression of 9-desaturase in carp by transcriptional and post-translational mechanisms. *Science* 271, 815–818.

Tsai, H.J. (2000) Electroporated sperm mediation of a gene transfer system for finfish and shellfish. *Molecular Reproduction and Development* 56, 281–284.

Tsai, H.J., Lai, C.H. and Yang, H.S. (1997) Sperm as a carrier to introduce an exogenous DNA fragment into the oocyte of Japanese abalone (*Haliotis diversicolor suportexta*). *Transgenic Research* 6, 85–95.

Tseng, F.S., Tsai, H.J., Liao, I.C. and Song, Y.L. (2000) Introducing foreign DNA into tiger shrimp (*Penaeus monodon*) by electroporation. *Theriogenology* 54, 1421–1432.

Tsoi, S.C., Ewart, K.V., Penny, S., Melville, K., Liebscher, R.S., Brown, L.L. and Douglas, S.E. (2004) Identification of immune-relevant genes from Atlantic salmon using suppression subtractive hybridization. *Marine Biotechnology* 6, 199–214.

Udvadia, A.J. and Linney, E. (2003) Windows into development: historic, current, and future perspectives on transgenic zebrafish. *Developmental Biology* 256, 1–17.

Uzbekova, S., Chyb, J., Ferriere, F., Bailhache, T., Prunet, P., Alestrom, P. and Breton, B. (2000) Transgenic rainbow trout expressed sGnRH-antisense RNA under the control of sGnRH promoter of Atlantic salmon. *Journal of Molecular Endocrinology* 25, 337–350.

Venugopal, T., Anathy, V., Pandian, T.J. and Mathavan, S. (2002) Molecular cloning of growth hormone encoding cDNA of an Indian major carp *Labeo rohita* and its expression in *E. coli* and zebrafish. *General Comparative Endocrinology* 125, 236–247.

Wang, R., Zhang, P., Gong, Z. and Hew, C.-L. (1995) Expression of the antifreeze protein gene in transgenic goldfish (*Carassius auratus*) and its implication in cold adaptation. *Molecular Marine Biology and Biotechnology* 4, 20–26.

Wang, Y., Hu, W., Wu, G., Sun, Y., Chen, S., Zhang, F., Zhu, Z., Feng, J. and Zhang, X. (2001) Genetic analysis of 'all-fish' growth hormone gene transferred carp (*Cyprinus carpio* L.) and its F_1 generation. *Chinese Science Bulletin* 46, 1174–1177.

Williams, T.D., Gensberg, K., Minchin, S.D. and Chipman, J.K. (2003) A DNA expression array to detect toxic stress response in European flounder (*Platichthys flesus*). *Aquatic Toxicology* 65, 141–157.

Winn, R.N. (2001) Bacteriophage-based transgenic fish for mutation detection. US Patent (publication # US6307121). Available at European Patent Office website: http://ep.espacenet.com

Wu, G., Sun, Y. and Zhu, Z. (2003) Growth hormone gene transfer in common carp. *Aquatic Living Resources* 16, 416–420.

Wu, S., Hwang, P., Hew, C. and Wu, J. (1998) Effect of antifreeze protein on cold tolerance in juvenile tilapia (*Oreochromis mossambicus*) and milkfish (*Chanos chanos*). *Zoological Studies* 37, 39–44.

Yamamoto, A., Hen, R. and Dauer, W.T. (2001) The ons and offs of inducible transgenic technology: a review. *Neurobiology of Disease* 8, 923–932.

Yoshizaki, G., Takeuchi, Y., Sakatani, S. and Takeuchi, T. (2000) Germ cell-specific expression of green fluorescent protein in transgenic rainbow trout under control of the rainbow trout vasa-like gene promoter. *International Journal of Developmental Biology* 44, 323–326.

Yoshizaki, G., Takeuchi, Y., Kobayashi, T. and Takeuchi, T. (2003) Primordial germ cell: a novel tool for fish bioengineering. *Fish Physiology and Biochemistry* 28, 453–457.

Zbikowska, H.M. (2003) Fish can be first – Advances in fish transgenesis for commercial applications. *Transgenic Research* 12, 379–389.

Zeng, Z., Hu, W., Wang, Y., Zhu, Z., Zhou, G., Liu, S., Zhang, X., Luo, C. and Liu, Y. (2000) The genetic improvement of tetraploid fish by pCAgcGHc-transgenism. *High Technology Letters* 10, 6–12.

Zeng, Z., Shan, T., Tong, Y., Lam, S.H. and Gong, Z. (2005) Development of estrogen-responsive transgenic medaka for environmental monitoring of endocrine disrupters. *Environmental Science and Technology* 39, 9001–9008.

Zhang, P., Hayat, M., Joyce, C., Gonzalez-Villasenor, L.I., Lin, C.M., Dunham, R.A., Chen, T.T. and Powers, D.A. (1990) Gene transfer, expression and inheritance of pRSV-rainbow trout-GH cDNA in the common carp *Cyprinus carpio* (Linnaeus). *Molecular Reproduction and Development* 25, 3–13.

Zhong, J., Wang, Y. and Zhu, Z. (2002) Introduction of human lactoferrin gene into grass carp (*Ctenopharyngodon idellus*) to increase resistance against GCH virus. *Aquaculture* 214, 93–101.

Zhu, Z., Li, G., He, L. and Chen, S. (1985) Novel gene transfer into the fertilized eggs of goldfish (*Carassius auratus* L. 1758). *Journal of Applied Ichthyology* 1, 31–34.

4 Gene Construct and Expression: Information Relevant for Risk Assessment and Management

Z. Gong, N. Maclean, R.H. Devlin, R. Martinez,
O. Omitogun and M.P. Estrada

Introduction

Although concerns over the environmental risks posed by transgenic organisms are wide ranging, many begin at the basic level of the transgene itself: how it is constructed, integrated and expressed in the transgenic organism. Some of these concerns include questions about the impacts of a transgene integrating into an existing functional gene, unexpected expression of a gene product (protein) after transgenesis or genes from a viral vector recombining with those of an endogenous viral sequence to create a novel pathogen. Understanding some fundamental concepts about transgene locus structure and expression provides the foundation to begin an environmental risk assessment of a transgenic organism. This chapter introduces and explores the main questions about transgene loci and their expression that are relevant for assessing possible environmental effects of transgenic aquatic organisms, particularly transgenic finfishes. The information presented in this chapter is useful for assessing a pre-existing transgenic organism that a particular party wishes to introduce for widespread use. The information can also provide guidance at an early stage of research and development of a new transgenic organism destined for aquaculture or other large-scale application.

The transgene locus is the actual physical location of the integrated transgene in a chromosome; and a variety of genetic elements could be integrated, including the protein coding region (exons), promoter, introns (non-coding regions of DNA) and transcriptional termination signals. If plasmid DNA is also introduced, the transgene locus may also contain an antibiotic resistance gene and its promoter in addition to the plasmid backbone sequence. The exact process of genomic integration is still not understood, and the integration locus and final structure of integrated transgenes remain unpredictable. Although regulation of transgene expression is relatively well known, the expression pattern of the inserted transgene may not always be predictable

 Environmental Risk Assessment of Genetically Modified Organisms: Vol. 3. Methodologies for Transgenic Fish (eds A.R. Kapuscinski *et al.*)

because of the potential chromosomal effect, which can vary across different sites of transgene integration. Because each transgene integration event is unique, risk assessment of the transgene must be done on a case-by-case basis. Basic molecular genetics information, such as the type of vector used, is needed to inform all risk assessments. Such information provides the basis for a systematic scientific assessment of the risks associated with the transgene. As the process of transgenesis is essentially the same in animals and plants, the risk concerns and assessments with regard to the transgene constructs and integration locus are also similar (Andow *et al.*, 2004; Grossi-de Sa *et al.*, 2006).

Transgene Constructs and Transgenesis

As with any prokaryotic or eukaryotic gene, a transgene construct consists of three basic and essential components: a promoter, a protein coding region and a transcriptional termination signal (terminator). The chapter presents a brief summary of the main components of a transgene construct, followed by a more detailed discussion of the features of transgene integration and expression relevant for risk assessment.

Elements of a transgene construct for transgenic fish

The promoter is critical to determine in which tissues and developmental stages the transgene is expressed, as well as its level of expression. The minimal promoter harbours a transcription initiation site and provides a physical site for assembly of the transcriptional machinery. Promoters also often include additional sequence elements, which provide binding sites (named *cis*-elements or enhancers if the sites are for transcriptional activators) for specific transcription factors. The combination of the minimal promoter and enhancer regions provides a fully functional promoter for use in transgenic fish. The protein coding region is the core of the transgene construct, and is transcribed into mRNA, which serves as a template for protein synthesis. For example, the growth hormone (GH) coding region of transgenes, which has been inserted into a number of aquaculture fish species (see Chapter 3, this volume), expresses GH in certain tissues, leading to increased rates of growth in transgenic fish. The last component of the construct, the transcriptional terminator, consisting of a polyadenylation signal and a subsequent T-rich region, signals where to terminate transcription of mRNA.

An example of a transgene construct for transgenic fish is shown in Fig. 4.1. It contains a liver-specific promoter from an ocean pout antifreeze protein gene (*opAFP*), a salmon GH cDNA and a termination signal from the same *opAFP* gene (Du *et al.*, 1992a). This construct has been successfully used in generating a number of fish species with enhanced growth rate (Du *et al.*, 1992b; Devlin *et al.*, 1995a; Tsai *et al.*, 1995; Rahman *et al.*, 1998).

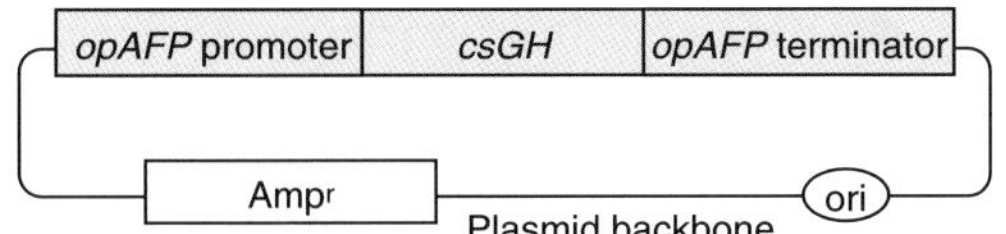

Fig. 4.1. Diagram of the *opAFP–csGH* construct used in the generation of growth hormone (GH)-transgenic fish. The construct consists of a liver-specific promoter derived from the ocean pout antifreeze protein gene (*opAFP*), chinook salmon growth hormone (*csGH*) cDNA and *opAFP* transcription termination signal (*opAFP* terminator) in a plasmid backbone that contains an ampicillin resistance gene (Amp^r) and a DNA replication origin (ori) for prokaryotic cells. Prior to introduction of the construct into transgenic fish, the plasmid backbone, including the antibiotic resistance gene, can be removed by restriction enzymes (e.g. Devlin *et al.*, 1995a). The construct has been used successfully in several salmonid species (Atlantic salmon, coho salmon, chinook salmon, rainbow trout and cut-throat trout; Du *et al.*, 1992b; Devlin *et al.*, 1995a,b), loach (Tsai *et al.*, 1995) and tilapia (Rahman *et al.*, 1998). The Aqua Bounty AquAdvantage™ transgenic salmon currently under review by the US Food and Drug Administration were produced with this construct.

DNA replication using a plasmid vector system

Once the transgene construct of interest is assembled *in vitro*, it generally is multiplied in bacterial cells using a plasmid DNA vector. This vector is usually comprised of at least a replication origin and an antibiotic resistance gene. The former is for amplification of plasmid DNA in bacterial cells, and the latter is required to facilitate the use of an antibiotic to select for bacteria that contain the DNA construct. The plasmid backbone, including the replication origin and the antibiotic resistance gene, is not required for the production of transgenic fish and can be easily removed before the construct is introduced into fish cells (e.g. Devlin *et al.*, 1995a).

Delivery of the transgene construct

Once a transgene construct is made, the next step is to introduce it into fish. Generally, fish eggs are targeted for gene delivery, and the most popular approach is to use a fine glass needle to directly inject DNA into the eggs. This procedure is called microinjection, and is generally performed on fertilized eggs (or occasionally in unfertilized eggs or oocytes). So far, the majority of transgenic fish have been produced by microinjection. Several other methods of transgene delivery have also been reported for the generation of transgenic fish, including electroporation, gene bombardment, liposome-mediated gene transfer and use of retroviruses (for further discussion of methods of transgene delivery, see Chapter 3, this volume).

Integration of the transgene

In most cases, the injected foreign DNA is eventually lost. However, in a small percentage (generally $<10\%$) of injected eggs, the foreign DNA is integrated into the host genome and thus permanently retained in the transgenic fish (for a review, see Gong and Hew, 1995). When the foreign DNA is integrated into the chromosomes of cells destined to become germ cells (eggs or sperm cells), it may be transmitted to the next generation, and thus a stable transgenic line may be established.

However, exogenously introduced DNA is generally not present in all of the cells in the founder fish, yielding mosaic transgenic individuals. Mosaicism is a result of transgene integration delayed until after a few rounds of cell division, so that only a fraction of embryonic cells actually have integrated transgene copies. In adult transgenic fish, each founder fish has a distinct pattern of foreign DNA persistence in tissues examined. In one study, for example, tissues showed a similar frequency of foreign DNA retention (Gross *et al.*, 1992).

Multiple copies of the transgene are frequently integrated into the genome at the same chromosomal locus. This is due to the formation of transgene concatemers, or repetitions of the DNA sequence, shortly after microinjection. In addition, transgene integration into multiple sites on different chromosomes is also possible. The event of DNA integration into the host genome is poorly understood, and the integration site is also unpredictable.

Inheritance of the transgene

Following DNA injection, individuals must be screened for the presence of the transgene. This is generally performed by use of the polymerase chain reaction (PCR). Transgenic founders are then crossed with a wild-type individual to determine whether the transgene is passed to the F_1 generation. The transmission rate from the founder generation to the F_1 generation is variable, generally ranging from near zero to less than 50% for a transgene with a single integration locus. The variation in transmission rates is due to the fact that founder fish are mosaic, and the foreign DNA is unevenly integrated in their gonadal tissue. The fraction of sperm or eggs containing the transgene in a transgenic founder varies and is generally below the Mendelian ratio of 50% expected for a single insertion event (e.g. Nam *et al.*, 1999; Gong *et al.*, 2002). However, transmission rates can be higher than 50%, which indicates the presence of multiple transgene inserts in the founder fish, which can be discerned by Southern blot analysis of F_1 progeny (Gaiano *et al.*, 1996).

However, in backcrosses with the non-transgenic strain, the transmission rate from transgenic F_1 to F_2 offspring is usually in a typical Mendelian ratio. This means that 50% of the offspring from a cross between a transgenic F_1 and wild-type fish will be transgenic. This is because the transgenic F_1 individual is a hemizygote, i.e. bears the transgene on only one homologous chromosome. In all future generations, the transmission rate is strictly based on Mendelian genetics; thus, the transgene has been stably integrated into the host genome, and a stable line of transgenic fish is established from the F_1 generation (e.g. Nam *et al.*, 1999; Gong *et al.*, 2002; Udvadia and Linney, 2003). So far, stable transmission of transgenes in Mendelian fashion has been observed for more than six generations in many of the transgenic fish developed for aquaculture purposes (see Chapter 3, this volume).

Expression of Transgenes

Once a stable transgenic line is generated, the transgene expression pattern is an important factor in determining phenotypic changes in the transgenic organism. It is important to understand the process of transgene expression before one can assess the potential benefits and risks of any phenotypic changes.

Importance of gene promoters

An important determinant in transgene expression is the use of a suitable gene promoter to drive expression of the target gene. Choice of a suitable promoter will help ensure that the target gene is expressed in the correct place (cells, tissues and organs), at the correct time (developmental stages), at the appropriate physiological level and under a particular condition (i.e. in the presence or absence of controlling environmental factors). Promoters can be constitutive or inducible. Constitutive promoters drive the downstream gene to be expressed at all times, whereas inducible promoters activate the gene only under specific conditions (i.e. certain temperatures or the presence of specific chemicals and hormones). In terms of tissue expression, there are ubiquitous promoters and tissue-specific promoters; the former drive the expression of genes in all types of tissues or cells, while the latter drive the expression of genes only in certain tissues or cells. For example, a liver-specific promoter will restrict gene expression to liver cells.

When transgenic fish technology was developed in the 1980s, fish molecular biology was in its infancy and the availability of DNA sequences derived from fish species was quite limited. Thus, essentially all early transgenic fish studies used heterologous promoters from non-fish origins, such as mice, humans and viruses. Commonly used promoters included the mouse metallothionein gene promoter, the simian virus (SV40) early gene promoter and the Rous sarcoma virus (RSV) promoter/enhancer. Although many groups reported successful generation of GH-transgenic fish bearing constructs with viral promoters and showing significant growth enhancement, viral promoters could raise questions when used in transgenic foodfish. The viral promoters used are generally ubiquitous promoters. Although there is no evidence to indicate harmful effects of viral promoters on the health of transgenic fish, people may have concerns due to uncertainty regarding their potential to recombine with viruses in the fish's cells or surrounding aquatic environment, or to turn on cancer genes in the fish genome.

From the early 1990s, fish promoters have been used in developing transgenic fish. Fish gene promoters, such as the carp β-actin promoter (Liu *et al.*, 1990), *opAFP* (Du *et al.*, 1992a) and salmon metallothionein-B promoter (Devlin *et al.*, 1994) have proven successful. Both β-actin and metallothionein-B promoters are strong promoters, with broad application in transgenic fish development. These fish promoters were used to construct 'all-fish' (or even 'all-salmon') transgene constructs (Liu *et al.*, 1990; Du *et al.*, 1992b; Devlin *et al.*, 1994). In recent years, researchers have tended to use gene components all from the same species as the transgenic host, generating what are called 'autotransgenic' fish. It is believed that genes from the host species are more equivalent to genes already present in the genome of the recipient (Hwang *et al.*, 2003), and that using them avoids concerns over the insertion of genes from other organisms (such as viruses and bacteria).

Faithful expression and unexpected expression of the transgene

Faithful expression of transgenes in fish can be directed by a homologous promoter (from the same fish species) or a heterologous promoter (from another species) provided the promoter is sufficiently long to include all important gene regulatory elements (Long *et al.*, 1997; Zeng *et al.*, 2005a). Faithful expression occurs when transgenes show the same temporal and spatial expression pattern as the endogenous gene from which the promoter is derived. For most genes showing a simple expression pattern (i.e. they are expressed in only one or a few cell types), short promoters (usually <2 kb) from these genes are sufficient to provide correct tissue specificity (e.g. Gong *et al.*, 2001; Ju *et al.*, 2003). Most fish promoters have strong functional conservation in related fish species. For example, liver-dominant gene expression driven by the *opAFP* promoter was observed in GH-transgenic Atlantic salmon, chinook salmon, coho salmon, rainbow and cut-throat trout, loach and tilapia (Du *et al.*, 1992b; Devlin *et al.*, 1995a; Tsai *et al.*, 1995; Rahman *et al.*, 1998).

Although traits targeted or phenotypes expected have been obtained in most reported transgenic fishes (see Chapter 3, this volume), occasionally, changes in non-targeted traits may occur (Devlin *et al.*, 2006). Transgene expression may not depend only on the promoter used and may also be affected by neighbouring regulatory elements in the host chromosome where the transgene is inserted. Thus, ectopic expression (gene expression in a tissue where it is not normally expressed) of the transgene could occur outside of the cells in which the promoter is normally active (e.g. Zeng *et al.*, 2005b). Sometimes the transgene may not be expressed because it is inserted into a heterochromatin region where genes can be transcriptionally silenced.

Pleiotropic effects from over-expression of a transgene can also affect non-target traits. For example, over-expression of the GH gene in many GH-transgenic fish results not only in enhanced growth (the targeted trait), but also in a broad range of pleiotropic effects (e.g. morphological changes in the skeleton and gut, behavioural changes in feeding and swimming and changes in other endocrine hormones (Devlin *et al.*, 1995b; Ostenfeld *et al.*, 1998)). Researchers can eliminate transgenic fish with undesirable traits during screening and characterization of the multiple lines of transgenic individuals produced (see Chapter 6, this volume, for further discussion of pleiotropic effects).

Stability and silencing of transgene expression

Data on stability of transgene expression are essential information for assessing environmental risks of transgenic fish on a case-by-case basis. Stable, long-term transgene expression and multiple generations of expected phenotypes have been observed. For example, fluorescent transgenic zebrafish, or Glofish, R., have exhibited stable transgene inheritance and fluorescent phenotype for over ten generations (Gong *et al.*, 2003; Gong *et al.*, 2006). Transgene silencing, which may be a defence mechanism developed by the host organisms against abnormal gene expression or alien transcription units, has been reported in some transgenic plants and mammals after a few generations (Henikoff, 1998; Lorence and Verpoorte, 2004). Studies also report observations of gene silencing in several transgenic fish lines (e.g. Nam *et al.*, 1999; Uzbekova *et al.*, 2003).

Transgene expression in different genetic backgrounds

Sometimes the same transgene construct may have different effects on transgenic fish phenotypes in different genetic backgrounds. The best documented example is provided by GH-transgenic rainbow trout lines derived from a wild strain and a domestic strain. The slow-growing wild strain was stimulated to grow three to ten times faster by transgenesis, but there was little or no growth enhancement with the fast-growing domestic strain (Devlin *et al.*, 2001). (See Chapters 5 and 6, this volume, for a detailed discussion of the effects of genetic background and environmental factors on transgene expression and resulting phenotypes.)

Overview of Main Concerns in Transgenesis and Transgene Expression

Unintended genotypic effects

Introducing DNA randomly into a genome is a potentially mutagenic event, depending on whether the transgene is integrated into a functional gene and whether genetic material besides the transgene itself is inserted; this has been observed in some lines of transgenic mice (NRC, 2002) and could also occur in transgenic fish. Moreover, transgene promoters and enhancers could activate other genes adjacent to or even some distance from the site of integration. These actions have the potential to cause unexpected genotypic and even phenotypic changes in the transgenic organism (NRC, 2002). Such changes, if they affect ecologically important traits, could increase uncertainty in conducting an environmental risk assessment (see Chapter 7, this volume, for a discussion of the types of uncertainty encountered during risk assessment and how to address them).

Transgene insertion in host genomes

The transgene may be integrated at multiple sites in a fish genome. One potential unintended effect from this is insertional mutagenesis, a mutational effect on a host gene due to transgene insertion. Fish genomes contain only a small percentage of their total DNA as coding or regulatory sequences, and the frequency of interruption of a functional gene or regulatory element is low. However, the potential for interrupting functional genes has been demonstrated by a large-scale screening of insertional mutagenesis in zebrafish that used a highly efficient viral vector for transgene integration (Gaiano *et al.*, 1996).

Targeted integration of transgenes into specific genome locations would reduce variability in transgene expression levels resulting from position effects (differential expression of the transgene based on its position in the chromosome). Targeted integration would make it easier to select individuals with desired levels of expression and minimum pleiotropic effects. Gene targeting technology (i.e. insertion of DNA into a desired chromosomal locus by homologous recombination) was developed in transgenic mice. This involved inserting a genetic construct into embryonic stem cells *in vitro*, selecting from among the transformed cells for the desired insertion site after homologous recombination, then making chimeric embryos using modified stem cells, and finally generating an entire transgenic mouse. However, such gene targeting technology has not yet been established in fish species despite efforts by several fish research groups (e.g. Hong *et al.*, 1998; Fan *et al.*, 2004). If ever achieved in fish, targeted integration would reduce phenotypic variability, an important source of scientific uncertainty in an environmental risk assessment (see Chapter 7, this volume).

Creation of new pathogens

Viral delivery systems (e.g. retroviral vectors) have been attempted in fish (Gaiano *et al.*, 1996; Lu *et al.*, 1997), and they may pose the risk of creating a novel pathogen. The viral delivery system may contain sequences that are capable of recombining with endogenous or exogenous viruses to create new types of viruses. In particular, viruses continually modify their genetic make-up and evolve, allowing for their prolonged persistence and virulence in the environment. Concerns about creating novel pathogens would be overcome by avoiding the use of viral vectors in transgenesis.

Mobile genetic elements

Mobile genetic elements, also known as transposons, can move from one chromosomal site to another and have been used by genetic engineers to improve transgene integration (Ivics *et al.*, 1997; Kawakami, 2004). The use of transposons may raise questions about unintended movements of DNA within the host genome, as well as horizontal gene flow between different species. There is some evidence of conservation of transposon-like sequences across species, leading to the hypothesis that these elements may have been distributed by transfer of genes via transposon-like elements such as the *mariner* and *Tc1* (Plasterk *et al.*, 1999; Krasnov *et al.*, 2005). In the process of generating transgenic fish, the transposon elements and specific transposases (or their mRNAs) are generally co-injected into fish eggs. Normally, these transposon elements become immobile after integration into the host genome because of the lack of a specific transposase. Thus, generally these integrated transposon elements will not move around within transgenic fish genomes or to other species unless the transposase is reintroduced. However, the effect of the transposon construct on the stability of the genome and genotype may need further examination.

Other DNA elements

In addition to viral vectors and transposons, DNA elements of bacterial origin are sometimes introduced into transgenic fish. They include the plasmid backbone, antibiotic resistance genes and reporter genes (which encode a protein that can be readily assayed). Antibiotic resistance genes are discussed in the next section on phenotypic changes. Reporter genes, such as the *lacZ* gene from *Escherichia coli*, have been used in some transgenic fish studies as a convenient tool to test the function of the promoter involved in transgene expression (Rahman and Maclean, 1999). Concerns about the use of such bacterial elements include the possibility that they might enter the food chain, and pose adverse effects to fish health. However, many bacteria like *E. coli* are common organisms of the human gut, and thus these bacterial elements are naturally present. Potential hazards associated with such bacterial elements can be addressed in the risk assessment process (Chapter 1, this volume), but it is preferable to avoid introducing bacterial elements to transgenic fish by removing them prior to integration into the genome and thereby avoid any concerns that might be associated with such sequences.

Unintended phenotypic effects

Unexpected phenotypic effects could sometimes result from transgenesis. Phenotypic side effects can include physiological and morphological abnormalities (Devlin *et al.*, 1995b), increased susceptibility to diseases (Jhingan *et al.*, 2003) and unexpected gene by-products, as discussed below. Reliable data on phenotypic side effects are necessary to inform an environmental risk assessment, particularly when trying to predict ecological effects from transgenic phenotypes. Chapter 6 discusses the importance of detailed knowledge about phenotypic effects for predicting ecological effects of transgenic fish before their introduction into nature. The ability to gather the necessary data to inform this component of a risk assessment hinges upon knowing at a molecular level whether the transgene is expressed as anticipated (i.e. does it become active at the desired developmental stage and in the desired tissue), whether it is producing the correct phenotypic effect on the targeted trait, as well as whether it is producing unintended side effects. Gaps in this kind of information will increase the uncertainty in an environmental risk assessment (see Chapter 7, this volume). The discussion below addresses some of these effects, along with ways to detect and mitigate them.

Activation of Other Genes in Host Genome

The enhancer region of the integrated transgene construct could activate the expression of neighbouring genes in the host's genome. It is also possible that a transgene's promoter, integrated randomly into the genome of the transgenic fish, could lead to the activation of a nearby host gene, although the authors are unaware of any published documentation of this event. Alteration of expression of genes at genome sites far removed from a transgene has been reported in cell lines (Muller *et al.*, 2001), but it is not known if this occurs in whole transgenic animals (NRC, 2002).

Side effects from promoters

To minimize the uncertainty surrounding phenotypic side effects, careful selection of suitable promoters is very important. Ubiquitous promoters allow the transgenes to be expressed in all cells at all times. In most cases, this is undesirable, as certain gene products constantly produced in unexpected places could be physiologically damaging. Thus, tissue-specific promoters might provide the most suitable expression because they minimize transgene expression in unnecessary places and may reduce potential pleiotropic effects that are a source of uncertainty during the risk assessment process.

Gene by-products

If the inserted transgene DNA forms a new reading frame either with endogenous host genes or by recombination with transgene DNA sequences, it is possible that a new and unexpected protein product could be produced. The transgene transcript and product for each insertion event should be examined to ensure expression of the intended sequence. Furthermore, if organisms are engineered to produce or secrete new substances such as hormones or toxins, they could pose new sources of environmental risk. Thus the nature of the protein product of the gene must be carefully considered when developing a transgene and conducting environmental risk assessments.

Inclusion of antibiotic resistance genes and drug resistance

Most transgene constructs are amplified in bacterial cells using a selectable marker gene that confers antibiotic resistance. If the whole plasmid DNA is introduced into the host, the antibiotic resistance gene will also be integrated into the fish genome. However, antibiotic resistance genes could be activated if they are driven by a suitable promoter or if they recombine with host genes or parts of the transgenic sequence. Should this happen, concerns could be raised about increasing antibiotic resistance in the environment. The use of antibiotic resistance genes has also led to the concern that these genes may escape or transfer to sensitive bacterial strains when the transgenic organism is introduced into the environment, further contributing to the increasing problem of antibiotic resistant diseases. To date, there has been no evidence of antibiotic resistance genes moving from transgene constructs in a transgenic fish to bacterial strains in the fish's environment or drug resistance developing from transgenic organisms. Nevertheless, these concerns can be easily mitigated by removing the plasmid backbone containing the antibiotic resistance gene before introducing the foreign DNA into fish cells.

Pleiotropic and secondary effects

Transgenesis can unexpectedly influence phenotypes because of pleiotropic effects from the transgene. For example, in growth enhanced transgenic fish some secondary phenotypes have been observed, including cranial deformities (Devlin *et al.*, 1995b; Ostenfeld *et al.*, 1998), alteration of muscle fibre structure (Hill *et al.*, 2000; Pitkanen *et al.*, 2001) and altered surface area of gill filaments (Stevens and Sutterlin, 1999). These changes may also affect mobility and swimming performance traits (Farrell *et al.*, 1997), which may indirectly alter a fish's ability to migrate, evade predators or capture evasive prey. It is important to determine if there are pleiotropic effects expressed in the transgenic fish line, because such effects may contribute to environmental risks (see Chapter 6, this volume).

Sometimes, new traits of transgenic fish may have adverse ecological effects. For example, some transgenic fish with increased disease resistance could prove to be pathogen carriers, resulting in increased numbers of animals able to carry the pathogen without individual consequence. The carrier status of transgenic fish can be assessed by pathological assessments, molecular detection of pathogens and by cohabitation trials with native fish.

General Strategies for Characterizing Transgene Locus and Expression

In the two previous volumes of *Environmental Risk Assessment of Genetically Modified Organisms* focusing on transgenic plant cases, four general issues were identified for characterizing transgene loci and expression: (i) transgene design; (ii) analysis of transgene locus structure; (iii) analysis of transgene expression; and (iv) transgene transmission (Andow *et al.*, 2004; Grossi-de Sa *et al.*, 2006). These general strategies are also applicable in this volume on environmental risk assessment of transgenic fish.

Transgene design

Proper design and characterization of transgene constructs can help address some concerns raised during an environmental risk assessment of transgenic fish. As discussed in previous sections, the main concerns about construct design include the selection of suitable promoters, the retention of antibiotic resistance genes after transgenesis and the use of viral vectors and transposons for improving transgene integration efficiency. It is important to adequately characterize gene expression driven by the promoter, including expression under different environmental conditions.

Analysis of transgene locus structure

Concerns about the transgene locus site include the following: (i) foreign DNA could interrupt endogenous host genes and cause loss-of-function mutations; (ii) foreign DNA could affect adjacent host genes and either increase or decrease their expression; (iii) the transgene could be expressed ectopically (expressed in an unexpected manner) under promoters of adjacent host genes; conversely, endogenous host genes could be expressed ectopically via interactions of transgenic promoters or other DNA elements; and (iv) transgene rearrangements during integration could create spurious open reading frames (ORFs), which might produce unintended gene products. Most of these possibilities could be addressed by cloning, sequencing and characterizing the transgene locus and its flanking regions. The number of integration sites may be determined by conventional Southern blot hybridization, classical breeding and by chromosome *in situ* hybridization. The complete characterization of the transgene locus structure, including both the transgene and the flanking segments in the host genome, can provide information on the integrity and number of copies of the inserted DNA at given loci, the number of transgene loci and potential gene rearrangement in the host genome (Uh *et al.*, 2006; Yaskowiak *et al.*, 2006). In addition, when the number of inserts in a strain is large, an approximate copy number can be estimated relative to another control gene by quantitative real-time PCR assays.

Analysis of transgene expression

Characterizing the transgenic line's phenotype on a case-by-case basis is essential for risk assessment. In addition to observing the phenotype of the whole organism, scientists can test for expression of the protein encoded by the transgene in expected tissues based on the original transgene construct design such as the specificity of the promoter and the function of the structural gene. Unexpected expression can be detected by applying various molecular detection methods to different tissues. Protein products from the transgene should be characterized to ascertain whether the correct protein is produced. This can be achieved by SDS polyacrylamide gel electrophoresis to confirm the protein's molecular weight, followed by Western blot analysis to confirm its identity. Alternatively, other antibody-based methods can be used, such as enzyme-linked immunosorbent assay (ELISA), immunoprecipitation and immunocytochemistry.

Simple examination of the transgene locus may not uncover the unintended activation of non-target genes. Investigating changes in expression of the neighbouring genes as an indicator of changes in non-target gene expression may be helpful, but it could be a blind and tedious approach. Since chances of possible activation of neighbouring genes are low, a better strategy may be to perform analyses for unexpected expression of non-target genes only if an abnormal phenotype appears. This could be done by systematically comparing major morphological and physiological traits of transgenic fish and non-transgenic individuals derived from the same founder line, under the same contained rearing conditions. In the future, DNA microarray and proteomic technologies may be adequately developed to feasibly employ them to examine the extent of phenotypic changes at the transcriptomic and proteomic level. In particular, analysing molecular changes in certain biological pathways in transgenic fish may help predict potential risks.

Analysis of transgene expression should also account for possible temporal changes in expression during the animal's life cycle. Transgene expression may change diurnally, seasonally and by developmental stage, depending on the regulatory sequences utilized in the transgene construct and on the host's cellular and organismal response to an altered state of gene expression. Furthermore, variability in transgene expression could contribute to genotype by environment interactions. These complexities must be kept in mind when conducting an environmental risk assessment of a transgenic fish, both at the transgene and the larger ecosystem level (Chapter 6, this volume).

Transgene transmission

The inheritance and stability of each introduced trait should also be determined. Evaluation of cross-generational inheritance is necessary to ensure that the transgene is inherited as a normal Mendelian trait. The pattern and stability of transgene inheritance should both be demonstrated, not just the pattern of transgene expression. Because the initial transgene integration rate in embryos from injected eggs is low, very few of the first generation progeny will be transgenic. Such transgenic progeny are hemizygous, carrying the transgene on only one member of a pair of homologous chromosomes. However, once transgenic progeny fish are recovered, breeding should confirm the Mendelian inheritance of transgenes in future generations (see Zhang *et al.*, 1990; Nam *et al.*, 1999).

Recommendations for Minimizing Environmental Risks

Because of the variability in the process of producing transgenic fish and the influence of environmental factors on transgene stability and expression, it is impossible to reach one general, encompassing conclusion about whether the transgene constructs themselves, their genomic integration and expression as proteins pose environmental risks. Every transgenic line is different even if the same transgene construct is used; this is due to the fact that the integration locus, transgene copy number, transgene expression and potential recombination of transgenes differ between lines and environments. The risk assessment data obtained from one transgenic line may not be applicable to other transgenic lines. Thus, it is necessary to assess individual lines on a case-by-case basis. Nevertheless, the recommendations listed below should address the more common questions relevant for environmental risk assessment of most of transgenic fish. These recommendations should be helpful to guide analysis of an existing transgenic line of animals or to guide early stages of new DNA construct design.

Recommendations for generation of transgenic fish

1. Remove unnecessary DNA elements (e.g. antibiotic resistance genes, plasmid sequences) prior to production of transgenic fish. This will prevent unnecessary

DNA from being integrated into the transgenic fish's genome and avoid the associated concerns.
2. Avoid using transposable elements and viral vectors. This can eliminate a major mechanism for unexpected movements of genetic material within a transgenic fish's genome or to other unrelated species.

Recommendations to avoid unexpected genotypic and phenotypic effects

1. Test hemizygous and homozygous transgenics for possible undesirable phenotypes, screening for individuals whose modified traits may pose additional ecological effects (see Chapter 6, this volume).
2. Carefully select lines of transgenic fish to optimize copy number and minimize undesirable phenotypes. Again, this will help to minimize the creation of altered phenotypes with potential adverse ecological effects (see Chapter 6, this volume).
3. Ensure that there is no interruption of endogenous host genes by cloning, sequencing and characterizing the gene integration site. Evidence of no interruption lowers the risk of unanticipated phenotypic effects in the transgenic individual.
4. Ensure correct transgene transcription by Northern blot hybridization or reverse transcriptase-polymerase chain reaction (RT-PCR) and correct protein products by Western blot or radioimmunoassay.
5. Carefully analyse gene products produced by transgenic fish to avoid production of potentially toxic or ecologically disruptive substances.
6. Test transgene expression in stable lines under different environmental conditions and life stages relevant to the case in hand (see Chapters 1, 5 and 6, this volume). This will allow selection of individuals more likely to exhibit stable and appropriate transgene expression under the variety of environmental conditions that a transgenic individual may encounter during its lifetime. It will also facilitate the culling of individuals expressing unintended phenotypic traits that might pose environmental hazards.

Chapter Summary

Environmental risk assessment of transgenic organisms begins at the most basic molecular level, with the transgene construct itself. How the transgene is constructed, integrated into the host genome and expressed are some of the most basic concerns raised when conducting an environmental risk assessment of transgenic organisms. Due to the complexity of these molecular processes, it is impossible to predict the exact risks posed by transgene constructs, but it is possible to determine some of the most important issues on which to focus analysis. This chapter reviews basic transgenic technology and presents major concerns surrounding the molecular structure and expression of transgenes in fish, and it also identifies strategies and methodologies for addressing them. The recommendations presented at the end of this chapter outline general strategies to help determine whether a gene construct poses an environmental risk, and they also inform the reader of ways to avoid such risks beginning at the earliest stages of transgene construct design.

Although these strategies provide a good roadmap for assessing and reducing the environmental risks of transgene construction, integration and expression, they are not all-encompassing. Due to the great variability in transgene design and integration into genomes of targeted organisms and the difference between each event, care should be taken to avoid broad generalizations about the safety or risk of transgenes in aquatic animals. Instead, each case should be considered separately and the risks assessed accordingly.

The transgene's construction, integration and expression are influenced by many factors, including the environment in which the organism resides. Thus, an environmental risk assessment process must include other important components beyond looking at the transgenic construct. These additional components are addressed in detail in subsequent chapters in this book.

References

Andow, D.A., Somers, D.A., Amugune, N., Aragao, F.J.L., Ghosh, K., Gudu, S., Magiri, E., Moar, W.J., Njihia, S. and Osir, E. (2004) Transgene locus structure and expression of Bt maize. In: Hilbeck, A. and Andow, D.A. (eds) *Environmental Risk Assessment of Genetically Modified Organisms, Volume 1: A Case Study of Bt Maize in Kenya*. CAB International, Wallingford, UK, pp. 83–118.

Devlin, R.H., Yesaki, T.Y., Biagi, C.A., Donaldson, E.M., Swanson, P. and Chan, W.K. (1994) Extraordinary salmon growth. *Nature* 371, 209–210.

Devlin, R.H., Yesaki, Y.T., Donaldson, E.M., Du, S.J. and Hew, C.L. (1995a) Production of germline transgenic Pacific salmonids with dramatically increased growth performance. *Canadian Journal of Fisheries and Aquatic Sciences* 52, 1376–1384.

Devlin, R.H., Yesaki, T.Y., Donaldson, E.M. and Hew, C.L. (1995b) Transmission and phenotypic effects of an antifreeze GH gene construct in coho salmon (*Oncorhynchus kisutch*). *Aquaculture* 137, 161–169.

Devlin, R.H., Biagi, C.A., Yesaki, T.Y., Smailus, D.E. and Byatt, J.C. (2001) Growth of domesticated transgenic fish. *Nature* 409, 781–782.

Devlin, R.H., Sundstrom, L.F. and Muir, W.M. (2006) Interface of biotechnology and ecology for environmental risk assessments of transgenic fish. *Trends in Biotechnology* 24, 89–97.

Du, S.J., Gong, Z., Tan, C.H., Fletcher, G.L. and Hew, C.L. (1992a) The design and construction of 'all fish' gene cassette for aquaculture. *Molecular Marine Biology and Biotechnology* 1, 290–300.

Du, S.J., Gong, Z., Fletcher, G.L., Shears, M.A., King, M.J., Idler, D.R. and Hew, C.L. (1992b) Growth enhancement in transgenic Atlantic salmon by the use of an 'all fish' chimeric growth hormone gene construct. *Biotechnology* 10, 179–187.

Fan, L., Crodian, J. and Collodi, P. (2004) Culture of embryonic stem cell lines from zebrafish. *Methods in Cell Biology* 76, 151–160.

Farrell, A.P., Bennett, W. and Devlin, R.H. (1997) Growth-enhanced transgenic salmon can be inferior swimmers. *Canadian Journal of Zoology* 79, 335–337.

Gaiano, N., Allende, M., Amsterdam, A., Kawakami, K. and Hopkins, N. (1996) Highly efficient germ-line transmission of proviral insertions in zebrafish. *Proceedings of the National Academy of Sciences USA* 93, 7777–7782.

Gong, Z. and Hew, C.L. (1995) Transgenic fish in aquaculture and developmental biology. *Current Topics in Developmental Biology* 30, 177–214.

Gong, Z., Ju, B. and Wan, H. (2001) Green fluorescent protein (GFP) transgenic fish and their applications. *Genetica* 111, 213–225.

Gong, Z., Ju, B., Wang, X., He, J., Wan, H., Sudha, P.M. and Yan, T. (2002) Green fluorescent protein (GFP) expression in germ-line transmitted transgenic zebrafish under a stratified epithelial promoter from *Keratin8*. *Developmental Dynamics* 223, 204–215.

Gong, Z., Wan, H., Tay, T.L., Wang, H., Chen, M. and Yan, T. (2003) Development of transgenic fish for ornamental and bioreactor by strong expression of fluorescent proteins in the skeletal muscle. *Biochemical and Biophysical Research Communications* 308, 58–63.

Gross, M.L., Schneider, J.F., Moav, N., Moav, B., Alvarez, C., Myster, S.H., Liu, Z., Hallerman, E.M., Hackett, P.B., Guise, K.S., Faras, A.J. and Kapuscinski, A.R. (1992) Molecular analysis and growth evaluation of northern pike (*Esox lucius*) microinjected with growth hormone genes. *Aquaculture* 102, 253–273.

Grossi-de Sa, M.F., Lucena, W., Souza, M.L., Nepomuceno, A.L., Osir, E.O., Amugune, N., Hoa, T.T.C., Hai, T.N., Somers, D.A. and Romano, E. (2006) Transgene expression and locus structure of Bt cotton. In: Hilbeck, A. and Andow, D.A. (eds) *Environmental Risk Assessment of Genetically Modified Organisms, Volume 2: Methodologies for Assessing Bt Cotton in Brazil*. CAB International, Wallingford, UK, pp. 93–107.

Henikoff, S. (1998) Conspiracy of silence among repeated transgenes. *BioEssays* 20, 532–535.

Hill, J.A., Kiessling, A. and Devlin, R.H. (2000) Coho salmon (*Oncorhynchus kisutch*) transgenic for a growth hormone gene construct exhibit increased rates of muscle hyperplasia and detectable levels of differential gene expression. *Canadian Journal of Fisheries and Aquatic Sciences* 57, 939–950.

Hong, Y., Winkler, C. and Schartl, M. (1998) Production of medakafish chimeras from a stable embryonic stem cell line. *Proceedings of the National Academy of Sciences USA* 95, 3679–3684.

Hwang, G.L., Azizur, R.M., Abdul, R.S., Sohm, F., Farahmand, H., Smith, A., Brooks, C. and Maclean, N. (2003) Isolation and characterisation of tilapia beta-actin promoter and comparison of its activity with carp beta-actin promoter. *Biochimica et Biophysica Acta* 1625, 11–18.

Ivics, Z., Hackett, P.B., Plasterk, R.H. and Izsvak, Z. (1997) Molecular reconstruction of Sleeping Beauty, a *Tc1*-like transposon from fish, and its transposition in human cells. *Cell* 91, 501–510.

Jhingan, E., Devlin, R.H. and Iwama, G.K. (2003) Disease resistance, stress response and effects of triploidy in growth hormone transgenic coho salmon. *Journal of Fish Biology* 63, 806–823.

Ju, B., Chong, S.W., He, J., Wang, X., Xu, Y., Wan, H., Tong, Y., Yan, T., Korzh, V. and Gong, Z. (2003) Recapitulation of fast skeletal muscle development in zebrafish by transgenic expression of GFP under the *mylz2* promoter. *Developmental Dynamics* 227, 14–26.

Kawakami, K. (2004) Transgenesis and gene trap methods in zebrafish by using the *Tol2* transposable element. *Methods in Cell Biology* 77, 201–222.

Krasnov, A., Koskinen, H., Afanasyev, S. and Molsa, H. (2005) Transcribed *Tc1*-like transposons in salmonid fish. *BMC Genomics* 6, 107.

Liu, Z.J., Moav, B., Faras, A.J., Guise, K.S., Kapuscinski, A.R. and Hackett, P.B. (1990) Development of expression vectors for transgenic fish. *Biotechnology* 8, 1268–1272.

Long, Q., Meng, A., Wang, H., Jessen, J.R., Farrell, M.J. and Lin, S. (1997) *GATA-1* expression pattern can be recapitulated in living transgenic zebrafish using *GFP* reporter gene. *Development* 124, 4105–4111.

Lorence, A. and Verpoorte, R. (2004) Gene transfer and expression in plants. *Methods in Molecular Biology* 267, 329–350.

Lu, J.K., Burns, J.C. and Chen, T.T. (1997) Pantropic retroviral vector integration, expression, and germline transmission in medaka (*Oryzias latipes*). *Molecular Marine Biology and Biotechnology* 6, 289–295.
Muller, K., Heller, H. and Doerfler, W. (2001) Foreign DNA integration: genome-wide perturbations of methylation and transcription in the recipient genomes. *Journal of Biological Chemistry* 276, 14271–14278.
Nam, Y.K., Noh, C.H. and Kim, D.S. (1999) Transmission and expression of an integrated reporter construct in three generations of transgenic mud loach *Misgurnus mizolepis*. *Aquaculture* 172, 229–245.
NRC (2002) *Animal Biotechnology: Science-based Concerns*. National Academies Press, Washington, DC.
Ostenfeld, T.H., McLean, E. and Devlin, R.H. (1998) Transgenesis changes body and head shape in Pacific salmon. *Journal of Fish Biology* 52, 850–854.
Pitkanen, T.I., Xie, S.Q., Krasnov, A., Mason, P.S., Molsa, H. and Stickland, N.C. (2001) Changes in tissue cellularity are associated with growth enhancement in genetically modified arctic char (*Salvelinus alpinus* L.) carrying recombinant growth hormone gene. *Marine Biotechnology* 3, 188–197.
Plasterk, R.H., Izsvak, Z. and Ivics, Z. (1999) Resident aliens: the *Tc1/mariner* superfamily of transposable elements. *Trends in Genetics* 15, 326–332.
Rahman, M.A. and Maclean, N. (1999) Growth performance of transgenic tilapia containing an exogenous piscine growth hormone gene. *Aquaculture* 173, 333–346.
Rahman, M.A., Mak, R., Ayad, H., Smith, A. and Maclean, N. (1998) Expression of a novel piscine growth hormone gene results in growth enhancement in transgenic tilapia. *Transgenic Research* 7, 357–369.
Stevens, E.D. and Sutterlin, A. (1999) Gill morphometry in growth hormone transgenic Atlantic salmon. *Environmental Biology of Fishes* 54, 405–411.
Tsai, H.J., Tseng, F.S. and Liao, I.C. (1995) Electroporation of sperm to introduce foreign DNA into the genome of loach (*Misgurnus anguilicaudatus*). *Canadian Journal of Fisheries and Aquatic Sciences* 52, 776–787.
Udvadia, A.J. and Linney, E. (2003) Windows into development: historic, current, and future perspectives on transgenic zebrafish. *Developmental Biology* 256, 1–17.
Uh, M., Khattra, J. and Devlin, R. (2006) Transgene constructs in coho salmon (*Oncorhynchus kisutch*) are repeated in a head-to-tail fashion and can be integrated adjacent to horizontally-transmitted parasite DNA. *Transgenic Research* 15, 711–727.
Uzbekova, S., Amoros, C., Cauty, C., Mambrini, M., Perrot, E., Hew, C.L., Chourrout, D. and Prunet, P. (2003) Analysis of cell-specificity and variegation of transgene expression driven by salmon prolactin promoter in stable lines of transgenic rainbow trout. *Transgenic Research* 12, 213–227.
Yaskowiak, E., Shears, M., Agarwal-Mawal, A. and Fletcher, G. (2006) Characterization and multi-generational stability of the growth hormone transgene (*EO-1α*) responsible for enhanced growth rates in Atlantic salmon. *Transgenic Research* 15, 465–480.
Zhang, P., Hayat, M., Joyce, C., Gonzalez-Villasenor, L.I., Lin, C.M., Dunham, R.A., Chen, T.T. and Powers, D.A. (1990) Gene transfer, expression and inheritance of PRSV-rainbow trout-GH cDNA in the common carp *Cyprinus carpio* (Linnaeus). *Molecular Reproduction and Development* 25, 3–13.
Zeng, Z., Liu, X., Seebah, S. and Gong, Z. (2005a) Faithful expression of living color reporter genes in transgenic medaka under two tissue-specific zebrafish promoters. *Developmental Dynamics* 234, 387–392.
Zeng, Z., Shan, T., Tong, Y., Lam, S.H. and Gong, Z. (2005b) Development of estrogen-responsive transgenic medaka for environmental monitoring of endocrine disrupters. *Environmental Science and Technology* 39, 9001–9008.

5 Approaches to Assessing Gene Flow

A.R. Kapuscinski, J.J. Hard, K.M. Paulson, R. Neira, A. Ponniah, W. Kamonrat, W. Mwanja, I.A. Fleming, J. Gallardo, R.H. Devlin and J. Trisak

Introduction

Potential gene flow from transgenic individuals to wild relatives is a major pathway through which transgenic fish might affect natural populations. Introgression of transgenes into wild populations has the potential to depress adaptation, alter (and likely reduce) genetic diversity within and between wild populations and consequently reduce their probability of continued existence. Environmental risk assessment should thus assess the probability and consequences of gene flow whenever a transgenic fish line has relatives in accessible aquatic ecosystems. However, gene flow is not a prerequisite for potential environmental effects resulting from escapes of transgenic fish. Chapter 6 addresses additional processes through which escaping transgenic fish can alter natural aquatic ecosystems.

This chapter focuses on ways to assess the likelihood of gene flow *before* approved use and actual entry of transgenic fish into the environment. (Chapter 9, this volume, addresses methods of monitoring gene flow *after* approval of a specific use or environmental entry of transgenic fish.) This chapter is relevant for proposed uses of transgenic fish that could lead to accidental escape from an aquaculture operation, ornamental fish aquarium, other confined systems or live fish markets. Assessing gene flow would also be needed to evaluate proposals for purposeful release of transgenic fish into natural water bodies, which is being contemplated as a novel way to control invasive fish species (Thresher *et al.*, 1999; Kapuscinski and Patronski, 2005).

In this chapter, the term 'relative' refers to any fish with which the transgenic fish can interbreed, and use of this term applies to cases in which wild relatives are either a native or non-native species that have established a feral population in the aquatic ecosystem. Wild relatives could belong to the same species as the transgenic fish, or to a closely related species. It is important to note that, although the chapter focuses on gene flow from transgenic fish, its assessment approach, and much of the supporting scientific knowledge, also applies to assessing gene flow from selectively bred farmed fish to wild relatives.

This chapter presents a step-by-step approach for assessing gene flow based on a fault tree of the chain of events necessary for introgression of transgenes into a wild population (Fig. 5.1). This chain of events is divided into two main parts: events contributing to the probability of entry and events contributing to the probability of introgression (Eq. 5.1). The chapter then discusses how to partition entry and introgression into sub-components (Eqs 5.2–5.5). A complete risk assessment would estimate the probabilities for different magnitudes of each event (see Chapter 1, this volume), for instance, by estimating probabilities of introgression that result in transgene frequencies ranging from 0% to 100% in the wild population.

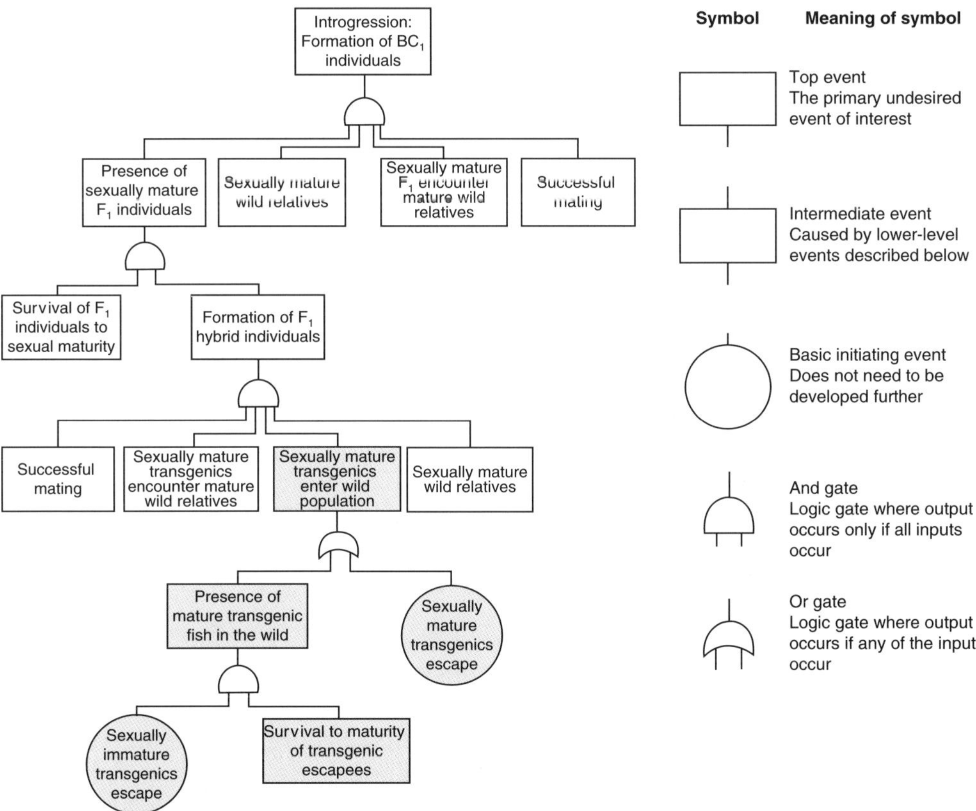

Fig. 5.1. Fault tree of events leading to introgression of transgenes from escaping transgenic fish into the gene pool of wild relatives. (Modified from figure developed by K. Hayes.) The fault tree should be read from the bottom to the top. Grey backgrounds: events leading to entry of sexually mature transgenic fish into a breeding population of wild relatives. White backgrounds: post-entry events leading to introgression. F_1: first-generation hybrids from transgenic and wild fish matings; BC_1: first generation of backcrosses between F_1 hybrids and wild relatives. Formation of BC_n individuals would occur via the same events shown in the penultimate row but involving BC_1, BC_2, etc. instead of F_1 individuals. Additional lower-level events could contribute to events in rectangular boxes but they have not been shown for brevity.

The chapter addresses the importance of assessing how both the transgene and overall genetic background of the escaping transgenic fish may affect the likelihood and consequences of gene flow to wild relatives. It also identifies possible changes in genetic and species diversity resulting from introgression of transgenes or from the genetic background of the transgenic line. It is important to consider both sources of genetic change because, as discussed in Chapter 3, many lines of transgenic fish are hemizygous (bearing the transgene on only one homologous chromosome); if hemizygous transgenic fish enter and breed with a wild population, some of their descendants will not carry the transgene but will carry much of their genetic background. Changes in genetic and species diversity are a subset of the broader set of potential ecological consequences of gene flow. Chapter 6 provides a more complete discussion of other possible ecological effects stemming from gene flow to wild relatives.

This chapter also identifies the main data needs for assessing gene flow and briefly describes possible modelling studies, field studies (involving non-transgenic fish) and laboratory experiments (some of which involve transgenic fish) to obtain needed data. It stresses the value of conducting laboratory experiments under different environmental conditions, and with different populations bearing the same transgene, to understand how genotype-by-environment (G × E) interactions and genetic background, respectively, might affect the likelihood and fate of gene flow to wild relatives. This discussion also highlights the impracticality of testing for all G × E interactions and for all effects of genetic background on transgene expression. Figure 5.2 summarizes the pathway for conducting an assessment of gene flow, the main types of baseline data needed and the major types of empirical information required about the transgenic fish line.

Estimating Gene Flow

Identification of wild fish populations with which escaping transgenic fish could interbreed is the first step in gene flow assessment.

Wild populations vulnerable to interbreeding with escaped transgenics

Gene flow estimations should identify all the wild populations into which the transgenes might introgress. Relevant wild populations are those in accessible aquatic ecosystems, defined as the ecosystems into which escaping transgenics might enter and encounter wild relatives. Various steps in gene flow assessment require information on population structure of conspecifics (individuals of the same taxonomic species), as explained in several places below.

Transgenics might interbreed, not only with conspecifics, but also with closely related species. Hybridization between related fish species occurs commonly at low frequencies (e.g. Hubbs, 1955; Philippart and Ruwet, 1982), and is more common among freshwater fishes (Campton, 1987). Hybridization can be favoured when one species (e.g. transgenic escapees) is more abundant than wild individuals of the other species. Some interspecific hybrids are fertile. Assessing gene flow therefore requires knowing whether the transgenic line involves a species that can

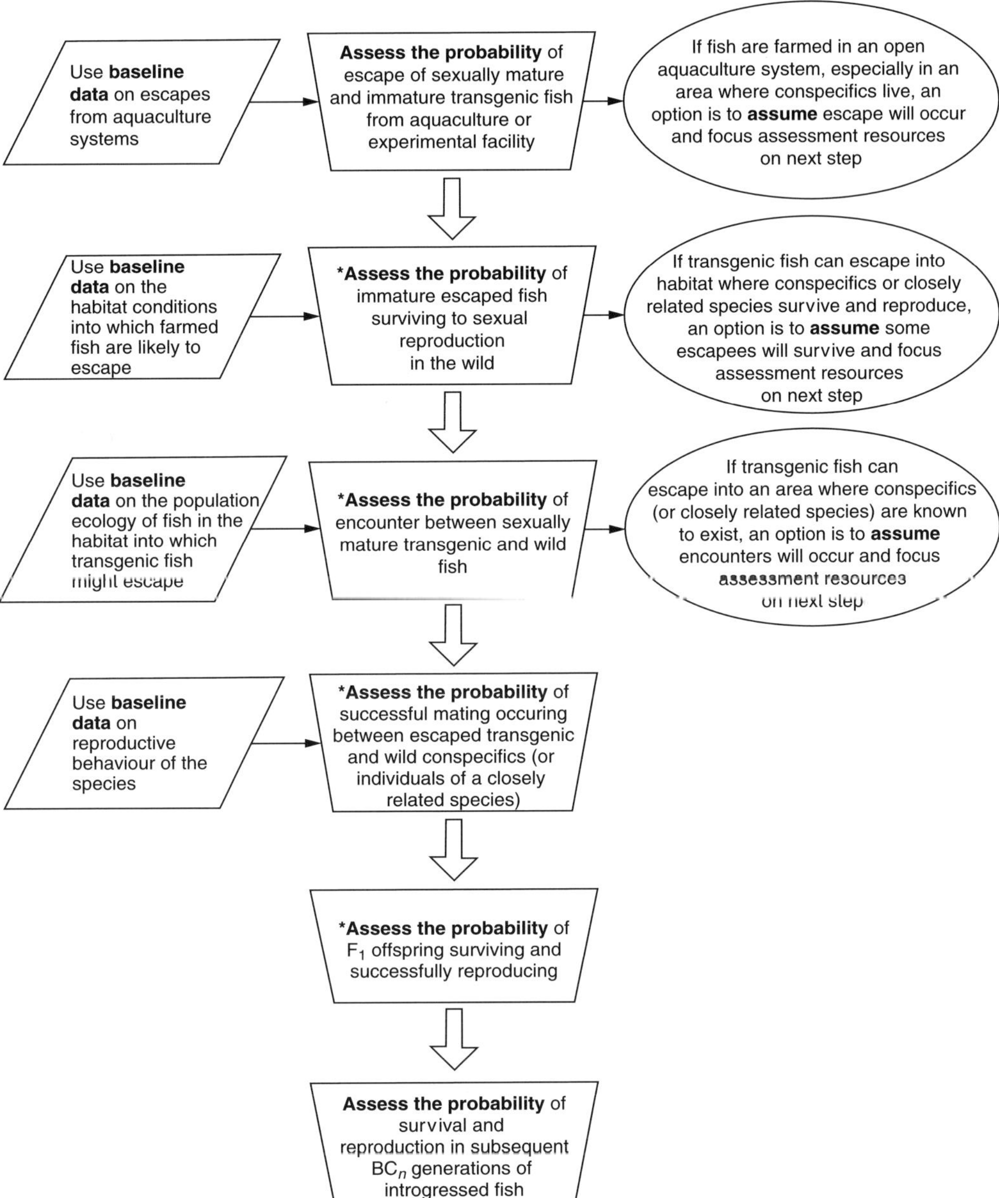

Fig. 5.2. Pathway for conducting an assessment of gene flow. * denotes an assessment step that requires empirical information on traits of the genetically modified organism (GMO) that are specifically related to this step; see text for discussion of specific traits. Empirical data can be obtained via confined experiments or studies as further discussed in the chapter.

produce fertile hybrids by interbreeding with closely related species. If such information is missing, traditional fish survey methods can determine whether there are any close relatives in natural waters (see Chapter 9, this volume), and laboratory mating experiments can determine viability and fertility of hybrids.

Assessing components of gene flow

Assessing gene flow involves determining the likelihood and magnitude of gene flow and evaluating its potential consequences relative to a desired condition, which can be identified using deliberative processes discussed in Chapter 2. Below, we present conceptual relationships of components of gene flow in the form of simple equations. These equations should be used to guide a step-wise assessment, not necessarily for quantitative estimation. Following the logic of these equations will help risk assessors consider components of gene flow and evaluate the importance of gene flow for a particular risk assessment case.

Gene flow is the outcome of two major processes: *entry* of sexually mature transgenic fish into a population of wild relatives in an aquatic ecosystem; and *introgression* or incorporation of the transgenic genotype into the gene pool of wild relatives by interbreeding between transgenic and wild fish, and back-crossing of the resulting hybrids to wild relatives. The probability of gene flow from transgenic to wild relatives can thus be estimated as:

$$\text{Probability of gene flow} = \text{probability of entry} \times \text{probability of introgression} \quad (5.1)$$

It is further possible to partition entry and introgression into sub-component probabilities (Eqs 5.2–5.5), each of which should be easier to estimate than trying to assess entry or introgression as a single variable. To facilitate the chapter's discussion, the variables in Eqs 5.2–5.5 are presented as being independent. However, the authors recognize that these variables, and their sub-components, may be dependent on and related to each other in complex ways. Furthermore, each of these variables can depend on a number of additional variables, for which data are often unavailable. For instance, many of these variables depend on whether the environment provides suitable physico-chemical conditions for transgenic escapee survival and reproduction. Also, variables involved with survival, sexual maturation and reproduction may depend on G × E interactions, an issue introduced later in this chapter and further addressed in Chapter 6. On a case-by-case basis, risk assessors should explicitly incorporate relevant dependencies between these variables (e.g. via conditional probabilities), or adopt mathematical techniques that account for possible dependencies (see Chapter 7, this volume).

Estimation of Entry Potential

Events that can lead to the entry of sexually mature transgenic individuals into a population of wild relatives are shown in Fig. 5.1 (denoted by grey boxes). Based on this chain of events, the probability of entry can be estimated as:

$$\text{Probability of entry} = P(\text{escape}_{\text{mature}}) + (P[\text{escape}_{\text{immature}}] \times P[\text{survival to sexual maturity}]) \quad (5.2)$$

Before estimating probabilities on the right-hand side of Eq. 5.2, risk assessors can determine whether immature and mature wild relatives of the transgenic fish line have entered the accessible environments previously (e.g. as aquaculture escapees) and established populations. An affirmative answer, based on solid empirical data for the area, suggests that the transgenic fish line has a high probability of entry. In this case, assessors may wish to make this assumption and focus their limited resources on assessing introgression (Eq. 5.3), an option indicated in Fig. 5.2.

Probability of escape

Estimating the probabilities of escape of immature and mature transgenic fish requires information on the pattern of escapes from the same aquaculture system and geographical location where the transgenic fish might be introduced. The probability of escape will nearly always be a positive value because escapes of farmed fish are inevitable from most aquaculture systems. Some species of transgenic fish could be produced within highly contained systems (see Chapter 8, this volume). This will add to the cost of aquaculture and would likely not be accepted by resource-poor farmers.

Estimating the scale of escape

Risk assessors will also need information on the scale of escapes. Many countries lack quantitative estimates of farmed fish escapes. Instead, they tend to have qualitative observations, which can still help inform predictions of transgenic fish escapes. For instance, the common occurrence of farmed hybrid catfish in Thai flood plains is attributed to fish escapes after frequent flooding of aquaculture ponds (Senanan *et al*., 2004). Ideally, risk assessors need more quantitative data, collected over adequate temporal and spatial scales, to predict the scale of transgenic fish escapes. A later section of this chapter suggests possible field studies to establish such baseline data.

Propagule pressure

One way to characterize the scale of escapes is in terms of propagule pressure, which is a composite measure of the frequency of escape events and the number of escapees per event in a specified region (Lockwood *et al*., 2005). Previous research on fishes (Kolar and Lodge, 2002; Marchetti *et al*., 2004) suggests that increased propagule pressure would increase the probability of entry (Eq. 5.2) and hybridization (Eq. 5.4). Risk assessors would ideally anticipate the temporal and spatial pattern of escapes from all facilities in the watershed or coastal areas that might rear the transgenic fish.

Drawing on experience with escapes from existing fish farms

Escapes from aquaculture operations vary greatly depending on the environment and type of rearing units. They range from infrequent, even rare, escapes of large numbers per event to continuous, frequent escapes of small numbers per event. In either case, propagule pressure may be high enough to result in a high probability of entry.

Cage culture systems located in areas with relatively strong water currents are most prone to infrequent but large-scale escapes of transgenic fish (e.g. Anonymous,

2002). Aquaculture production using cages and pens suspended in rivers or coastal waters is prevalent in developing and developed countries; large numbers of fish can escape from such systems due to storm damage, human negligence and damage by animals such as river otters and seals. Many aquaculture systems in developing countries are in low-lying wetlands, consisting of or connected to breeding grounds of potential wild relatives of transgenics. Wild relatives in such wetlands would likely be exposed to frequent, small-scale escapes of transgenic fish from numerous operations in an area. For instance, if transgenic tilapias are adopted in extensive fish farming systems prevalent in much of the tropics, their escape into surrounding waters is inevitable (see examples in Box 5.1). In countries

Box 5.1. Escapes of farmed tilapia from extensive aquaculture systems.

A starting point for estimating the probability of entry of transgenic fish (Eq. 5.2) is to search for evidence of feral populations of the same or related species that resulted from prior escapes of unmodified farmed fish. Evidence from Uganda and Thailand, described below, shows that tilapia farmed beyond their native range have escaped and established feral populations in the receiving environment. (Tilapia is used here as the common name for a large group of species belonging to the tribe Tilapiini, Family Cichlidae, Order Perciformes.) This evidence suggests that any transgenic tilapia produced in similar aquaculture systems in these geographical areas would also escape and breed with feral relatives. Credible rejection of this possibility would require solid scientific evidence of complete disruption of a key survival or reproductive trait of the transgenic fish.

Escapes from aquaculture ponds in Uganda

Subsistence production, with little regulatory oversight, has characterized aquaculture in Uganda for the last 50 years. Aquaculture is practised in small ponds, usually located within wetlands, or by the edges of water bodies (reservoirs, rivers and lakes), for easy exchange of water into and out of the ponds. In most cases the ponds lack diversion channels to guard against flooding and stormwater collection during high rains. Consequently, Nile tilapia (*Oreochromis niloticus*) and other introduced tilapiine species have ended up in most waters in the country through farm escapes, flooding, stormwaters or deliberate release of farmed species into the wild (Mwanja and Mwanja, 2002; Mwanja *et al.*, 2007). Prior to the 1960s, Nile tilapia, *O. leucostictus* and *Tilapia zilli* were restricted to Lake Albert and the Albert Nile in Uganda because of a barrier at the rapids at Muchison Falls between Lakes Kyoga and Albert. These species have been extensively introduced throughout the country for aquaculture purposes, and they now occur in all major water bodies and many wetlands.

Currently, there are nearly 20,000 tilapia aquaculture ponds of an average of 400 m^2 scattered throughout the country, with the highest concentration within the Lake Kyoga catchment, followed by the Lake Victoria Basin. These two lake bodies are very important for fisheries and biodiversity in the region. Indeed, the fisheries of Lake Victoria and Kyoga are currently dominated by introduced species, including Nile perch, which is the mainstay of the two fisheries.

Continued

Box 5.1. *Continued*

A new kind of business, described by the Ugandan Department of Fisheries Resources as 'emerging commercial fish farmers', has recently emerged. These farmers are undertaking aquaculture at a much more intensive level than subsistence farmers, with a sole motive of profit. Their ponds are much bigger, better constructed and, to a fair degree, better protected against floods and stormwaters than those of subsistence farmers. However, intensive aquaculture practices used by these commercial fish farmers (e.g. high stocking rates, expanded pond areas and drive for higher returns) still present risks for farmed species to enter the wild. The ponds of these commercial fish farms are also located in areas quite prone to flooding, and the effects of stormwater, coupled with ineffective regulation and monitoring of these farms (despite a fairly good regulatory framework for aquaculture), could combine to increase the 'propagule pressure' of escaping farmed fish (Lockwood *et al.*, 2005).

There has not yet been an adequate study on the escape and unintentional release of farmed fish in Uganda. Mwanja and Mwanja (2002) only described occurrence of wild species in culture systems (ponds) and farmed species in surrounding waters (wetlands) around the culture systems; they did not assess the rate of escape of farmed fish into the wild. However, occurrence of the farmed strains where they are not native in the wild is an indication of entry of the farmed forms into the wild, either through deliberate introduction by farmers or escape from the ponds. In a recent study by Mwanja *et al.* (2007), many farmers reported fish escapees into the wild, and they mentioned that they have found cultured species in natural wetlands around their ponds. The poor water distribution system (pipes, canals, etc.) discourages farmers from locating aquaculture ponds in upland areas less prone to flooding and fish escapes; it is too expensive for them to transport water to such upland sites.

Escapes from cage aquaculture in Thailand

Although the tribe Tilapiini is not indigenous to Thailand, several species have been introduced, mostly for aquaculture use. *O. mossambicus* and *T. zillii* were among the first species, introduced in 1949, followed by *T. rendalli* in 1955, *O. niloticus* in 1965 and *O. aureus* in 1970. Selectively bred genetically improved farmed tilapia (GIFT), a strain of *O. niloticus*, was introduced during 1994–1996 from the Philippines (ADB, 2005). The GIFT strain is now widely used for aquaculture in Thailand. *Sarotherodon melanotheron* was introduced for aquatic weed control in reservoirs. A hybrid red tilapia has been also introduced for aquaculture.

Among the various species introduced, only *O. niloticus* and hybrid red tilapia are presently used for aquaculture production in Thailand. There have been reports of relatively low levels of tilapia production in natural reservoirs (ranging from 0.08 to 10.09 kg/ha/year). The size and depth of these reservoirs are believed to limit the production of tilapia, especially *O. niloticus*, and production per unit area bears a negative exponential relationship to size of the water body (De Silva *et al.*, 2004). Indeed, tilapia require shallow water to build nests when spawning. However, in four Thai reservoirs (Mae Chang, Chulaphon, Kwan Phayao and Lam Takhong), the contribution of *O. niloticus* exceeded 20% of the total fish landings (De Silva *et al.*, 2004). Of all the introduced tilapia species,

Continued

Box 5.1. *Continued*

three, *O. mossambicus*, *O. niloticus* and *S. melanotheron*, have currently established feral populations in Thailand's natural water bodies. Introgressive hybridization among the three species was reported in the high salinity area in Phechaburi province (Karnasuta *et al.*, 1999). Aquaculture of saltwater tolerant *O. mossambicus* was most common in this area and nearby coastal provinces before the introduction of *O. niloticus*. Interspecific hybridization among tilapia species has also been reported elsewhere when the species are introduced beyond their natural range (Macaranas *et al.*, 1986). Thus, any future farming of transgenic *O. niloticus* could also lead to introgression of transgenes into feral populations of all three tilapia species.

Tilapias are cultured in both earthen ponds and cages. Cages usually give somewhat better quality of fish, presumably due to better water quality, but their production cost is also two to three times higher than that of ponds. Culture sites for cage culture are in big reservoirs and rivers. National tilapia aquaculture production is about 100t/year, 85% of which is in ponds and 5% in cages. In 2003, there were about 1000 registered tilapia cage farmers located in several major rivers throughout the country, especially the Mun, Kok, Chaopraya and Mekong Rivers. There are few quantitative estimates of feral tilapia abundance, although a 2002 fishery survey of major rivers in the north-east of Thailand (e.g. the Pong, Chi and Mun Rivers) reported tilapia comprised 0.04% of all species found (Dumrongtripob, 2002). Extensive tilapia cage culture is present in these rivers but their biophysical characteristics, such as steep banks, deep water and strong water currents, might make it harder for tilapia to establish feral populations. Although several tilapias can adapt to fast-flowing rivers, they tend to remain in shallow inshore water more fit for their reproduction and feeding behaviour (Trewavas, 1982). Such biophysical impediments to successful reproduction could likewise reduce or prevent introgression of transgenes in spite of entry by escaping transgenic fish.

like Bangladesh, India and Cambodia, certain areas undergo periodic flooding, providing more chances for farmed transgenic fish to escape from aquaculture facilities. Above-ground units, such as concrete raceways and fibreglass tanks, may be located above flood-prone areas; but they still have possible routes of escape, depending on the unit's water management, exposure to predators and aquaculture practices (Scientists' Working Group on Biosafety, 1998).

However, it is important to recognize that geographic variations within many developing countries influence the location of aquaculture ponds, and there are many ponds located far from any natural bodies of water. The potential scale of transgenic fish escapes into natural waters containing wild relatives will be much lower from these isolated facilities.

Survival of immature escapees to sexual maturity

Compared to sexually mature escapees, immature escapees must survive longer in the wild and complete the process of sexual maturation before they can encounter and mate with wild relatives (Fig. 5.1 and Eq. 5.2). Their survival to sexual maturity requires suitable environmental conditions. Thus, risk assessors need to determine whether accessible ecosystems meet the escapees' needs for food, habitat and environmental cues for reproductive development. (Some of this information can be obtained while conducting assessments of potential ecological consequences, as described in Chapter 6, this volume.) Prior presence of conspecific or closely related species of the transgenic fish in the accessible ecosystems indicates that a suitable environment exists for survival and reproduction of these species, thus increasing the chances of the transgenics to survive to maturity. In such cases, assessors may wish to assume this survival to maturity and focus on assessing introgression (Eq. 5.3), an option indicated in Fig. 5.2. However, it is important to determine whether transgenic fish will reach maturity in a manner similar (e.g. at the same age) to wild relatives. Confined experiments to obtain such data should be conducted under conditions mimicking the natural environment as closely as possible (see Chapter 6, this volume).

Survival rate from immature to mature stage

Survival rates of immature transgenic fish to sexual maturity may differ from their wild relatives because of effects of the transgene or genetic background, especially if a highly selected line was used to produce the transgenics. For example, survival rates of artificially propagated and hybrid strains of salmonid fishes are often lower than those of naturalized salmonids, and this trend is consistent with reduced predator avoidance behaviour by these groups (Negus, 1999; Fleming *et al.*, 2000; McGinnity *et al.*, 2003). Laboratory experiments can compare predator avoidance behaviour and other survival traits of transgenic individuals against wild individuals (see Chapter 6, this volume).

Baseline data on survival rates of wild conspecifics or closely related species can facilitate predicting survival rates of immature transgenics to maturity in the wild. If such baseline data are not available, investigators can apply the mark–recapture method to estimate survival rates of fish populations in the accessible ecosystem (method described in Chapter 9, this volume). An alternative approach for estimating survival rate is a tag recovery study, in which individual dead, tagged fish are recovered by fishermen, who return the tag to managers of the study (Williams *et al.*, 2002).

Ability of immature escapees to complete process of sexual maturation

Empirical data on sexual maturation traits of the transgenic line, with an emphasis on any differences from wild relatives, will help assess the ability of transgenic escapees to reach sexual maturity in accessible ecosystems. For instance, some transgenic lines bearing growth hormone (GH) gene constructs exhibit decreased age of sexual maturity (e.g. Devlin *et al.*, 1995, 2004a; Muir and Howard, 2001), although this might not be the case under restricted food conditions in the wild.

Baseline data on sexual maturation of conspecifics or closely related species inhabiting similar aquatic ecosystems can help predict the ability of transgenic escapees to complete sexual maturation. A direct source of information is observation of the stage of reproductive development via one of the common methods (Wootton, 1990; Anderson and Neumann, 1996; Estay *et al.*, 1998). An indirect source of information is size or age structure, when there is good knowledge of correlation with the stage of sexual development. When gathering baseline data, assessors should beware of the potential for surprise. For instance, in spite of assumptions that immature pink salmon (*Oncorhynchus gorbuscha*) could not survive in freshwater, the Laurentian Great Lakes experienced population explosions of this species two decades after 21,000 juveniles were flushed down the drain of a hatchery (Emery, 1981; Kwain and Lawrie, 1981).

Estimation of Introgression Potential

The next major step in assessing gene flow is to estimate the probability of introgression of transgenes from any transgenic fish successfully entering a wild population, with entry defined by Eq. 5.2. Although the concept of introgression implies longer-term persistence of the transgene in a wild population, we focus here on assessing introgression to the first backcross generation (BC_1). The BC_1 generation is created when first-generation hybrids (F_1 progeny from mating between transgenic escapees and wild parents) successfully interbreed with the wild population. Persistence of transgenes to the BC_1 generation is necessary, although not sufficient, for persistence of transgenes over the longer term. In many cases, it will be difficult to assess gene flow beyond this point due to limits in scientific information and methods, as well as practical limits in funds, time and facilities for conducting relevant tests. Nevertheless, the methodologies and limitations presented in this chapter are relevant for assessors who choose to assess gene flow through additional generations of backcrossing.

Risk assessors can break down the task of assessing introgression by first determining the potential for first-generation hybridization (to yield F_1 progeny), and then assessing the potential for advanced generation backcrosses (BC_1 through BC_n), at least through the BC_1 generation. Thus,

$$\text{Probability of introgression} = P(F_1 \text{ hybridization}) \times P(BC_1 \text{ backcrossing}) \quad (5.3)$$

It is further helpful to divide the task into the sub-component probabilities in Eqs 5.4 and 5.5:

$$P(F_1 \text{ hybridization}) = P(\text{encounter}) \times P(\text{escapee} \times \text{wild mating}) \quad (5.4)$$

$$P(BC_1 \text{ backcrossing}) = P(F_1 \text{ offspring survival to maturity}) \times P(\text{encounter}) \times P(F_1 \times \text{wild mating}) \quad (5.5)$$

Further introgression would require BC_1 fish to reach sexual maturity and successfully interbreed with wild fish and so on.

Probability of first-generation hybridization

The probability of producing F_1 hybrid progeny from interbreeding between wild and transgenic parents depends on two primary probabilities: (i) sexually mature transgenics encounter sexually mature wild relatives; and (ii) the two types mate successfully (Fig. 5.1). In general, these elements will differ between situations involving conspecific individuals and those involving individuals from related species. Transgenic individuals are more likely to mate successfully with wild fish of the same species because conspecifics are generally genetically, phenotypically and behaviourally similar. However, the risk of introgression when the transgenic line and recipient population are different species may not be negligible, especially when these species are closely related.

Travel of escapees to areas where they may encounter wild relatives

Dispersal of transgenic fish to places where they may encounter wild relatives depends on: (i) the distance between fish farms and aquatic habitats where wild relatives reside; and (ii) distribution patterns of the escaping fish. The easiest way to estimate distances between aquaculture operations and relevant aquatic habitats is to use geographic information system (GIS) databases. Otherwise, one can map aquaculture facilities in relation to relevant habitats, using records on aquaculture facilities and information collected in prior fisheries management assessments or ichthyology field studies. Data gaps can be filled by a census of aquaculture operations and surveys of aquatic habitats. Predicting distribution patterns of escaping transgenics also requires basic information on movement behaviour and preferred habitats of different life stages of the unmodified species. It is also important to determine if the transgenic line has any special advantage or disadvantage conferred by dispersal-related traits (see Chapter 6, this volume). Such dispersal behaviour information is also relevant to estimating the probability of encounter (Fig. 5.1 and Eq. 5.4).

The dispersal of transgenic escapees will be species-specific and depends on factors such as the escapee's life stage, survival in the new habitat, the species' innate migratory and dispersal behaviour and presence of stimuli for dispersal. Eggs and larval stages might be dispersed further because they are easily carried by water currents; however, mortality rates would be higher compared to juveniles and adults. There are big gaps in baseline information on distances travelled by escapees of most farmed fish species (see section below on data needs).

Encounter between mature transgenic escapees and mature wild relatives

Hybridization between sexually mature transgenic individuals and wild fish is most likely when sufficient numbers of transgenic individuals (of the appropriate physical and behavioural phenotype) enter wild populations when, and where, natural breeding is occurring. Conditions must also be conducive for successful encounter of the mature transgenics and wild adults. Reproductive-stage transgenics that escape near wild breeding populations pose a much higher risk of encounter than transgenic juveniles escaping far from wild breeders. For example, Butler and Watt (2003) found that Atlantic salmon escapes from farms in western Scotland contributed an average of 9% of caught salmon in nearby rivers, compared with 2% of catches (due to straying escaped salmon) in rivers away from farms.

In the absence of scientific evidence to the contrary, risk assessors can reasonably assume that sexually mature transgenics, after entering a wild population's habitat, have no special disadvantage in encountering wild mates, and thus they can focus resources on assessing the next step: successful mating between transgenics and wild relatives (Fig. 5.2). This assumption is reasonable given the scant evidence that current transgenes alter the seasonal or spatial patterns of fish breeding (e.g. Bessey *et al.*, 2004; Howard *et al.*, 2004); although they may decrease age at maturity as observed in GH-transgenic fish (e.g. Devlin *et al.*, 1995, 2004a; Muir and Howard, 2001).

Factors other than phenotypic effects of the transgene may affect the probability of encounter. Depending on the origin of the founders used to create the transgenic line, breeding time – which has a significant genetic component – may differ from that of local wild populations, thus limiting encounter. Moreover, the domestication involved in producing cultured lines of fish may alter breeding times (reviewed in Fleming and Petersson, 2001). Similarly, for fish that 'learn' the location of breeding grounds through juvenile experience (i.e. returning to natal areas for breeding), artificial rearing will impede such capabilities and inhibit spatial overlap with wild breeding populations. To assess whether these factors are relevant to the transgenic fish in question, information is needed about the transgenic line's genetic background, domestication history and importance of juvenile experience to reproductive behaviour.

Mating between transgenic escapees and wild relatives: information needed on mating systems

Reproductive-stage transgenic individuals need to show appropriate physiological, morphological and behavioural characteristics before they can mate successfully with wild relatives. These characteristics may include appropriate timing of sexual maturation, morphology and coloration, as well as courtship and agonistic behaviour. Successful mating between transgenic and wild relatives also depends on the presence of appropriate environmental cues for courtship and breeding behaviour. For many species, the types of environmental cues and their importance to mating success often vary with breeder density and habitat quality. Risk assessors need comparative data on transgenics and wild relatives regarding their reproductive characteristics and responses to environmental cues. Chapter 6 discusses studies for generating such data.

Assessing the probability of mating requires data on mating system traits. The mating system includes the number of mates and how they are obtained. Mates can be obtained through competition for mates and resources, courtship and mate choice. In some species, the mating system also extends to the form and duration of parental care, which affects early survival of F_1 progeny (see Eq. 5.5). These traits may vary across transgenic lines and should be measured for each line. In some cases, there is no evidence that the transgene itself alters competitive ability; rather, the environmental effect of captive rearing greatly impairs the performance of both transgenic and non-transgenic fish in a similar fashion (Bessey *et al.*, 2004). In other cases, transgenic fish are clearly superior to non-transgenics, particularly males, mainly due to their larger size (Howard *et al.*, 2004). There is, however, no indication that currently used transgenes impair breeding behaviour, either in terms of courtship or spawning synchronization between the sexes (Bessey *et al.*, 2004; Howard *et al.*, 2004). If size-assortative mating is predominant in the species under examination, and transgenic fish are larger than non-transgenics, this could constrain interbreeding, but is unlikely to prevent it. In fishes that are sequential hermaphrodites, where body size or social structure affects sex change (e.g. sea breams, wrasses, groupers; see Shapiro, 1984), GH-transgenes could have profound effects on sex ratios and thus the mating system.

If highly domesticated transgenic fish escape at or close to sexual maturity, they may be ill-equipped to mate in the wild due to effects of their captive rearing environment. For example, they may have been fed artificial diets and developed under higher rearing densities; both of these artificial conditions can alter their reproductive characteristics compared to wild relatives. The ability of domesticated fish to mate in the wild can be limited by active mate choice or by competition between individuals of the same sex for potential mates. For instance, in some species, cultured males have difficulty mating successfully because they do not express the appropriate phenotype or behaviour conducive to mating or competing for mates (Fleming and Gross, 1993; Fleming *et al.*, 1996; Berejikian *et al.*, 1997; Weir *et al.*, 2004). A risk assessment should therefore consider whether domesticated transgenic fish display reproductive traits that disrupt their ability to mate with wild relatives.

In summary, risk assessors need reliable information on breeding overlap and mating system characteristics of the transgenics compared to wild relatives before they can estimate the probability of mating. Limited evidence (to date) on a few lines of growth-enhanced transgenic fish suggests that growth-hormone transgenes do not inhibit interbreeding and, in some cases, may even promote it (e.g. de la Fuente *et al.*, 1999; Howard *et al.*, 2004). For these particular transgenes and species-specific mating systems, it appears that transgene introgression into wild populations would be reduced more by the consequences of artificial rearing and associated domestication than by the action of the transgene itself. Thus, for fish species such as salmon, which aquaculturists propagate via fully artificial means, the concomitant domestication can reduce the potential for introgression. This is less likely to occur in species such as tilapia, which are allowed to breed naturally in aquaculture operations.

Mating between transgenic escapees and wild relatives: demographic information needed on wild populations

Assessing the probability of mating also requires population dynamics information on the wild populations with which the transgenics could interbreed. Such information includes the number of wild individuals and the proportion of the total abundance that is reproductively mature. Methods for estimating the abundance of different life stages of fish in the wild are detailed in Chapter 9.

Probability of BC_1 backcrossing

The probability of first-generation backcrosses depends on three primary probabilities (Eq. 5.5 and Fig. 5.1): (i) F_1 offspring survive to sexual maturity; (ii) F_1 and sexually mature wild relatives encounter each other; and (iii) the two types mate successfully. Assessment of F_1 offspring survival to sexual maturity is discussed below. Previously discussed approaches (used to assess events leading to the F_1 offspring) can be used to assess the other two probabilities.

F_1 offspring survival to maturity

The likelihood that viable F_1 hybrid progeny will result from matings between transgenic and native individuals, and survive to the point of successful reproduction, is influenced by all the factors that determine post-zygotic success. These factors include fertilization success, gamete number and quality, parental care (in species where this is applicable), successful foraging and predator avoidance behaviour, as well as appropriate growth and reproductive schedules. Growth rate and size at maturity can affect gamete quality and number, particularly in females (Jonsson *et al.*, 1996). For example, Bessey *et al.* (2004) found decreased gamete size and increased gamete number in female, growth-enhanced transgenic coho salmon; but found no detectable changes in gamete quantity or quality in transgenic males. As parental care exists in 21% of bony fish families, including a number of species used to produce transgenic lines (e.g. tilapia – *Oreochromis* spp. and channel catfish – *Ictalurus punctatus*), the quality of this care can directly influence embryo and larval survival. However, the authors are unaware of studies to look for such effects associated with transgenes.

Net fitness methodology

Recognizing the difficulty of predicting the effect of each factor influencing introgression of transgenes, Muir and Howard (1999, 2001) have proposed a net fitness methodology that focuses on estimating fitness traits that span the entire life cycle of a fish species and thus represent the outcome of all these contributing factors. Under certain assumptions, the net fitness methodology predicts the transgene frequency in the F_1 hybrid generation and subsequent generations of backcrossing to the wild population. The methodology assumes that the sum of an organism's fitness can be quantified by measuring six component traits covering the entire life cycle of the species and that these data can be applied at a population level. These 'fitness components' include: juvenile viability, age at sexual maturity, fertility, fecundity, mating advantage and adult viability. Figure 5.3 briefly describes the data to be collected for each of these fitness traits. The methodology involves two major steps: (i) collection of fitness trait data on real transgenic individuals and their wild relatives in confined tests; and (ii) input of these data into a mathematical model that quantifies the joint effect of all six fitness traits to predict the fate of the transgene and the size of the introgressed population over multiple generations. Box 5.2 summarizes the three general classes of transgene fate predicted by the model.

The net fitness methodology calculates the relative fitness of the transgenic and wild genotypes based on data collected for these six fitness components. For small fish with a relatively short generation time, and that are capable of breeding in confinement, these data will be relatively easy to collect in confined tanks or ponds (see Chapter 8, this volume). However, for fish with long generation times, complex life histories and that have not been bred in captivity, collecting even the data needed to estimate these six fitness components can become quite difficult. For example, in tilapia species, adult viability would probably be the hardest

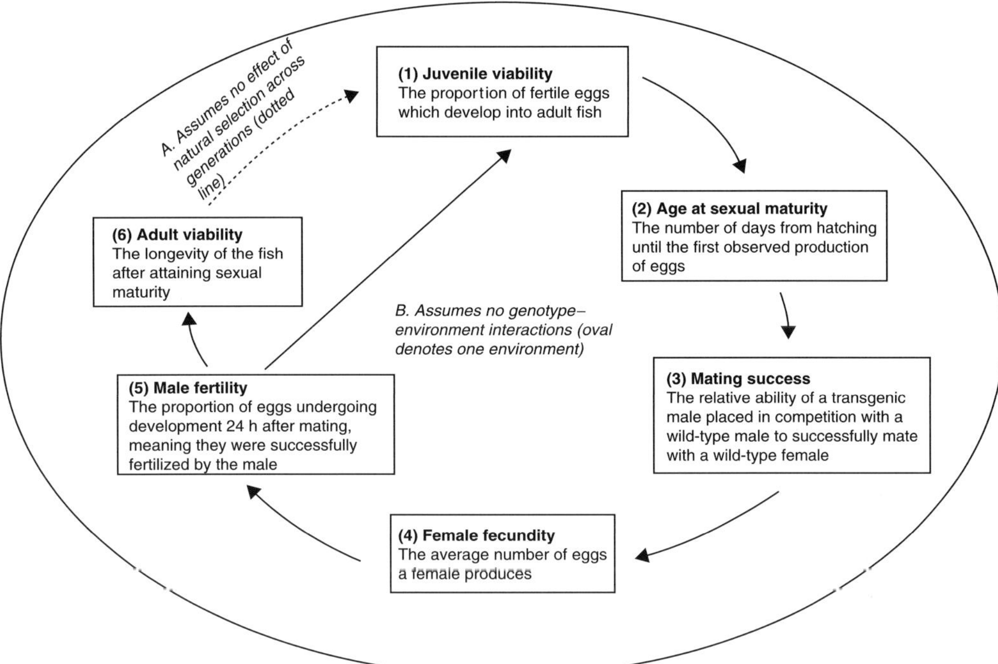

Fig. 5.3. The net fitness methodology involves measuring six fitness traits that cover critical points in a species' entire life cycle and then entering the trait data into a mathematical model that predicts transgene fate. The current version of the methodology assumes: (A) no evolution of fitness traits in response to natural selection across generations after initial transgene escape and (B) no G×E interactions affect the fitness traits. The feasibility and costs of testing these assumptions for a given case depend on the biology of the fish species and available scientific, technical and laboratory resources.

Box 5.2. Results from net fitness methodology simulations.

The net fitness methodology predicts transgene frequency and population size after a specified number of generations (Fig. 5.3). These predictions are based on relative fitness values among transgenic and wild genotypes for six traits spanning an individual's life cycle and a starting population size. The predictions can be characterized by the following scenarios (Muir and Howard, 1999, 2002; Howard *et al.*, 2004):

- **Purge scenario (lowest risk)**: the transgene is purged after the initial escape of transgenic fish, with no effect on initial wild population size. This scenario occurs when the net fitness of the transgenic fish is lower than that of its wild relatives.
- **Spread scenario (invasion risk)**: the transgene spreads through a population of wild relatives, with no impact on the size of the introgressed population. This occurs when the net fitness of transgenic fish is equivalent to or higher than that of its wild relatives.
- **Trojan gene scenario (extinction risk)**: the transgene initially spreads through a population of wild relatives, then triggers a decline in the size of the introgressed population. This occurs when the transgene has an antagonistic effect on different fitness traits.

Continued

Box 5.2. *Continued*

The Trojan gene scenario could result in the transgene being driven into the wild population because of an initial fitness advantage conferred by the transgene, but ultimately causing a decline in the introgressed population size due to a second fitness disadvantage. For example, under the simplifying assumptions of this deterministic model, simulations suggested that if a growth-enhanced transgenic line exhibits a mating advantage by large males, and if progeny of transgenic–wild crosses exhibit a moderate viability disadvantage, the antagonistic effect of both traits would trigger the Trojan gene effect over multiple generations (Muir and Howard, 1999, 2002). Figure 5.4A and B below illustrate the trade-off between increased mating advantage and decreased viability of transgenic fish, such that different combinations of these values predict different scenarios. In these graphs, the contours indicate the number of generations to extinction. A wide range of values for mating advantage and viability traits in transgenic fish was input into the model, while holding all other fitness traits equal to those of the wild fish.

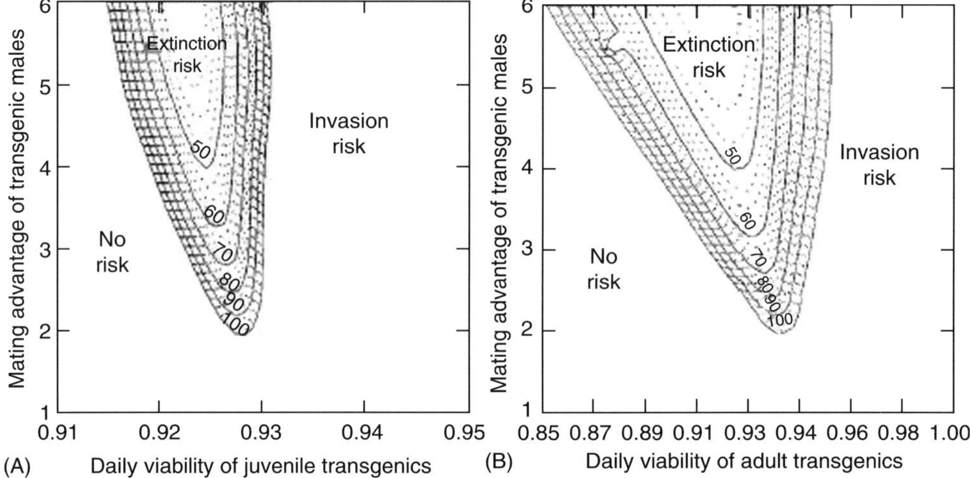

Fig. 5.4. (A) Extinction and invasion risks as a result of a trade-off between the transgene increasing male mating advantage and reducing juvenile viability. For comparison, wild male mating advantage is 1 and juvenile viability for wild individuals is 93.1%. (Reprinted from Muir and Howard, 2002, with permission from Springer Science and Business Media.) (B) Extinction and invasion risks as a result of a trade-off between the transgene increasing male mating advantage and reducing adult viability. For comparison, wild male mating advantage is 1 and adult viability for wild type is 95%. (Reprinted from Muir and Howard, 2002, with permission from Springer Science and Business Media.)

fitness trait to measure in captivity. When faced with such challenges, risk assessors can prioritize the transgenic line's most important fitness traits to measure based on a sensitivity analysis (see also Chapter 7, this volume). Specifically, sensitivity analysis of the current version of the net fitness model determined that the model's predictions are particularly sensitive to age at sexual maturity and juvenile viability measurements (Muir and Howard, 2001).

Estimation of fitness components will be most useful to risk assessors if they are based on data collected on the actual transgenic line and wild populations under consideration; such estimates will capture any important effects of genetic background, thus greatly reducing a potential source of incertitude in the assessment. (See below and Chapter 6, this volume, for further discussion of genetic background, and see Chapter 7, this volume, for approaches to addressing this kind of scientific uncertainty.) The analyst must specify the number of fishes in the wild population, the number of transgenic escapees and the number of generations for which to run the simulation model used during the second step of the net fitness methodology.

Validation of the net fitness model's predictions has not yet been published in the peer-reviewed literature. Validation involves comparing the model's prediction of transgene fate to the actual fate of the transgene in an experiment in which transgenic fish are released into a population of wild relatives. The model should be run using a complete set of data collected on real transgenic fish and non-transgenic conspecifics. The experiment should be in indoor laboratories or well-confined ponds or artificial streams under environmental conditions which allow the fish to survive and reproduce on their own. Transgene fate should be measured over multiple generations. It will be easiest to conduct such empirical tests using a model fish species, such as Japanese medaka, with short generation times and ease of natural reproduction in captivity.

Any future use of this methodology in a real risk assessment should explicitly acknowledge its limitations. The current mathematical model is deterministic, so it does not include stochastic (random) effects. Modelling experts can readily remove the limitation of the deterministic mathematical model by adding appropriate algorithms for stochasticity; one team is in the process of doing so (W.M. Muir, Purdue University, Indiana, 2005, personal communication). Genotype-by-environment interactions will limit the predictability of results for the real ecosystems at issue (Devlin *et al.*, 2004b), but this problem can be somewhat reduced by repeating fitness measurements (Fig. 5.3) under different biotic and abiotic conditions known to contribute to G × E interactions in the unmodified species. Chapter 6 discusses this issue in detail. Finally, the methodology presently assumes no change in fitness trait values over multiple generations in response to natural selection. This assumption may be unrealistic in light of the present knowledge (e.g. Endler, 1995); effects of this assumption on the ultimate fate of introgressed genes could be most easily tested via sensitivity analyses or other uncertainty analysis methods (see Chapter 7, this volume).

F_1 offspring survival to maturity: gamete incompatibility and outbreeding depression

Two additional genetic processes should be considered when assessing the likelihood of F_1 offspring surviving to maturity. First, gametic incompatibility (hybrid incompatibility) could reduce the viability of first-generation hybrid offspring. Gamete incompatibility will be primarily an issue if transgenic escapees hybridize with a wild population of a closely related species (e.g. in tilapias), and such incompatibility will be more likely due to existing species barriers than the transgenes themselves (Bolnick and Near, 2005). Laboratory experiments can be conducted to test for gametic incompatibility of transgenic embryos and offspring.

Second, F_1 offspring can exhibit outbreeding depression, the decline of a key fitness trait such as embryonic or juvenile viability upon interbreeding of two different populations (Hallerman, 2003). This may result from a breakdown of each population's different co-adapted gene complexes, i.e. sets of alleles that work together to confer high fitness. This may or may not occur in F_1 individuals, as each has a complete set of alleles from each parent which may mask low levels of outbreeding depression. Thus, detection of a fitness disadvantage in first-generation hybrids could reflect a large genetic incompatibility through interactions between alleles at different loci. Outbreeding depression in transgenic F_1 hybrids could be due to maladaptive transgenes or other alleles in the genetic background of the transgenic line. Testing for outbreeding depression in captivity is difficult due to the inherent differences between nature and confined lab or semi-natural testing units. It is still an important task because if outbreeding depression is detected in captivity, it is likely to occur in the wild. However, even if it is not detected in captivity, it may still occur in the wild. Confined testing for outbreeding depression will be easiest at early life stages, relatively easy at later life stages and quite difficult at later life stages. Also, confined testing will be easier for species with short generation times and ease of reproduction in captivity.

Long-term introgression of transgenes and genetic background into wild populations

Introgression of transgenes into the BC_1 generation and further generations of backcrossing (BC_n) depends ultimately on a balance between migration of the transgene into the population (gene flow) and natural selection against transgenic phenotypes. If transgenic phenotypes are sufficiently similar to wild phenotypes, natural selection against transgenic individuals will be relatively weak and introgression may occur. Assuming escapes of hemizygous transgenic fish, half of the F_1 progeny (from matings between transgenic and wild adults) will carry the transgene, and the other half will only carry background genes from the transgenic line. Natural selection against descendants carrying the transgene could be either greater or less than natural selection against those carrying only the genetic background; either consequence could alter the likelihood of transgene introgression into a wild population. Eventually, in a stable environment, migration and selection will reach equilibrium, and the transgene will persist at a frequency that is determined by the selection coefficient (i.e. a measure of the intensity of selection) against it (Hedrick and Velasco, 1999; Tufto, 1999; Frankham *et al.*, 2002). Mutation and random genetic drift (principally in small populations) can affect the final transgene frequency as well.

It is possible that selection against transgenic individuals resulting from introgression will initially be weak but become stronger as introgression proceeds. Outbreeding depression may become more pronounced in BC_1 and later generations, compared to the F_1 generation, owing to the unmasking upon segregation[1] of unfavourable genetic combinations through deleterious recessive alleles. These unfavourable genetic combinations may or may not involve the transgenic loci. Outbreeding depression of this type has been observed in species of salmonids and largemouth bass (e.g. Gharrett *et al.*, 1999; Philipp *et al.*, 2002; McGinnity *et al.*, 2003; Thrower *et al.*, 2004; McClelland *et al.*, 2005), but to date few such multigenerational studies have been undertaken.

If the transgene alters the first age at maturity, this can increase or decrease the rate of introgression. For instance, the reduced age at maturity observed in certain growth-enhanced transgenic lines (Muir and Howard, 2001; Devlin *et al.*, 2004a) could speed the rate of introgression through multiple rounds of hybridization and backcrossing.

Potential Consequences of Transgene Introgression for Genetic Diversity and Persistence of Wild Populations

It is critically important to conserve genetic and phenotypic diversity within and among wild populations of a species because this diversity determines how populations will respond to current and future environmental variation (Frankham *et al.*, 2002). Phenotypic diversity in life history traits reflects evolutionary processes that have affected variation in fitness within populations and subsequent local adaptation that can lead to genetic divergence among populations (Roff, 1992, 2002; Stearns, 1992). Variation in life history traits is therefore an expression of how genetic diversity is distributed within and among populations. Gene flow from introduced transgenic organisms can alter the distribution of genetic diversity within a recipient wild population and among populations. Because genetic diversity ultimately determines a population's ability to adapt to changing conditions, such a change in how genetic diversity is partitioned within and among populations can affect extinction risk.

Hybridization and introgression of introduced genetic material from transgenic escapees (transgenes plus genetic background) pose several genetic risks to wild populations, ranging from changing allelic frequencies to the loss of population genetic structure and variation. Hybridization of introduced populations with wild fish, followed by introgression of their genetic backgrounds into wild populations, is known to have caused loss of local adaptation to the natural environment (Rhymer and Simberloff, 1996). Alternatively, hybridization and introgression might promote the spread of introduced populations by increasing their genetic variation, which may later increase their degree of local adaptation. Under this alternative scenario, introgression would break up existing co-adapted gene complexes in the wild population, and only later might natural selection favour new, co-adapted gene complexes. In either case, hybridization and introgression may eventually lead to extinction of the wild populations if

[1] Separation into different gametes, hence into different offspring, of the two members of the pair of alleles possessed by the parent.

the introduced population is large or repeatedly released. Therefore, it is important that such gene flow be detected as early as possible (see Chapter 9, this volume) to prevent or reduce changes in native genetic diversity.

Introgression of transgenes may change genetic diversity at other loci within a native population. If the transgene alters fitness, natural selection may have broader effects on the genome, depending on the loci linked to the transgene (Gepts and Papa, 2003). For example, if the transgene confers higher fitness by increasing reproductive success, alleles at loci linked to it may be subject to a selective sweep that increases their frequency. If, on the other hand, the transgene reduces fitness, linked loci will be subject to background selection (Charlesworth *et al.*, 1993, 1995). In both cases, genetic diversity in the wild population is altered and may be reduced, even if transgenic individuals have lower fitness than wild individuals.

If the transgene is maladaptive in the wild, natural selection will eventually purge it from an introgressed population. This is the most benign outcome for conservation of genetic diversity, but it may not always be impact free. Purging of maladaptive transgenic fish is not instantaneous and requires a number of generations depending on the degree of natural selection against the transgenic phenotypes. Persistence of maladaptive transgenic individuals during this generational time lag can heighten extinction risk if the introgressed wild population is already in decline.

If a gene flow assessment predicts the Trojan gene scenario (Box 5.2), the predicted population decline is itself an ecological change. This decline should be assessed for potential cascading ecological effects, a topic taken up in Chapter 6. If the affected wild population is native (hence a centre of origin for the species), there will be loss of unique genes. If the affected population is already threatened or endangered, the Trojan gene effect would increase the chance of extinction. The loss of an entire population, in turn, might reduce the resilience of the aquatic biological community, for instance through simplification of the food web, unless the community contains other species that serve a similar ecological function (Olden *et al.*, 2004). All of these possible consequences are important issues to discuss during multi-stakeholder deliberation (see Chapter 2, this volume) in order to ascertain their broader societal implications.

Data Needs to Estimate Gene Flow

Estimating gene flow between the transgenic fish and its related forms in the wild requires knowledge of the organismal biology and ecology, population parameters (including population size or abundance, population structure, fecundity, recruitment of juveniles to the population, mortality and other life history traits), and community ecology for the wild relatives of the transgenic fish. This information is important especially for estimating the probability of encounter and mating between the transgenic fish and its related wild forms.

Table 5.1 summarizes the data needs discussed throughout this chapter. For certain data gaps, field and laboratory experiments will be needed. The

Table 5.1. Examples of types of data, studies and scientific expertise needed to assess gene flow from transgenic fish to wild relatives.

Description of data need	Types of studies[a] (generally from simplest to most complex)	Studies (may) require expertise in
Data to estimate entry potential		
What is the rate of escape, from existing aquaculture or experimental facilities ('propagule pressure')?	• Field studies to detect and quantify escapees • Mandatory self-reporting of escapes by relevant facilities (requires infrastructure for enforcement) • Mark–recapture studies • Use of molecular genetics markers • Mixed-stock analysis • Video surveillance	• Fisheries assessment methods • Molecular genetics methods, such as PCR-based detection of specific genes
What is the pattern of escapes from existing aquaculture facilities?	• Field studies to detect escapees • Molecular lab studies, especially when genetic markers are the only way to differentiate cultured and wild fish • Use of telemetry systems	• Fish population dynamics and field assessment methods • Life history of the species in question • Spatial (GIS) modelling
What proportion of immature transgenic escapees are likely to survive to sexual maturity in the natural environment?	• Mark–recapture field experiments • Laboratory experiments to determine survival rates relative to wild-type • Mixed-stock analysis	• Life history of the species in question • Fish population dynamics and field assessment methods • Fish ecology
Data to estimate introgression potential		
Do transgenic escapees disperse in a spatial and temporal pattern and in a phenotypic state that make them likely to find available mates?	• Field sampling for presence of escapees at critical times and places vis-à-vis the native population • Laboratory experiments and spatial modelling	• Life history of the species in question • Fisheries assessment methods • Spatial (GIS) modelling
Are transgenic escapees likely to mate with wild conspecifics (or to hybridize with closely related species) in the natural environment?	• Laboratory studies of mating behaviours of transgenic fish • Field sampling to determine what environments are suitable for reproduction	• Life history of the species in question, especially of mating behaviours and breeding in captivity • Fisheries assessment methods
Are F_1 or BC_n progeny likely to survive and reproduce successfully in the natural environment?	• Laboratory experiments in which matings between transgenic and wild fish can be controlled	• Life history of the species in question, especially of mating behaviours and breeding in captivity • Genetics and breeding programmes
What is the relative net fitness of transgenic fish, compared to a selected captive or wild population?	• Laboratory experiments in which transgenic and comparative strains of fish can be bred and measured for fitness components (fecundity, fertility, age at sexual maturity, mating advantage, juvenile viability, adult viability)	• Life history of the species in question, especially as it might guide prioritizing the most important fitness component traits to examine

Continued

Table 5.1. *Continued*

Description of data need	Types of studies[a] (generally from simplest to most complex)	Studies (may) require expertise in
What is the spatial distribution of populations of wild conspecifics, or closely related species, in the accessible ecosystem?	• Field sampling for presence of wild fish • Telemetry studies	• Fish systematics (ichthyology) for correct identification of fish species in the wild • Fish behavioural ecology • Fisheries assessment methods • Population genetics techniques and analysis
How many reproductively active wild conspecifics, or closely related species, live in the accessible ecosystem?	• Field sampling for direct estimation of abundance of wild fish • Mark–recapture studies	• Fish population dynamics and field assessment methods
Other desirable data		
How might transgenic fish's phenotype be expressed in a variable natural environment?	• Laboratory experiments in which fish can be exposed to manipulations of environmental variables contributing to survival and reproductive success in the wild (e.g. variable density, natural food or other simulations of natural habitat features)	• Fish behaviour • Fish genetics • Life history of the species in question, especially as it might guide prioritizing the most important environmental variables
What is the population genetic structure of the wild populations?	• Field sampling wild fish to collect tissue • Laboratory analysis of genetic structure of population (allozyme to DNA-marker studies)	• Population genetics techniques and analysis
How will the genetic background of the transgenic and wild strains affect the probability of introgression?	• Laboratory experiments in which matings between transgenic and wild fish from different strains can be controlled	• Life history of the species in question, especially of mating behaviours and breeding in captivity • Genetics and breeding programmes

[a]Any studies using transgenic fish should be well confined to prevent the escape of transgenic fish into the wild (see Chapter 8, this volume).

discussion below addresses two kinds of data needs in further detail. Chapter 10 discusses capacity building needed to facilitate baseline ecological data collection and access.

Baseline data needs

Effective estimation of gene flow between transgenics and wild relatives depends on baseline data on the relevant populations of wild relatives in accessible ecosystems. Preferably such data are collected long before a country considers adopting transgenic fish. Table 5.1 indicates which baseline data are needed to estimate specific parameters in Eqs 5.2, 5.4 and 5.5. These attributes include reproductive traits, survival traits and population ecology of the wild relatives. It is preferable to have baseline data from multiple-year classes of the wild populations, from different seasons, and after any major environmental shock (such as major pollution discharges, dam construction, invasive species spread) likely to lead to cascading effects on existing wild populations. Baseline data used to predict gene flow *before* adopting transgenic fish in aquaculture operations are also needed to inform monitoring after adoption of transgenic fish. This chapter introduces some methods for baseline data collection; Chapter 9 provides additional guidance.

Distinguishing escapees from wild relatives in accessible ecosystems

The ability to differentiate aquaculture escapees from wild relatives in the accessible ecosystems is crucial for collecting accurate baseline data necessary to estimate the probability of escape in Eq. 5.2 (Table 5.1). In some cases, it may be possible to distinguish aquaculture escapees solely by external differences (e.g. Craik *et al.*, 1987; Fleming *et al.*, 1994; Lund *et al.*, 1997; Hard *et al.*, 2000; Aparicio *et al.*, 2005; von Cramon-Taubadel *et al.*, 2005). More technically demanding methods with potential to differentiate farmed fish from wild fish include scale analysis (Lund *et al.*, 1989), carotenoid pigments (Craik *et al.*, 1987; Lura and Sægrov, 1991), stable isotope profiles of otoliths (ear bones of fish) (Dempson and Power, 2004) and induced thermal signatures on otoliths. However, genetic marking offers a potentially much more powerful technique for marking farmed fish (Crozier, 2000) because the mark is permanent (see Chapter 9, this volume, for detailed discussion of use of genetic markers).

Use of markers, regardless of type, would have to be widely adopted – and probably have to be required by law – to truly allow estimation of baseline rates of escape from aquaculture operations. The few cases of mandatory tagging involve sea cage farming of salmon in the USA and Iceland (Naylor *et al.*, 2005). Tagging of cultured fish would be hard to implement on a wide scale in many developing and developed countries. Interestingly, recent scientific publications (Senanan *et al.*, 2004; Hassanien and Gilbey, 2005; Mohindra *et al.*, 2005) indicate that some developing countries have an active research programme on genetic markers, and it might be feasible to develop genetic markers to quantify escapees.

Wild population genetic structure

Knowledge of the population structure of wild relatives of transgenic fish is essential for predicting the likely patterns of transgene movement. Accurate information on the spatial distribution of conspecific or closely related wild populations is essential (Roderick and Navajas, 2003). It is important to determine the degree of genetic differentiation between the transgenic and wild populations, and population structure of the wild populations, *prior* to the introduction of transgenics into aquaculture systems. All of these data needs can be met by application of genetic markers combined with modern statistical methods, as summarized below under possible field studies.

Data needed to understand genotype-by-environment interactions

Genotype-by-environment interactions are widespread in natural populations and can have profound consequences for the fate of transgenic individuals released into wild environments. The most important of these consequences is the unpredictability of the transgenic phenotype in the wild. Traits measured on transgenic individuals in captivity may not accurately reflect the phenotypes of their progeny in natural environments (Bessey *et al.*, 2004; Devlin *et al.*, 2004b; Sundström *et al.*, 2004). The challenge of assessing G × E interactions for transgenic fish before they enter natural ecosystems is discussed in detail in Chapter 6.

Confined experiments can manipulate environmental variables (e.g. by varying density, water flow patterns or abundance of natural food) to test the influence of G × E interactions on fitness-component traits of transgenic fish. This will improve understanding of how G × E interactions may affect mating of transgenic fish with wild relatives. For example, for GH-transgenic coho salmon, food availability was found to strongly influence the growth and survival of transgenic and non-transgenic fish differently, and such results provide evidence for significant G × E effects on fitness (Devlin *et al.*, 2004a). Combined with mating trials in captivity, such experiments would help to identify conditions under which transgenic individuals are most likely to successfully encounter and interbreed with wild individuals. Such experiments cannot remove all incertitude about effects of G × E interactions on gene flow because they cannot fully replicate natural environments. (See Chapter 7, this volume, for strategies to address such incertitude.)

Possible Field Studies and Experiments to Address Data Needs

This section discusses the design and construction of possible field studies and confined experiments for collecting some of the data mentioned in the previous section. A few overriding considerations apply to all experiments designed to address questions about gene flow: (i) who should design the experiments; (ii) appropriate transgenic and control groups for experiments; (iii) who should collect the data; (iv) confinement of experiments involving transgenic fish; and (v) timing of the experiments.

First, field studies and experiments should be designed by scientists who are experts in the biology, ecology and behaviours of the species in question. Knowledge of the variance in the parameter being measured is needed so that experiments can be statistically robust and detect expected differences. (See Chapter 9, this volume, for discussion of issues related to sample size and designing statistically robust studies.) Ideally, confined experiments should involve real transgenic fish from the strain of interest. Proper control groups for such experiments depend on the situation at hand. For transgenic fish lines produced from wild-type individuals, the meaningful control is the source parental line. This line often differs from the currently farmed unmodified line and may even be a wild population. Some transgenic fish lines developed for aquaculture, however, may originate from or have been crossed with a farmed population exhibiting the best available characteristics. In this situation, the farmed population is the proper control. Regardless of the ancestry of the transgenic fish line, some experiments may need a second control consisting of the wild population with which transgenic escapees might interbreed.

Once experiments for comparing life history trait values of transgenic and the wild strain have been designed, most of these data are relatively simple to collect. These experiments can be carried out by a group of scientists who are not necessarily experts on the species. It is important to have well-confined facilities for housing and rearing the test fish, including the transgenic strain (see Chapter 8, this volume, for confinement options). Much of this research will take place in the context of a pending policy decision. In other words, transgenic fish will likely be proposed for a particular application, and scientists will collect data to help decision makers determine whether a particular transgenic fish should be approved for that application. The time-sensitive nature of many proposals underscores the benefits of collecting baseline data far in advance of any transgenic fish proposals. However, even when decision makers impose deadlines, tools such as the net fitness methodology can be used to help prioritize the most important life history characteristics to investigate.

Field studies to address entry potential

Baseline data on escapees

Estimating entry potential of transgenic lines is much easier if there are baseline data on patterns of escape and dispersal behaviour of unmodified lines from existing aquaculture operations. One approach to establishing baseline escape data is to implement mandatory reporting of escapes of currently farmed, non-transgenic strains from existing aquaculture operations. Such reporting recently became required in some salmon cage-farming regions (Anonymous, 2001a,b; Naylor *et al.*, 2005). Site specific markers for individual hatcheries and farms, under exploration in the USA and Norway (Naylor *et al.*, 2005), would enhance such reporting. Enforcing such marking requirements would be unfeasible in most developing counties due to the limited regulatory power and staff in governmental fisheries departments. An alternative approach could be to estimate total escapees using readily available information on the location and size of existing aquaculture units, as well as the frequency and damage levels of events that facilitate escapes from these operations. The scale of escapes could be estimated from transect analysis, starting from the aquaculture operation, through the main watercourse and along the drainage leading out of the aquaculture operation. A scientist familiar with statistical field sampling methods should design the sampling protocol, which can then be carried out by technicians trained in traditional fisheries assessment methods (Murphy and Willis, 1996). A slightly different approach would be direct sampling (preferably stratified sampling) of the vicinity around many aquaculture operations in an area. Both sampling approaches require distinguishing escapees from wild relatives. Alternatively, the total number of escapees could be deduced by comparing fish farm records for numbers of fish stocked and numbers harvested, provided the farmers maintain accurate records on fish mortalities and the assessors can access such records.

The pathways of movement and the rate of travel are key components to determining the likelihood that escapees will encounter native breeders. Radio-tagging and monitoring of farmed fish by tracking them directly or through pre-tuned automatic listening stations located at strategic points within catchments would make it possible to collect some baseline data (Butler *et al.*, 2005). Also, some simple experiments might involve mark–recapture studies of members of the wild population, especially of adults during the breeding season, in the vicinity of potential transgenic escape sites. Visibly marking or tagging individuals and recovering them in a systematic way, guided by a sufficiently powerful sampling design, can help elucidate spatial patterns of movement, rates of movement and, potentially, survival (Seber, 1982). The sampling design depends critically on the biology of the organism, including whether it is migratory or not, particularly during the reproductive period. For example, a sampling design for fish migrating along river channels might involve a series of transects radiating from the potential release site, with sampling sites located at periodic intervals or focused near known mating aggregations. By contrast, a design for non-migratory fish might require only a simple array of sampling sites spread over suitable habitat in the area surrounding the point of release, with the size of the area guided by knowledge of the fish's home range.

Population genetic studies for assessing the extent and genetic impacts of transgene flow

Assessing the geographical extent of potential introgression of transgenes and the genetic impacts of this introgression on wild populations requires two types of baseline information: (i) the genetic population structure of the wild populations prior to any escapes of transgenic fish; and (ii) the degree of genetic differentiation between wild populations and the transgenic line. Such baseline information is also needed to monitor changes in population genetic structure *after* any approval to culture transgenic fish in a specific place (see Chapter 9, this volume).

A broad sampling of wild fish populations across regions is needed to identify wild population subdivisions that may reflect adaptation to local conditions. Hence, a hierarchical sampling scheme is recommended (Shaklee and Currens, 2003). Prior understanding of the ecology of the target species (e.g. preferred habitats for reproduction), ecosystem variation and barriers to movement of individuals within a water system is paramount for designing an adequate sampling scheme to determine the population genetic structure of the species. Sampling design should be guided by experts on population genetics and the biology and ecology of the local species. The sampling scheme, including number of individuals sampled, is also dictated by the kind of genetic marker used.

Once there is information about the population structure of wild relatives, mixed-stock analysis offers one way to determine the degree of genetic differentiation between the transgenic line and these wild populations (Millar, 1987; Pella and Masuda, 2001). For example, Hansen (2002) used such methods to detect different degrees of genetic introgression of domesticated trout into wild brown trout populations in two rivers in Denmark. Kalinowski (2004) gives guidance on mixed-stock analysis.

A variety of software packages exist for evaluating genetic constitutions and population structure, as well as for mixed-stock analysis. Among those for population genetic analyses are ARLEQUIN (Excoffier *et al.*, 2005), FSTAT (Goudet, 2002), GENEPOP (Raymond and Rousset, 1995) and TFPGA (Miller, 1997). Examples of software programs for mixed-stock analysis include GMA (Kalinowski, 2003) and SPAM (Debevec *et al.*, 2000).

Confined experiments to assess influence of genetic background on transgene flow

An important issue to consider when assessing gene flow is how the genetic background of escaping transgenic fish will affect the traits influencing transgene flow in natural populations. Most transgenic fish under development are designed to have special attributes, such as faster growth or disease resistance, to enhance productivity of an aquaculture system. As discussed in Chapter 3, two common objectives of fish genetic engineering are to reduce costs and produce a better-quality end product. Therefore, a highly selected and high-performance fish breed would be the most probable target for transgenesis. Consequently, the transgene will not be transmitted alone during introgression into a wild population; it will introgress along with an important set of loci with very high frequency of alleles advantageous for the selected traits, resulting in substantial changes in allele frequencies (including introduction of new alleles) in the wild population. McGinnity *et al.* (2003) showed that, under wild conditions, offspring of escaped farmed salmon and of crosses between farmed and wild fish (F_1, F_2 and BC_1 groups) had reduced survival rates compared to wild salmon. Devlin *et al.* (2001) showed that genetic background can also affect transgene expression, as a fish line highly selected for fast growth did not differ much in size-at-age from a growth-enhanced transgenic fish line of the same species. These two studies suggest that selectively bred fish and transgenic fish with similar changes in a trait (e.g. enlarged adult size) might show similar initial changes in fitness. However, very different multigeneration effects are possible because of differences in the genetic basis of the trait: monogenic control (by allelic variation at a single locus) of size in the transgenic line versus polygenic control (by allelic variation at a multiple loci) in the selectively bred line.

The genetic background into which a transgene is inserted will affect its effect on a recipient population because genes interact in their effect on the phenotype. Evidence that different genetic backgrounds can modify transgene expression has been found in many taxa of animals including mice (McGowan *et al.*, 1989; Elliot *et al.*, 1995; Opsahl *et al.*, 2002), fruit flies (Pal-Bhadra *et al.*, 1997; Aagaard *et al.*, 1999) and fish (Devlin *et al.*, 2001). Furthermore, modelling of changes in fitness of transgenic fish under differing genetic backgrounds (Devlin *et al.*, 2005) indicates that even small changes in fitness arising from variation in genetic background can override a Trojan gene effect resulting from invasion of the transgene. Multigeneration effects of genetic background on the fate of the transgene can be quite profound. For example, one long-term study evaluated the effects of a transgene on correlates of fitness over 13 generations of selection for 8-week body weight in two transgenic lines of mice, both carrying a GH-transgene (*oMtla-oGH*), but of very different genetic backgrounds – one selected for body weight and the other non-selected. Mice carrying this particular construct had reduced litter size and reproductive effort (Clutter *et al.*, 1996; Siewerdt *et al.*, 1998). Genetic response accumulated in the transgenic lines via changes in polygenes affecting growth and not by increased transgene frequency, and the transgene was eventually lost (Siewerdt *et al.*, 2000).

Collectively, the above studies indicate that genetic background is likely to have a major effect on the flow and fate of transgenes in fish populations. The effect of natural selection on the transgene within a particular genetic background can have unpredictable effects in wild gene pools. Novel environments can produce unpredictable results – even a dramatic increase in frequency of a transgene – if selection favours a particular genetic background. Selection need not even act on the trait directly affected by the transgene; natural selection operating on any other trait influenced by or linked to the transgene will affect its frequency as well.

One approach to evaluating the importance of variation in genetic background is to assess the survival and phenotypes of individuals experimentally cross-bred from different candidate transgenic lines and wild populations in confinement. The kind of experiment outlined below could involve crosses between wild fish and transgenic lines founded from a number of already cultivated populations. The next section's discussion uses a number of quantitative genetic concepts and terms; readers may want to consult a quantitative genetics text (e.g. Falconer and Mackay, 1996) for further detail.

Experimental design to measure effect of genetic background
For brevity, the discussion here considers only two types of transgenic lines (t) with contrasting genetic backgrounds: a cultured fish line with little or no prior selective breeding (Wt), and a highly selected line of the same species (St). A possible experimental design is to make the following crosses (Fig. 5.5):

- Reciprocal crosses between a wild population (W) and Wt;
- Reciprocal crosses between W and St;
- Pure lines of W, Wt and St.

Ideally, the experimental design would allow the measurement of genetic components of phenotypic variation, in order to estimate genetic parameters that may influence gene flow. This requires marking fish individually to keep track of the pedigree. It also requires using a mating design that provides good estimates of genetic components of variance. A hierarchical mating design (e.g. North Carolina design 1; Falconer and Mackay, 1996) may be a prudent choice.

The number of generations of crosses required for the experiment depends on the information needed. If investigators want to assess the probability of introgression, the experiment should entail at least two generations, F_1 crosses and second-generation crosses (BC_1 and F_2). If F_1 progeny show greatly depressed survival or reproduction – assuming they are reared under excellent environmental conditions for the species – then the study may not need to assess backcrosses.

Measure phenotypes of targeted transgenic trait and fitness-component traits
Data should be collected on variation in the transgenic trait and fitness-component traits to understand how genetic background might affect gene flow. Inference about the effect of genetic background on the targeted transgenic trait can be obtained by comparing its expression between the two transgenic lines (Wt and St) and between their crosses with the wild population (W). Inference about the effect of the genetic background on fitness can be drawn from comparing fitness-component traits of the wild population (W) with those of the two transgenic lines. If facilities or other experimental resources are limited, the assessment of fitness-component traits could be simplified. For example, a more limited assessment might involve only measuring,

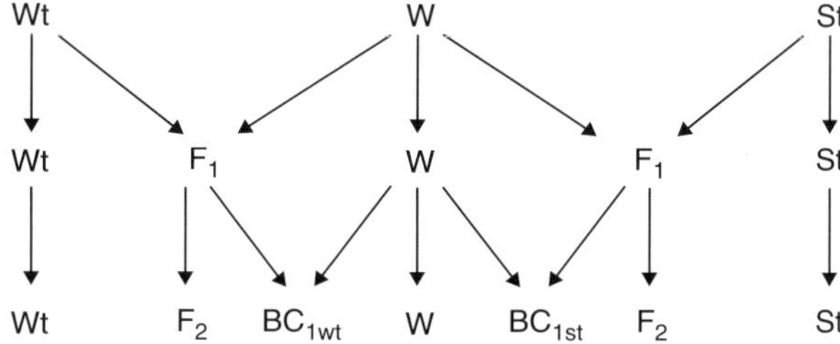

Fig. 5.5. Crosses of a wild population (W) that receives a transgene (t) from a population with a closely related, unselected background (Wt) or from a population of a very different, highly selected background (St). Genetic background experiments might examine F_1, F_2 and backcrosses to population W.

in a representative sample of individuals from each line, survival to a pre-adult stage known to be critical for viability in the wild. This life stage would differ on a case-by-case basis, depending on the biology of the species involved.

Data on the targeted transgenic trait and fitness-component traits in F_1 and F_2 hybrids and backcrosses can give information on whether the different genetic backgrounds alter expression of these traits due to dominance, heterosis, epistasis, outbreeding depression or maternal effects. In particular, backcrosses to the wild line permit determining whether there is outbreeding depression. Backcrosses to the wild line would also provide insight into whether different genetic backgrounds have a major effect on the fate of the transgene. If one of the examined genetic backgrounds appears to reduce the probability of introgression, this information should be used to guide the choice of transgenic line used for commercial development.

Benefit of incorporating genetic parameter estimation into the experiment

Data on how transgenesis changes quantitative genetic parameters may help to predict the effect of transgene introgression on the fitness of wild populations. For example, if expression of the transgene (e.g. for growth enhancement) is tied (through linkage or pleiotropy) to another gene of large effect on another trait (e.g. reproduction), then the genetic architecture underlying fitness in the wild populations could change quickly under selection. Hence, examination of genetic parameters should focus on comparing heritability and genetic correlations between fitness traits for the different genetic backgrounds. These parameters have been published for numerous economically important and fitness-component traits of fish, including different captive populations of salmon, trout, tilapia, carp, cod and many others (Tave, 1993; Hallerman, 2003). Heritability, the fraction of total phenotypic variation in a population due to the additive effect of genes, is used to predict response to selection. Genetic correlations are also important because they reflect pleiotropic effects and transient linkage between the selected transgenic trait and traits related to fitness, such as those affecting survival and reproductive success.

Expansion of experimental design: more generations and G × E interactions

Some key refinements to the experimental design can substantially improve inferences drawn from the data with respect to predicting the fate of the transgene in a wild population. The ability to add these refinements depends on available facilities, time and funding for such risk assessment studies. One important refinement is to include one or more additional environments in which to evaluate the variation in the pure and cross-bred lines. This can provide an idea of the importance of G × E interaction on phenotypic expression, which, in turn, affects potential introgression of the transgene in wild populations. Another valuable refinement is to evaluate additional second (or later) generations, including backcrosses to the wild population. This would permit assessment of potential fitness changes in advanced generations after the critical first generation of interaction between transgenic and native individuals, making it easier to predict the longer-term fate of the transgene. Finally, a study that overlays family structure within lines on to the line-cross analysis would provide estimates of genetic and environmental components of phenotypic variance; these core parameters can be used to predict how the lines would respond to different forms of selection.

Chapter Summary

Gene flow is a major pathway through which transgenic fish might affect natural populations. It can result from successful interbreeding between transgenic fish and individuals of the same species (conspecifics) or of a closely related species (fertile, interspecific hybrids). Although this book's focus is on risk assessment of transgenic fish, concepts and methodologies presented in this chapter also apply to gene flow from selectively bred lines to wild relatives. The chapter may therefore be of help to numerous groups around the world involved with governance, research or monitoring of aquaculture operations or fish stocking programmes. It is important to recognize that transgenic fish can have ecological effects even without occurrence of gene flow, and this is addressed in Chapter 6.

This chapter presents a systematic approach to assessing potential gene flow from transgenic individuals to wild fish *before* regulatory approval or actual entry into the environment. This approach involves assessing measurable end points in a chain of events that must occur to have incorporation of transgenes into a population of wild relatives (Fig. 5.1). Two major end points are: *entry* of sexually mature, fertile transgenic fish into a population of wild relatives in an aquatic ecosystem; and *introgression* or incorporation of the transgenic genotype into the gene pool of wild relatives by backcrossing of transgenic-wild fish hybrids to wild relatives. We present methodologies for partitioning the assessment of entry and introgression into sub-components (Eqs 5.2–5.5); each sub-component should be easier to assess than treating entry or introgression as a single variable.

It can still be a daunting task to collect the case-specific empirical data for all the sub-components of entry and introgression required for assessing gene flow. The chapter, therefore, provides a step-by-step strategy for deciding whether to accept a specific, worst-case assumption (e.g. that transgenic escapees will survive in the receiving environment), allowing a risk assessor to focus limited resources on assessing other sub-components (Fig. 5.2). This strategy provides risk assessors with flexibility while maintaining scientific quality of the assessment.

The chapter also presents key data needs for each step in assessing gene flow, as well as methodologies for their collection and the kinds of scientific expertise needed to obtain and analyse these data (Table 5.1). Baseline data needs – which do not require use of transgenic fish – include specific information about populations, species and habitats in environments that transgenic fish might enter. These include fitness traits and behaviours influencing the ability of transgenic fish to survive and encounter and mate with wild fish. We introduce approaches for measuring these traits in confined experiments but point out two important limitations of the resulting data.

Genotype-by-environment interactions pose the first limit when conducting confined experiments on transgenic fish for gene flow assessment. Because fish may express different phenotypes in different environments, data obtained in captivity may not accurately predict how they will perform in nature. Thus, it is important to conduct the kinds of confined tests presented in this chapter under different environmental conditions. Chapter 6 discusses how to prioritize the environmental factors that should be varied in risk assessment experiments. Effects of genetic background (encompassing the fish's entire genome) on expression of the transgene pose a second concern when testing transgenic fish for gene flow assessment. It is important to address genetic backgrounds of: (i) the aquaculture line into which the transgenes will be incorporated prior to commercial use; and (ii) the wild populations with which transgenic escapees might interbreed. We outline quantitative genetics studies, including a possible experimental design (Fig. 5.5) that can be conducted in the lab to measure genetic background effects.

The incorporation of transgenes into wild populations could lead to changes in frequencies of native alleles (variable forms of different genes), loss of genetic distinctiveness and loss of genetic variability. Such genetic changes can threaten the ability of wild populations to adapt and persist under current and future environmental conditions. This reduction in resilience is of particular concern for wild populations already in decline, in the species' centre of origin, or for species experiencing extreme rates of environmental change. Consequences of transgenes in natural populations depend on whether, and at what rate, the transgene is purged from or spreads throughout the wild population over the long term. Purging is most benign, though not completely risk-free. Transgene spread could displace native genotypes or, in a worst case under very specific conditions, trigger a crash of the wild population.

References

Aagaard, L., Laible, G., Selenko, P., Schmid, M., Dorn, R., Schotta, G., Kuhfittig, S., Wolf, A., Lebersorger, A., Singh, P.B., Reuter, G. and Jenuwein, T. (1999) Functional mammalian homologues of the *Drosophila* PEV-modifier Su(var)3–9 encode centromere-associated proteins which complex with the heterochromatin component M31. *The EMBO Journal* 18, 1923–1938.

Anderson, R.O. and Neumann, R.M. (1996) Length, weight, and associated structural indices. In: Murphy, B.R. and Willis, D.W. (eds) *Fisheries Techniques*, 2nd edn. American Fisheries Society, Bethesda, Maryland, pp. 447–482.

Anonymous (2001a) *Scottish Fish Farms Annual Production Survey 2000*. Fisheries Research Services, Scottish Executive Environment and Rural Affairs Department, Edinburgh, Scotland.

Anonymous (2001b) Reglamento ambiental para la acuicultura (RAMA), D.S. Minecon N 320-01. Available at: http://www.subpesca.cl/

Anonymous (2002) *Review and Synthesis of the Environmental Impacts of Aquaculture*. The Scottish Association for Marine Science and Scottish Executive Central Research Unit, Napier University, Edinburgh, Scotland.

Aparicio, E., Garcia, B., Araguas, R.M., Martinez, P. and Garcia-Marin, J.L. (2005) Body pigmentation pattern to assess introgression by hatchery stocks in native *Salmo trutta* from Mediterranean streams. *Journal of Fish Biology* 67, 931–949.

ADB (2005) *An Impact Evaluation of the Development of Genetically Improved Farmed Tilapia and Their Dissemination in Selected Countries*. Operations Evaluation Department, Asian Development Bank, Metro Manila, Philippines. Available at: http://www.adb.org/Documents/Books/Tilapia-Dissemination/IES-Tilapia-Dissemination.pdf

Berejikian, B.A., Tezak, E.P., Schroder, S.L., Knudsen, C.M. and Hard, J.J. (1997) Reproductive behavioral interactions between wild and captively reared coho salmon (*Oncorhynchus kisutch*). *ICES Journal of Marine Science* 54, 1040–1050.

Bessey, C., Devlin, R.H., Liley, N.R. and Biagi, C.A. (2004) Reproductive performance of growth-enhanced transgenic coho salmon. *Transactions of the American Fisheries Society* 133, 1205–1220.

Bolnick, D.I. and Near, T.J. (2005) Tempo of hybrid inviability in centrarchid fishes (Teleostei: Centrarchidae). *Evolution* 59, 1754–1767.

Butler, J.R.A. and Watt, J. (2003) Assessing and managing the impacts of marine salmon farms on wild Atlantic salmon in western Scotland: identifying priority rivers for conservation. In: Mills, D.H. (ed.) *Salmon at the Edge*. Blackwell Scientific, Oxford, UK, pp. 93–118.

Butler, J.R.A., Cunningham, P.D. and Starr, K. (2005) The prevalence of escaped farmed salmon, *Salmo salar* L., in the River Ewe, western Scotland, with notes on their ages, weights and spawning distribution. *Fisheries Management and Ecology* 12, 149–159.

Campton, D.E. (1987) Natural hybridization and introgression in fishes: methods of detection and genetic interpretation. In: Ryman, N. and Utter, F. (eds) *Population Genetics and Fisheries Management*. University of Washington Press, Seattle, Washington, DC, pp. 161–192.

Charlesworth, B., Morgan, M.T. and Charlesworth, D. (1993) The effect of deleterious mutations on neutral molecular variation. *Genetics* 134, 1289–1303.

Charlesworth, D., Charlesworth, B. and Morgan, M.T. (1995) The pattern of neutral molecular variation under the background selection model. *Genetics* 141, 1619–1632.

Clutter, A.C., Pomp, D. and Murray, J.D. (1996) Quantitative genetics of transgenic mice: components of phenotypic variation in body weights and weight gains. *Genetics* 143, 1753–1760.

Craik, J.C., Harvey, S.M., Jakupsstovu, S.H.I. and Shearer, W.M. (1987) Identification of farmed and artificially reared Atlantic salmon among catches of the wild salmon fishery of the Faroes. International Council for the Exploration of the Sea meeting, Copenhagen, Denmark.

Crozier, W.W. (2000) Escaped farmed salmon, *Salmo salar* L., in the Glenarm River, Northern Ireland: genetic status of the wild population 7 years on. *Fisheries Management & Ecology* 7, 437–446.

de la Fuente, J., Guillen, I., Martinez, R. and Estrada, M.P. (1999) Growth regulation and enhancement in tilapia: basic research findings and their applications. *Genetic Analysis: Biomolecular Engineering* 15, 85–90.

De Silva, S.S., Subasinghe, R.P., Bartley, D.M. and Lowther, A. (2004) Tilapias as alien aquatics in Asia and the Pacific: a review. FAO Fisheries Technical Paper. No. 453. FAO, Rome, Italy.

Debevec, E.M., Gates, R.B., Masuda, M., Pella, J., Reynolds, J.J. and Seeb, L.W. (2000) SPAM (version 3.2): Statistic program for analyzing mixtures. *Journal of Heredity* 91, 509–511. Software available at: http://www.cf.adfg.state.ak.us/geninfo/research/genetics/software/spampage.php

Dempson, J.B. and Power, M. (2004) Use of stable isotopes to distinguish farmed from wild Atlantic salmon, *Salmo salar. Ecology of Freshwater Fish* 13, 176–184.

Devlin, R.H., Yesaki, T.Y., Donaldson, E.M. and Hew, C.L. (1995) Transmission and phenotypic effects of an antifreeze/GH gene construct in coho salmon (*Oncorhynchus kisutch*). *Aquaculture* 137, 161–169.

Devlin, R.H., Biagi, C.A., Yesaki, T.Y., Smailus, D.E. and Byatt, J.C. (2001) Growth of domesticated transgenic fish. *Nature* 409, 781–782.

Devlin, R.H., Biagi, C.A. and Yesaki, T.Y. (2004a) Growth, viability and genetic characteristics GH transgenic coho salmon strains. *Aquaculture* 236, 607–632.

Devlin, R.H., D'Andrade, M., Uh, M. and Biagi, C.A. (2004b) Population effects of GH transgenic salmon are dependant upon food availability and genotype by environment interactions. *Proceedings of the National Academy of Sciences USA* 101, 9303–9308.

Elliot, J.I., Festenstein, R., Tolaini, M. and Kioussis, D. (1995) Random inactivation of a transgene under the control of a hybrid *hCD2* locus control region/Ig enhancer regulatory element. *The EMBO Journal* 14, 575–584.

Emery, L. (1981) Range extension of pink salmon into the lower Great Lakes. *Fisheries* 6, 7–10.

Endler, J.A. (1995) Multiple-trait coevolution and environmental gradients in guppies. *Trends in Ecology and Evolution* 10, 22–29.

Estay, F., Neira, R., Díaz, N., Valladares, L. and Torres, A. (1998) Gametogenesis and sex steroid profiles in cultured coho salmon (*Oncorhynchus kisutch*, W.). *Journal of Experimental Zoology* 280, 429–438.

Excoffier, L., Laval, G. and Schneider, S. (2005) Arlequin ver. 3.0: an integrated software package for population genetics data analysis. *Evolutionary Bioinformatics Online* 1, 47–50.

Falconer, D.S. and Mackay, T.F.C. (1996) *Introduction to Quantitative Genetics*, 4th edn. Prentice-Hall, London.

Fleming, I.A. and Gross, M.R. (1993) Breeding success of hatchery and wild coho salmon (*Oncorhynchus kisutch*) in competition. *Ecological Applications* 3, 230–245.

Fleming, I.A. and Petersson, E. (2001) The ability of released, hatchery salmonids to breed and contribute to the natural productivity of wild populations. *Nordic Journal of Freshwater Research* 75, 71–98.

Fleming, I.A., Jonsson, B. and Gross, M.R. (1994) Phenotypic divergence of sea-ranched, farmed and wild salmon. *Canadian Journal of Fisheries and Aquatic Sciences* 51, 2808–2824.

Fleming, I.A., Jonsson, B., Gross, M.R. and Lamberg, A. (1996) An experimental study of the reproductive behaviour and success of farmed and wild Atlantic salmon (*Salmo salar*). *Journal of Applied Ecology* 33, 893–905.

Fleming, I.A., Hindar, K., Mjølnerød, I.B., Jonsson, B., Balstad, T. and Lamberg, A. (2000) Lifetime success and interactions of farm salmon invading a native population. *Proceedings of the Royal Society, London B* 267, 1517–1524.

Frankham, R., Ballou, J.D. and Briscoe, D.A. (2002) *Introduction to Conservation Genetics*. Cambridge University Press, Cambridge.

Gepts, P. and Papa, R. (2003) Possible effects of (trans)gene flow from crops on the genetic diversity from landraces and wild relatives. *Environmental Biosafety Research* 2, 89–103.

Gharrett, A.J., Smoker, W.W., Reisenbichler, R.R. and Taylor, S.G. (1999) Outbreeding depression in hybrids between odd- and even-broodyear pink salmon. *Aquaculture* 173, 117–129.

Goudet, J. (2002). FSTAT, a program to estimate and test gene diversities and fixation indices 2.9.3.2. Institute of Ecology, University of Lausanne: Lausanne, Switzerland. Available at: http://www2.unil.ch/popgen/softwares/fstat.htm

Hallerman, E.M. (ed.) (2003) *Population Genetics: Principles and Applications for Fisheries Scientists*. American Fisheries Society, Bethesda, Maryland.

Hansen, N.M. (2002) Estimating the long-term effects of stocking domesticated trout into wild brown trout (*Salmo trutta*) populations: an approach using microsatellite DNA analysis of historical and contemporary samples. *Molecular Ecology* 11, 1003–1015.
Hard, J.J., Berejikian, B.A., Tezak, E.P., Schroder, S.L., Knudsen, C.M. and Parker, L.T. (2000) Evidence for morphometric differentiation of wild and captively reared adult coho salmon: a geometric analysis. *Environmental Biology of Fishes* 58, 61–73.
Hassanien, H.A. and Gilbey, J. (2005) Genetic diversity and differentiation of Nile tilapia (*Oreochromis niloticus*) revealed by DNA microsatellites. *Aquaculture Research* 36, 1450–1457.
Hedrick, W. and Velasco, A.L. (1999) Testing for inbreeding and outbreeding depression in the endangered Gila topminnow. *Animal Conservation* 2, 121–129.
Howard, R.D., DeWoody, J.A. and Muir, W.M. (2004) Transgenic male mating advantage provides opportunity for Trojan gene effect in a fish. *Proceedings of the National Academy of Sciences USA* 101, 2934–2938.
Hubbs, C.L. (1955) Hybridization between fish species in nature. *Systematic Zoology* 4, 1–20.
Jonsson, N., Jonsson, B. and Fleming, I.A. (1996) Phenotypically plastic response of egg production to early growth in Atlantic salmon, *Salmo salar. Functional Ecology* 10, 89–96.
Kalinowski, S.T. (2003) Genetic mixture analysis 1.0. Department of Ecology, Montana State University, Bozeman Montana 59717. Available at: http://www.montana.edu/kalinowski/GMA/GMA_Home.htm
Kalinowski, S.T. (2004) Genetic polymorphism and mixed-stock fisheries analysis. *Canadian Journal of Fisheries and Aquatic Sciences* 61, 1075–1082.
Kapuscinski, A.R. and Patronski, T.J. (2005) Genetic methods for biological control of non-native fish in the Gila River Basin. Contract Report to the U.S. Fish and Wildlife Service. University of Minnesota, Institute for Social, Economic and Ecological Sustainability, Minnesota Sea Grant Publication, St. Paul, Minnesota. Available at: http://www.seagrant.umn.edu/publications/F20
Karnasuta, J., Kamonrat, W. and Ngamsiri, T. (1999) Specific identification of tilapia in high salinity sea water in artemia culture area of Phetburi Coastal Aqaculture Station using enzyme electrophoresis. *Thai Fisheries Gazette* 52, 533–542.
Kolar, C. and Lodge, D. (2002) Ecological predictions and risk assessment for alien fishes in North America. *Science* 298, 1233–1236.
Kwain, W. and Lawrie, A.H. (1981) Pink salmon in the Great Lakes. *Fisheries* 6, 2–6.
Lockwood, J.L., Cassey, P. and Blackburn, T. (2005) The role of propagule pressure in explaining species invasions. *Trends in Ecology and Evolution* 20, 223–228.
Lund, R.A., Hansen, L.P. and Jarvi, T. (1989) Identification of reared and wild salmon by external morphology, size of fins and scale characteristics. *NINA Forskningsrapp* 1, 1–54.
Lund, R.A., Midtlyng, P.J. and Hansen, L.P. (1997) Post-vaccination intra-abdominal adhesions as a marker to identify Atlantic salmon, *Salmo salar* L., escaped from commercial fish farms. *Aquaculture* 154, 27–37.
Lura, H. and Sægrov, H. (1991) A method of separating offspring from farmed and wild Atlantic salmon (*Salmo salar*) based on different ratios of optical isomers of astaxanthin. *Canadian Journal of Fisheries and Aquatic Sciences* 48, 429–433.
Macaranas, J.M., Taniguchi, N., Pante, M.J.R., Capili, J.B. and Pullin, R.S.V. (1986) Electrophoretic evidence for extensive hybrid gene introgression into commercial *Oreochromis niloticus* (Linn.) stocks in the Philippines. *Aquaculture and Fisheries Management* 17, 249–258.
Marchetti, M.P., Moyle, P.B. and Levine, R. (2004) Invasive species profiling? Exploring the characteristics of non-native fishes across invasion stages in California. *Freshwater Biology* 49, 646–661.
McClelland, E.K., Myers, J.M., Hard, J.J., Park, L.K. and Naish, K.A. (2005) Two generations of outbreeding in coho salmon (*Oncorhynchus kisutch*): effects on size and growth. *Canadian Journal of Fisheries and Aquatic Sciences* 62, 2538–2547.

McGinnity, P., Prodöhl, P., Ferguson, A., Hynes, R., Maoilédigh, N.Ó., Baker, N., Cotter, D., O'Hea, B., Cooke, D., Rogan, G., Taggart, J. and Cross, T. (2003) Fitness reduction and potential extinction of wild populations of Atlantic salmon *Salmo salar* as a result of interactions with escaped farm salmon. *Proceedings of the Royal Society of London, Series B* 270, 2443–2450.

McGowan, R., Campbell, R., Peterson, A. and Sapienza, C. (1989) Cellular mosaicism in the methylation and expression of hemizygous loci in the mouse. *Genes and Development* 268, 1669–1676.

Millar, R.B. (1987) Maximum likelihood estimation of mixed stock fisheries composition. *Canadian Journal of Fisheries and Aquatic Science* 44, 583–590.

Miller, M.P. (1997) *Tools for Population Genetic Analyses (TFPGA) 1.3*. Department of Biological Sciences, Flagstaff, Arizona, USA. Available at: http://www.marksgeneticsoft ware.net/tfpga.htm

Mohindra, V., Narain, L., Punia, P., Gopalakrishnan, A., Mandal, A., Kapoor, D., Ponniah, A.G. and Lal, K.K. (2005) Microsatellite DNA markers for population-genetic studies of *Labeo dyocheilus* (McClelland, 1839). *Journal of Applied Ichthyology* 21, 478–482.

Muir, W.M. and Howard, R.D. (1999) Possible ecological risks of transgenic organism release when transgenes affect mating success: sexual selection and the Trojan gene hypothesis. *Proceedings of the National Academy of Sciences USA* 96, 13853–13856.

Muir, W.M. and Howard, R.D. (2001) Fitness components and ecological risk of transgenic release: a model using Japanese medaka (*Oryzias latipes*). *American Naturalist* 158, 1–16.

Muir, W.M. and Howard, R.D. (2002) Assessment of possible ecological risks and hazards of transgenic fish with implications for other sexually reproducing organisms. *Transgenic Research* 11, 101–104.

Murphy, B.R. and Willis, D.W. (1996) *Fisheries Techniques*, 2nd edn. American Fisheries Society, Bethesda, Maryland.

Mwanja, T.M. and Mwanja, W.W. (2002) Identification, characterization, and evaluation of farmed tilapiine species grown in eastern and central Uganda, Project Report. Makerere University Institute of Environment and Natural Resources, Kampala, Uganda.

Mwanja, W.W., Akol, A., Abubaker, L., Mwanja, M., Msuku, S.B. and Bugenyi, F. (2007) Status and impact of rural aquaculture practice on Lake Victoria Basin wetlands. *African Journal of Ecology* 45(2), pp. 165–174.

Naylor, R., Hindar, K., Fleming, I.A., Goldburg, R., Williams, S., Volpe, J., Whoriskey, F., Eagle, J., Kelso, D. and Mangel, M. (2005) Fugitive salmon: assessing the risks of escaped fish from net-pen aquaculture. *Bioscience* 55, 427–437.

Negus, M.T. (1999) Survival traits of naturalized, hatchery, and hybrid strains of anadromous rainbow trout during egg and fry stages. *North American Journal of Fisheries Management* 19, 930–941.

Olden, J.D., Poff, N.L., Douglas, M.R., Douglas, M.E. and Fausch, K.D. (2004) Ecological and evolutionary consequences of biotic homogenization. *Trends in Ecology and Evolution* 19, 18–24.

Opsahl, M.L., McClenaghan, M., Springbett, A., Reid, S., Lathe, R., Colman, A. and Whitelaw, C.B.A. (2002) Multiple effects of genetic background on variegated transgene expression in mice. *Genetics* 160, 1107–1112.

Pal-Bhadra, M., Bhadra, U. and Birchler, J.A. (1997) Cosuppression in *Drosophila*: gene silencing of alcohol dehydrogenase by *white-Adh* transgenes is polycomb dependent. *Cell* 90, 479–490.

Pella, J. and Masuda, M. (2001) Bayesian methods for stock-mixture analysis from genetic characters. *Fisheries Bulletin* 99, 151–167.

Philipp, D.P., Claussen, J., Kassler, T. and Epifanio, J. (2002) Mixing stocks of largemouth bass reduces fitness through outbreeding depression. *American Fisheries Society Symposium* 31, 349–363.

Philippart, J.C.L. and Ruwet, J.C.L. (1982) Ecology and distribution of tilapias. In: Pullin, R.S.V. and Lowe-McConnell, R.H. (eds) *The Biology and Culture of Tilapias. ICLARM Conference Proceedings* 7. International Center for Living Aquatic Resources Management, Manila, The Philippines, pp. 15–59.

Raymond, M. and Rousset, F. (1995) Genepop (Version 1.2): population-genetics software for exact tests and ecumenicism. *Journal of Heredity* 86, 248–249.

Rhymer, J.M. and Simberloff, D.S. (1996) Genetic extinction through hybridization and introgression. *Annual Review of Ecology and Systematics* 27, 83–109.

Roderick, G.K. and Navajas, M. (2003) Genes in new environments: genetics and evolution in biological control. *Nature Reviews Genetics* 4, 889–899.

Roff, D.A. (1992) *The Evolution of Life Histories: Theory and Analysis*. Chapman & Hall, London.

Roff, D.A. (2002) *Life History Evolution*. Sinauer Associates, Sunderland, Massachusetts.

Scientists' Working Group on Biosafety (1998) *Manual for Assessing Ecological and Human Health Effects of Genetically Engineered Organisms. Part One: Introductory Text and Supporting Text for Flowcharts. Part Two: Flowcharts and Worksheets*. The Edmonds Institute. Available at: www.edmonds institute.org. manual.html

Seber, G.A.F. (1982) *The Estimation of Animal Abundance and Related Parameters*, 2nd edn. Macmillan, New York.

Senanan, W., Kapuscinski, A.R., Na-Nakorn, U. and Miller, L.M. (2004) Genetic impacts of hybrid catfish farming (*Clarias macrocephalus x C. gariepinus*) on native catfish populations in central Thailand. *Aquaculture* 235, 167–184.

Shaklee, J.B. and Currens, K.P. (2003) Genetic stock identification and risk assessment. In: Hallerman, E.M. (ed.) *Population Genetics: Principles and Applications for Fisheries Scientists*. American Fisheries Society, Bethesda, Maryland, pp. 291–328.

Shapiro, D.Y. (1984) Sex reversal and sociodemographic processes in coral reef fishes. In: Potts, G.W. and Wooton, R.J. (ed.) *Fish Reproduction: Strategies and Tactics*. Academic Press, London. pp. 103–118.

Siewerdt, F., Eisen, E.J., Conrad-Brink, J.S. and Murray, J.D. (1998) Gene action of the *oMt1a-oGH* transgene in two lines of mice with distinct selection backgrounds. *Journal of Animal Breeding and Genetics* 115, 211–226.

Siewerdt, F., Eisen, E.J., Murray, J.D. and Parker, I.J. (2000) Response to 13 generations of selection for increased 8-week body weight in lines of mice carrying a sheep growth hormone-based transgene. *Journal of Animal Science* 78, 832–845.

Stearns, S.C. (1992) *The Evolution of Life Histories*. Oxford University Press, Oxford, UK.

Sundström, L.F., Lõhmus, M., Johnsson, J.I. and Devlin, R.H. (2004) Growth hormone transgenic salmon pay for growth potential with increased predation mortality. *Proceedings of the Royal Society of London-Biological Sciences* 271, S350–S352.

Tave, D. (1993) *Genetics for Fish Hatchery Managers*, 2nd edn. Van Nostrand Reinhold, New York.

Thresher, R.E., Hinds, L., Grewe, P., Hardy, C., Whyard, S., Patil, J., McGoldrick, D. and Vignarajan, S. (1999) *Reversible Sterility in Animals*. Australian Patent PG4884.

Thrower, F.P., Hard, J.J. and Joyce, J.E. (2004) Genetic architecture of growth and early life history transitions in anadromous and derived freshwater populations of steelhead (*Oncorhynchus mykiss*). *Journal of Fish Biology* 65, 286–307.

Trewavas, E. (1982) Tilapias: taxonomy and speciation. In: Pullin, R.S.V. and Lowe McConnell, R.H. (eds) *The Biology and Culture of Tilapias*. ICLARM Conference Proceedings 7.

International Center for Living Aquatic Resources Management, Manila, The Philippines, pp. 3–13.

Tufto, J. (1999) The wave of advance of introduced genes in populations of plants. In: Ammann, K., Jacot, Y., Simmonsenk, V. and Kjellson, G. (ed.) *Methods for Risk Assessment of Transgenic Plants: III. Ecological Risks and Prospects of Transgenic Plants, Where Do We Go from Here? A Dialogue Between Biotech Industry and Science*. Birkhauser Verlag, Basel, Switzerland.

von Cramon-Taubadel, N., Ling, E.N., Cotter, D. and Wilkins, N.P. (2005) Determination of body shape variation in Irish hatchery reared and wild Atlantic salmon. *Journal of Fish Biology* 66, 1471.

Weir, L.K., Hutchings, J.A., Fleming, I.A. and Einum, S. (2004) Dominance relationships and behavioural correlates of individual spawning success in farmed and wild male Atlantic salmon, *Salmo salar. Journal of Animal Ecology* 73, 1069–1079.

Williams, B.K., Nichols, J.D. and Conroy, M.J. (2002) *Analysis and Management of Animal Populations*. Academic Press, San Diego, California.

Wootton, R.J. (1990) *The Ecology of Teleost Fishes*. Chapman & Hall, London.

6 Assessing Ecological Effects of Transgenic Fish Prior to Entry into Nature

R.H. Devlin, L.F. Sundström, J.I. Johnsson, I.A. Fleming, K.R. Hayes, W.O. Ojwang, C. Bambaradeniya and M. Zakaraia-Ismail

Introduction

Researchers recognized the potential for adverse ecological consequences following the release of transgenic fish in the early years of research on this technology (Tiedje *et al.*, 1989; Kapuscinski and Hallerman, 1990). This recognition stimulated development and assessment of containment strategies and tools to assist in identifying environmental hazards and reducing hazard exposure via containment (Devlin and Donaldson, 1992; ABRAC, 1995; Scientists' Working Group on Biosafety, 1998). Although the literature has defined what needs to be assessed, the development and application of empirical methodologies for conducting such assessments is in its infancy, relying thus far on laboratory and modelling studies and not on field studies. This reflects the significant time and resources required to develop transgenic strains useful for controlled scientific investigations and to conduct ecological risk research under adequate confinement. It also reflects the lack of direct experience with how transgenic fish function in nature because, to our knowledge, none have escaped or have been introduced into aquatic ecosystems. Currently, ecologically relevant information about transgenic fish must come from contained laboratory observations or field data about related strains or species which may share characteristics with a transgenic strain of interest (Devlin *et al.*, 2006).

This chapter addresses methodologies to assess potential ecological effects of transgenic fish *before* they actually enter unconfined environments. It outlines a strategy for identifying the most important data and then discusses methodologies for generating them. The chapter also discusses: (i) types of information needed to accurately characterize potential receiving environments for the transgenic fish; (ii) phenotypic characteristics of transgenic fish that will influence their interaction with ecosystem processes and components; (iii) experimental approaches for assessing transgenic fish phenotypes and their ecological impact; and (iv) major sources of scientific uncertainty influencing empirical assessments. This chapter does not address whether potential ecological effects, revealed through scientific risk assessment studies, constitute harms; Chapters 1 and 2 discuss a multi-stakeholder process for determining harms and acceptable levels of risk.

Defining ecosystem and transgenic fish variables

A systematic approach to the process of assessing ecological effects of transgenic fish prior to their entry into nature is critical because the list of possible hazards and variables to measure will be so broad that it will impede conducting an effective and efficient risk assessment. It is thus imperative to narrow the scope of potentially affected ecosystem processes and components (abiotic and biotic), as well as the traits of the organism that may affect these processes and components, so that empirical studies can be designed to target the most relevant interactions.

Figure 6.1 and Box 6.1 present overviews of important steps to take when attempting to assess the ecological effects of transgenic fish before their entry into unconfined environments. Further explanation and elaboration of information required for these steps are provided below.

Single- versus multiple-generation effects of transgenic populations

Potential short-term ecological effects of transgenic fish on aquatic ecosystems will depend on how often, and how many, transgenic fish enter the ecosystem and on the magnitude of the effects induced by these fish (see Chapter 5, this volume). Potential long-term ecological effects can be indirect, through effects on keystone species, or direct and sustained depending on the successful reproduction of transgenic genotypes (see Chapter 5, this volume). Direct effects on the ecosystem from sterile transgenic fish may differ from those of reproducing transgenic strains. If a transgenic fish cannot reproduce, its impact, while still potentially damaging to aquatic ecosystems, will be limited by the lack of persistence of the transgene in nature provided the transgenic fish are not continually escaping into the receiving ecosystem. The consequences of such biologically confined transgenic fish in nature may be limited to direct effects of escaped individuals over the course of their lifetime. However, they could still alter the abiotic or biotic properties of their environment in permanent ways, especially if transgenic fish overexploit key resources. Hence, ecosystem resilience (i.e. the ability to return to a previous state following perturbation) will be an important factor in determining long-term effects of sterile transgenic fish (Table 6.1).

The potential ecosystem effects from the successful reproduction of transgenic fish will vary over time in response to changes in their abundance. Their abundance is partially determined by their survival and reproductive fitness relative to conspecifics and other ecosystem members (see Chapter 5, this volume, Box 5.2), as well as the continued rate of escape or release of transgenic individuals (less fit individuals can be maintained in the environment by continuous escape). Reduced fitness will, in most cases, result in eventual elimination of the

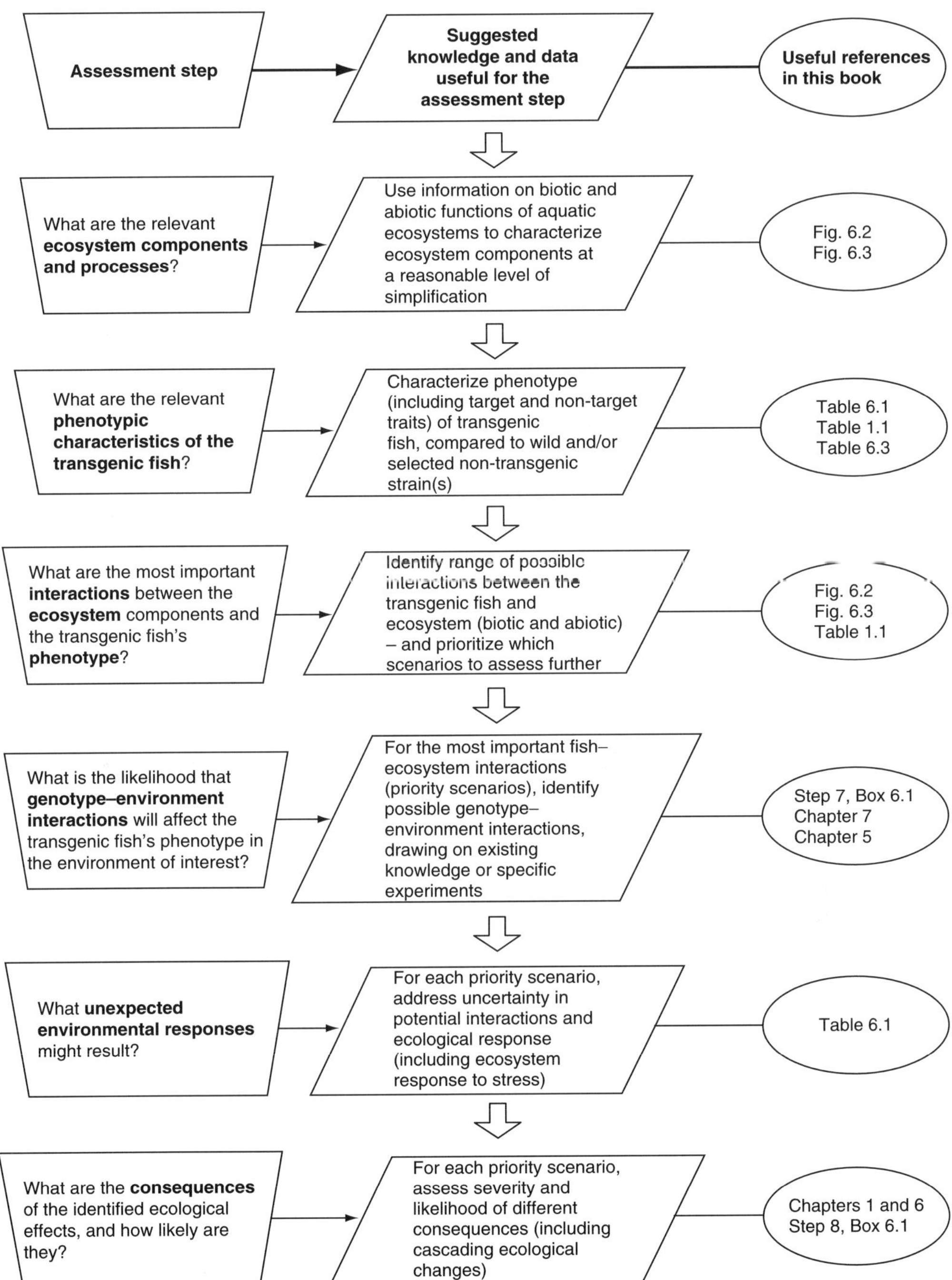

Fig. 6.1. Overview of steps and information required for ecological risk assessment of transgenic fish, with references to chapter and book sections. (Figure developed by K. Paulson.)

Box 6.1. Overview of important steps for assessing the ecological effects of transgenic fish before their entry into nature.

1. *Determine potential exposure of the ecosystem to transgenic fish*
Estimate the likelihood of escape for transgenic fish from culture facilities and aquaculture operations using methodologies from Chapter 5, which also gives guidance for assessing the next step: the transgenic individual's probability of survival and successful reproduction. One should consider whether ecosystem exposure to transgenic fish will vary over time or space. If the transgenic fish cannot reproduce successfully, then only single-generation direct effects of escaped individuals need be considered, unless carry-over effects from major shifts in ecosystem function are anticipated.

2. *Characterize the ecosystem*
Determine critical characteristics of potential receiving environments for transgenic fish. Consider relevant biotic and abiotic ecosystem components, as well as spatial and temporal variation in these parameters. Figs 6.2 and 6.3 show examples of such components.

3. *Determine potential ecosystem resources and services used and contributed by transgenic fish*
Refine potential ecological effects by determining which major biotic and abiotic resources and functions transgenic fish and relevant wild populations use from, and contribute to, the ecosystem (Table 6.1, columns 1–4) (see Birch *et al.*, 2004). This characterization process should be informed by expert knowledge of the species and potential receiving ecosystems, previously known biological actions of transgene products, transgenic fish phenotypes observed during their development, and data from studies on surrogate species with phenotypes similar to that of the transgenic strain. Qualitative analysis of expert and stakeholder conceptual models (see Chapter 1, this volume) will help identify critical interactions and optimum levels of complexity in this process. This step also identifies gaps in information which could lead to uncertainty in the risk assessment process. See below for discussion of major types of uncertainty associated with interpretation of empirical data, and see Chapter 7 for a discussion of how to address different types of uncertainty.

4. *Partition interactions into transgenic phenotypic traits and ecosystem variables*
Identify abiotic and biotic ecosystem components likely to interact with transgenic fish. These may include pathogens, prey, space for spawning, etc. (Table 6.1, column 4). Specific factors (i.e. transgenic fish phenotypes and environmental variables) that influence these interactions also need to be defined (Table 6.1, columns 5 and 6) in as much detail as possible. It is important to note that the same phenotype or environmental variable often may influence more than one fish–ecosystem interaction. Qualitative modelling and sensitivity analyses may help determine which variables are most important to assess empirically to determine potential ecological effects of the transgenic fish. These steps require considerable expert and relevant stakeholder knowledge, and they provide an important foundation for designing and conducting empirical evaluations.

5. *Define and prioritize hazards*
Compile a list of potential impacts (hazards) that may result from each transgenic fish–ecosystem interaction (e.g. Table 6.1, column 7) based on information

Continued

Box 6.1. *Continued*

derived in steps 1–4, using a variety of systematic hazard analysis tools (see Chapter 1, this volume). Hazard assessments should be structured around key developmental and reproductive stages of the organism (e.g. individual, genetic, population, community and ecosystem processes; Parker *et al.*, 1999) for all potential receiving environments. Hazards should be prioritized (ideally through a multi-stakeholder process), as discussed in Chapters 1 and 2. This allows the next steps of the risk assessment to focus on the highest priority hazards for the case in question.

6. *Design experiments to assess phenotypic traits and critical environmental variables*

Empirical assessments of transgenic fish phenotypes and effects of environmental variables should be conducted. Such experiments will help to estimate the likelihood and magnitude of the hazard (i.e. how the ecosystem could change in response to interactions between transgenic traits and environmental variables). Transgenic fish traits (Table 6.1, column 5) associated with prioritized hazards (Table 6.1, column 7) are summarized (Table 6.2, columns 1 and 2 show examples of intended and unintended phenotypes in the transgenic strain). In conjunction with ecosystem variables associated with prioritized hazards (Table 6.1, column 6), experiments are designed to assess the consequences of these traits (Table 6.2, column 3). Experiments should be designed using expert knowledge and considering the range of ecological conditions over which an effect needs to be detected. Table 6.3 provides examples of laboratory methods which have been used to assess phenotypes. Following laboratory assessments, a summary of empirical phenotypic characteristics should be compiled from all experiments examining a specific trait's response to environmental variables. These results should inform determination of whether the measured trait is increased, decreased or unaffected in the transgenic fish relative to wild relatives or other ecosystem members (Table 6.2, column 4). A major objective at this stage is to anticipate the response of the variable being tested, focusing on qualitative assessment relative to non-transgenic relatives or other species in the ecosystem. In many cases, the direction of the effect is all that can be confidently assessed. For example, would ecosystem entry of transgenic fish exhibiting increased foraging activity as well as elevated predation mortality lead to increased, decreased or unaltered predation on a prey species? Such relative and qualitative assessments may be more useful at this point than efforts to obtain specific quantitative data (e.g. number of prey consumed per hour).

7. *Identify factors contributing to uncertainty in empirical studies*

The validity or accuracy of studies can be influenced by a number of factors, including: (i) ability to extrapolate phenotypes to consequences in nature; (ii) pleiotropic effects; (iii) genotype-by-environment interactions (G × E); and (iv) evolutionary change caused by genetic selection. These factors may impact empirical assessments of phenotypic traits and their interaction with environmental variables, and they are discussed in detail in the text and in Chapters 5 and 7.

8. *Predict ecological consequences from empirical studies*

The results of empirical studies should be combined with expert and relevant stakeholder knowledge to assess whether the transgenic fish's phenotypic traits

Continued

Box 6.1. *Continued*

are likely to alter a biotic or abiotic process in the ecosystem (Figs 6.2 and 6.3). If they are, the potential consequences to the ecosystem should be determined (Table 6.2, column 5). The magnitude of the consequence may be further influenced by whether the transgenic trait is expected to affect the organism's fitness over time (Table 6.2, column 6), as discussed in Chapter 5. These final steps are critical because they determine the level of environmental risk posed by the organism. Accurate determinations are contingent upon the breadth of expertise utilized, as well as the availability of baseline information relating specific phenotypes to ecological effects.

transgene, but in the meantime it may still affect demographics of the wild conspecific population. Transgenes that enhance a fish's fitness compared to that of wild relatives will result in increasing abundance of transgenic fish in an ecosystem through time (see Table 6.2, column 6 and Chapter 5, this volume).

Potential ecological effects of transgenic fish also depend on whether they derive from a species that is indigenous or alien to the ecosystem. For example, a transgenic strain with reduced fitness relative to wild conspecifics might be rapidly eliminated from the ecosystem by intraspecific competition; therefore its ecological consequences may be of short duration. In contrast, the same transgenic strain entering an environment without conspecifics may become established if they successfully compete with other species; this situation is analogous to the successful introduction and establishment of non-transgenic invasive species. Thus, a risk assessment for transgenic fish requires evaluation of both absolute and relative consequences arising from its altered traits.

Characterization of Relevant Ecosystems

An important first step in assessing ecological effects of transgenic fish is to characterize the potential recipient ecosystem(s). An ecosystem functions as a biotic community living within its abiotic environment. Organisms interact with ecosystems by using and contributing resources and services within the ecosystem (Andow and Hilbeck, 2004; Birch *et al.*, 2004). Introducing new elements can disrupt these interactions, causing the ecosystem to shift to a new state, and alter species composition (e.g. changes in population sizes, local extirpations, species extinctions and altered ecosystem functions (Parker *et al.*, 1999)). Transgenic fish have the potential to be such a change-inducing element, depending on their altered phenotypes. As such, it is critical to anticipate the ecosystem processes and components which a transgenic fish strain might affect (Table 6.1, columns 1–4).

Characterization of the receiving ecosystem should include identification of the abiotic and biotic components with which transgenic fish are likely to interact (Table 6.1, column 4), using knowledge about the ecosystem's food webs and

Table 6.1. Examples of summary outputs from ecological hazard analysis of transgenic fish before their entry into nature[a].

Column 1	Column 2	Column 3	Column 4	Column 5	Column 6	Column 7
Ecosystem resources and functions	Type	Ecosystem components and attributes	Ecosystem subcomponents that interact with transgenic fish	Critical transgenic fish phenotypes that influence their interaction with the ecosystem	Environmental variables that influence the interaction of transgenic fish with the ecosystem	The hazard: how the ecosystem component could change
Required from the ecosystem by fish	Biotic	Food sources	Prey	Food consumption rate, competitive feeding ability, prey selection	Primary production, food availability and production, prey density and type	Change in biodiversity and population sizes of prey
		Mates	Wild relatives	Mating behaviour, aggression, secondary sex characteristics	Availability of mates, suitable spawning habitat	Changes in the number and behaviour of mates
	Abiotic	Feeding territory	Space	Movement, swimming ability, feeding motivation	Prey densities, pathways to new habitat	Invasion of novel habitats, novel prey species affected
		Chemical and physical properties	Oxygen, chemical compounds, water temperature and flow	Respiration and excretion	Water flow, temperature, food availability	Changes in water flow, excess toxic metabolites
Contributed to the ecosystem by fish	Biotic	Conspecific individuals	Population characteristics (genetics, demographics)	Breeding success, mate selection, foraging ability, disease resistance	Spawning substrate, selective forces (prey availability, etc.)	Changes in genetic diversity, number of conspecifics

Continued

Table 6.1. *Continued*

Column 1	Column 2	Column 3	Column 4	Column 5	Column 6	Column 7
Ecosystem resources and functions	Type	Ecosystem components and attributes	Ecosystem subcomponents that interact with transgenic fish	Critical transgenic fish phenotypes that influence their interaction with the ecosystem	Environmental variables that influence the interaction of transgenic fish with the ecosystem	The hazard: how the ecosystem component could change
		Predators	Transgenic fish predators	Susceptibility to predation, nutritional value	Predator density and type, habitat complexity, swimming ability of predators	Changes in the number of predators and predation rate on other species
		Pathogens	Pathogen abundance and distribution	Immune function, energy partitioning to immune function	Physical factors affecting immune function, distribution and type of pathogens	Change in pathogen numbers and exposure to other species
	Abiotic	Organic outputs	Faecal output, post-mortem decomposition, toxic compounds	Longevity, abundance, migration, feeding behaviour	Temperature, decomposer communities	Changes in adjacent trophic levels, toxic effects on other organisms
		Habitat restructuring	Stream bed structure	Body size, nest building, behaviours, feeding behaviour	Physical properties of habitat (gravel, weed substrates)	Changes in gravel structure, biotic productivity

[a]Steps include: determining ecosystem resources used and contributed by transgenic fish (columns 1–4), partitioning interactions into transgenic fish traits and environmental variables (columns 5, 6), and identifing hazards that could occur from these interactions (column 7). Hazard prioritization (not shown) would then occur via a multi-stakeholder process (see Chapter 1, this volume). Each cell exemplifies information needed for the assessment step. Each assessment should take into account that effects will differ at each developmental stage of the fish and between environments, necessitating case-by-case and iterative assessments.

Table 6.2. Example strategy for designing ecological risk assessment tests of traits identified in Table 6.1, column 5, for a hypothetical example of a growth-enhanced transgenic fish strain.[a]

Column 1	Column 2	Column 3	Column 4	Column 5	Column 6
Intended altered traits in transgenic strain	Potential unintended altered traits in transgenic strain	Design and execute experimental studies	Phenotypic effects in the transgenic fish	Predicted consequence in ecosystem	Magnitude of change when combined with fitness effects
Growth rate and feed conversion efficiency		Determine food consumption rate and efficiency, metabolic efficiency or prey selection	Increased appetite, food conversion	Increased consumption of resources, increased population size	Reduced survival from enhanced predation mortality
	Disease resistance	Examine pathogen challenge, transgenic fish pathogen carrier status	Decreased disease resistance, carries higher pathogen loads	Carrier transgenic individuals spread pathogen to susceptible species, reducing their population size	Decreased spread of pathogen due to increased predation mortality of carrier transgenic fish
	Temperature tolerance	Test the transgenic fish's thermal tolerance	Unaffected	Unaffected	No change anticipated
	Spawning ability	Examine spawning competition	Decreased	Decreased reproductive output	Reduced or increased depending on relative number of transgenic and wild fish in breeding populations

[a]Traits in columns 1 and 2 of this table require analysis to assess their influence on environmental variables (Table 6.1, column 6). All traits and variables identified need to be examined during a range of developmental stages, in multiple environments and in combinations of these (Parker *et al.*, 1999). Risk assessment tests need to be performed on the individual, genetic, population, community and ecosystem levels. Each cell in this table will in most cases require a separate comprehensive assessment and written evaluation. The information in columns 5 and 6 of this table is derived by hazard analysis (Chapter 1, this volume, and Table 6.1) and fitness assessments (Chapter 5, this volume).

trophic levels. Once these components are identified, one can assess how potential abiotic and biotic alterations caused by transgenic fish may influence other ecosystem components and processes. Ideally, this assessment should be conducted using systematic hazard analysis tools and expert and relevant stakeholder knowledge, making sure to consider available and relevant baseline data. Qualitative analysis of experts and relevant stakeholders' conceptual models (see Chapter 1, this volume) will help identify important interactions.

Abiotic components and processes: physical parameters

Ecosystem characteristics may be divided conceptually into abiotic and biotic components and processes, recognizing that significant interaction exists between these categories. It is important to identify potentially affected abiotic components and processes during the hazard analysis stage of the risk assessment. These components include physical characteristics such as water depth, water flow, substrate and temperature, as well as chemical characteristics like dissolved oxygen, nitrate content, pH and salinity (Fig. 6.2). Such physical properties of water are key determinants of an aquatic ecosystem's biotic structure.

Changes in physical properties can result from alteration of the ecosystem's trophic structure (i.e. energy flow and food web relationships), and these changes can have cascading effects. For instance, while the addition or removal of fish in freshwater systems typically has the greatest impact on plankton communities, there are often cascading effects on nutrient availability and water chemistry (Hansson *et al.*, 1998). Such effects have been inferred from marine systems where reductions in large predators appear to have cascading effects on phytoplankton production and nitrate availability (Frank *et al.*, 2005). A transgenic fish's impact on physical properties may have similar cascading effects. Furthermore, transgenesis may alter the natural state of some strains' phenotypes so dramatically (e.g. mud loach or rainbow trout gigantism (Devlin *et al.*, 2001; Nam *et al.*, 2001)) that the transgenic fish might begin to use alternate trophic levels.

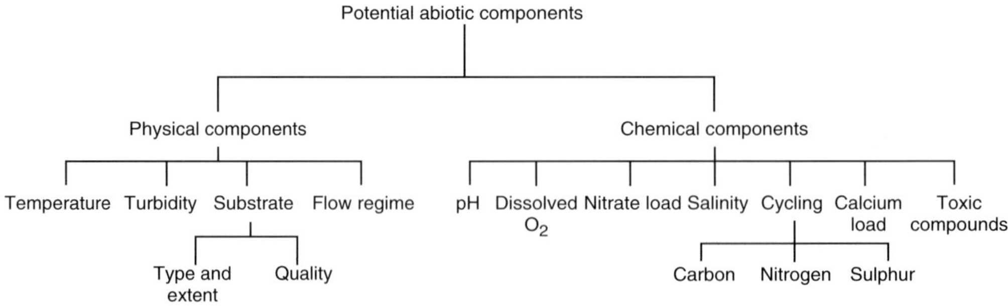

Fig. 6.2. Possible abiotic aquatic ecosystem components affecting or affected by transgenic organisms.

Transgenic fish which function as 'ecosystem engineers' (or displace ecosystem engineers in an aquatic habitat) could have important long-term ramifications. Some fishes, such as tilefish, grouper and salmon, are considered ecosystem engineers because they create, modify and maintain habitats (Coleman and Williams, 2002). Ecosystem engineering can affect the distribution and abundance of large numbers of organisms and significantly impact biodiversity. The potential for such ecosystem effects can be experimentally determined if specific behaviours or chemical outputs associated with engineering activities are altered in the transgenic fish strain (e.g. a fish's ability to displace substrate when creating nests or refuges).

Biotic components and processes: species and their interactions among trophic levels

Evaluating potential environmental effects of transgenic fish requires identification of key functional groups within the ecosystems, namely primary producers, consumers (herbivores, carnivores, omnivores) and decomposers. These functional groups interact through food webs involving multiple trophic levels. Some keystone species may have greater influence on trophic interactions than others, and their loss can transform ecological processes (e.g. Paine, 1969). Some fish species, in particular, exert 'top-down' regulation of aquatic ecosystem structure (e.g. on plankton or benthic communities) and ecosystem function (e.g. on primary and secondary production) (Leveque, 1997). Transgenic fish might affect biotic components through predation, competition, habitat alteration, inter- and intraspecific hybridization, as well as introduction of new parasites and diseases. Escape or intentional introduction of certain transgenic fish may therefore create trophic cascades (Leveque, 1997). A striking example of trophic cascades is in Lake Victoria where the depletion of haplochromines by Nile perch shows how a predator can affect other trophic levels (Goldschmidt *et al.*, 1993).

For example, one way that predatory effects of introduced transgenic fish may impact trophic structures is through intraguild predation, where they might compete with other species for prey at one stage of their ontogeny. Intraguild predation is known to affect trophic cascades, where an increase in the number of top predators could result in decreasing numbers of intermediate level predators, which then allows herbivores to increase, ultimately increasing plant grazing (Polis and Holt, 1992). In this regard, cannibalism has been shown to play a role in causing population level effects in laboratory experiments with growth hormone (GH) transgenic salmon (Devlin *et al.*, 2004b).

Characterization of Transgenic Fish Phenotypes Influencing Ecological Interactions

After characterizing the abiotic and biotic ecosystem components and processes of the ecosystem with which transgenic fish might interact, the next risk assessment step is to determine which phenotypic traits of the transgenic fish will influence these interactions (Table 6.1, column 5). Phenotype differences between the transgenic strain and the founder strain will be based primarily on the function of the inserted transgene, although other genetic and environmental factors will likely shape phenotype as well (see below and Chapter 5, this volume). Several kinds of phenotypic alterations can be predicted based on the intended function of the introduced transgene. The types of gene constructs and target traits anticipated for use in future generations of transgenic fish, crustaceans and molluscs are very broad (see Tables 1.1 in Chapter 1 and 3.1 and 3.2 in Chapter 3, this volume). Altered traits in a transgenic fish strain may be even more numerous due to pleiotropy, whereby expression of the transgene affects additional traits of the organism. Therefore, a transgenic fish line may have a number of modified traits influencing its ecological interactions. Rarely, some ecological effects may arise from indirect effects of bioactive compounds (e.g. gene products or consequent metabolites) produced within the transgenic organism (e.g. Dunham *et al.*, 2002; Sarmasik *et al.*, 2002), which may be actively secreted or passively released following its death.

Many example traits discussed in the next section involve growth-enhanced transgenic fish because they are the subject of the relevant published studies. Comparable studies on other kinds of transgenic fish have yet to be performed.

Competition and Predation

Transgenic fish which escape or are released into the wild are likely to compete with wild relatives and other species for ecosystem resources. Such competition can have a wide range of effects on survival and reproduction of wild populations, which in turn could affect the ecosystem's productivity and biodiversity (Persson, 2002; Polunin and Pinnegar, 2002).

All organisms survive through acquisition of resources from their ecosystem. It is difficult for two species (or genotypes) to use the same resources, if limiting, without affecting the population characteristics of the other. Major ecosystem resources for fish include biotic factors such as prey species and mates, and abiotic factors such as microenvironments with optimum temperatures and dissolved oxygen levels (see Figs 6.2 and 6.3).

Competition varies depending on species, developmental stage and environmental conditions. In exploitative competition, all individuals have access to the resource, and they all try to exploit it, but do not interact aggressively; interference competition occurs through the aggressive exclusion of competitors (Milinski and Parker, 1991). For example, salmon and tilapia often compete aggressively through interference (Einum and Fleming, 1997; Neat and Mayer, 1999), whereas most carp species exhibit exploitative competition (Weir and Grant, 2004). Altered behavioural traits in transgenic fish may affect competition directly (e.g. territoriality) or indirectly (e.g. feeding behaviour). Such changes should be considered when assessing phenotypic traits for ecological impacts and how their effects may vary under different resource levels and habitat complexities.

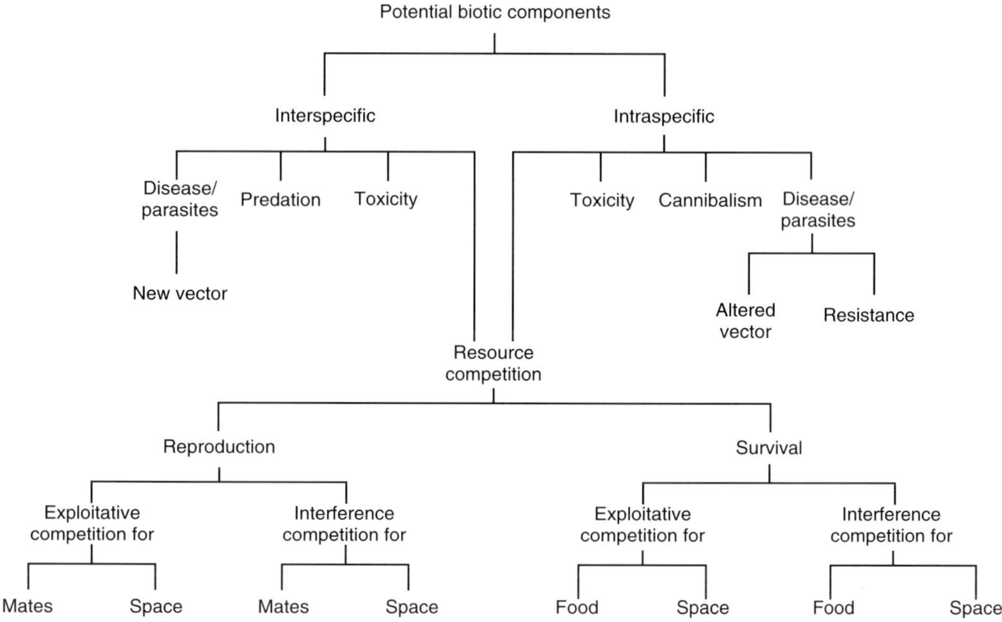

Fig. 6.3. Possible biotic aquatic ecosystem components affecting or affected by transgenic organisms.

Identifying competition and its consequences

Identifying whether competition is occurring in a natural system is often a difficult task. For example, two species with identical diets will compete only if the shared diet is limited in supply. Moreover, even if the abundance of a native species decreases after introduction of transgenic fish, this is not an unequivocal evidence of competition. The decline may be due to environmental factors unrelated to the presence of transgenic fish. This difficulty illustrates the importance of carefully designing controls in experimental studies. For example, if several similar streams are invaded by transgenic tilapia, changes in the abundance of native fish could be compared between affected and unaffected aquatic habitats, in the same area. If some parameter (e.g. abundance or growth rate of the native species) decreases in the streams with transgenic tilapia relative to the unaffected (i.e. control) ones, it is likely that the transgenic introduction had a negative effect on native species through competition (see Chapter 9, this volume, for further details on how to monitor changes in abundance). A good way of testing for this possibility under confined conditions is through experiments in mesocosms. Only by comparing aquatic systems possessing or lacking transgenics can one credibly infer that the presence of transgenics drove changes in key ecological indicators.

Choosing the appropriate method to study potential effects of competition depends on the specific traits of the transgenic species being assessed and the limiting factors in the affected ecosystem. These factors vary, but they include: food supply, nursery grounds and spawning habitat. In juvenile salmonids, for example, there is strong competition for food and shelter during the first weeks after emergence from the gravel bed, and only 1–5% of the fry survive their first summer (Elliott, 1994). The timing of emergence may, therefore, be important for predicting survival rates; indeed studies show that GH-transgenic coho salmon hatch and emerge from the gravel sooner than conspecifics (Devlin *et al.*, 2004a; Sundström *et al.*, 2005). Earlier emergence may confer a prior residence advantage to transgenic offspring emerging in the wild since territory owners generally win conflicts against intruders (Johnsson *et al.*, 1999). However, predation experiments suggest that predation is higher on early emergers, and reduced survival of transgenic offspring has been observed relative to the wild type when they emerge into naturalized habitats with predators (Sundström *et al.*, 2005). When assessing competition in confined studies using semi-natural habitats, the relative survival rates of transgenic and non-transgenic fish (and their offspring) can be monitored by repeated sampling of nursery areas; molecular genetic markers can be used to distinguish transgenic from non-transgenic offspring (see Chapter 9, this volume). Alternatively, groups of 100% hemizygous transgenic offspring can be generated for experimentation by crossing homozygous transgenic fish with non-transgenic fish. Similar approaches may be expanded to multi-species comparisons.

One way to obtain a preliminary indication of whether competitive interactions might occur is to assess similarity in resource use (Hurlbert, 1978) between invasive transgenic and wild specimens in mesocosms and other laboratory apparatus. The objective is to determine the degree to which the transgenic fish strain and wild specimens utilize the same range of resources (e.g. temperature, food particle size, spawning area). However, quantitative measures of resource use do not provide specific information about the mechanisms or effects of competitive interactions. Therefore, when possibilities to study the target ecosystem *in situ* are limited, competition experiments in the laboratory under semi-natural conditions, or in the wild using surrogate models (see below), may provide useful information (Johnsson and Björnsson, 2001; Sundström *et al.*, 2004c, 2005).

Expression of transgenes that affect growth and feeding behaviour is anticipated to influence competition for resources. GH-transgenic salmon, for example, are more active when feeding in order to realize their higher growth potential, making it important to determine whether such fish consume more food than non-transgenic genotypes under different conditions (Abrahams and Sutterlin, 1999; Devlin *et al.*, 1999; Sundström *et al.*, 2004b). Laboratory-based population dynamics studies suggest that the effect of transgenic fish on non-transgenic cohorts is highly dependent on the level of food availability (Devlin *et al.*, 2004b). High food supply levels did not affect growth and survival of non-transgenic salmon in the presence of transgenic salmon, whereas at normal levels of food availability (i.e. just enough to fully satiate non-transgenic populations), growth of GH-transgenic salmon declined to near non-transgenic levels. Additionally, their presence in populations suppressed the growth of non-transgenic fish, although the survival of the non-transgenic fish was unaffected. At very low food supply levels (i.e. extreme competition), transgenic salmon also depressed the growth of non-transgenic salmon. More importantly, all populations containing transgenic salmon crashed to extinction (or to a single individual) due to aggression and cannibalism from dominant transgenic salmon; pure populations of non-transgenic fish had high survival rates and continued to gain biomass.

Laboratory-based experiments cannot completely mimic the factors determining competition (Devlin *et al.*, 2006). For example, it is important to consider predation risk under heightened food competition because it is an important factor in determining feeding decisions (Lima and Dill, 1990). GH-transgenic coho salmon hide less in rocky substrates compared to normal fry when predators are present in semi-natural environments. Consequently, transgenic fry may suffer greater predation mortality than wild salmon. This effect is further exacerbated when food supplies are low, which presumably requires transgenic salmon to increase their foraging activity, thereby increasing their exposure to predation (Sundström *et al.*, 2004c). Incorporating density-dependent effects of competition into laboratory studies may also be important. For example, adding wild fish to a habitat will increase the density of fish, which might increase space competition; this can have effects on the non-transgenic fish's growth that need to be distinguished from effects caused by altered traits of the transgenic fish (Bohlin *et al.*, 2002; Sundström *et al.*, 2004a).

Transgenes affecting growth and feeding behaviour might also affect choice of prey species. In one study, Sundström *et al.* (2004b) varied the type of prey offered to GH-transgenic coho salmon and found that transgenic fish were more willing than the wild type to attack different prey types, including inedible artificial prey items. If transgenic fish consume a greater variety of prey types, and if their relative consumption of various food groups differs from that of wild genotypes, effects could cascade down the trophic chain, reorganizing the biotic community in ways which have not previously been observed with non-transgenic fish.

Competition effects on metabolism

A key factor influencing the ecological effects of an organism on an ecosystem is its ability to capture and efficiently utilize prey. Thus, it is critical to understand the metabolic efficiency with which transgenic fish utilize food resources. Growth-enhanced strains of transgenic fish may exhibit significant morphological and physiological changes affecting metabolism such as enzymes, gut structure, respiratory ability and enhanced feed conversion efficiency (Fu *et al.*, 1998; Krasnov *et al.*, 1999; Cook *et al.*, 2000a; Hill *et al.*, 2000; Martinez *et al.*, 2000; Leggatt *et al.*, 2003; Devlin *et al.*, 2004a; Stevens and Devlin, 2005; Raven *et al.*, 2006). If such changes alter metabolic efficiency, they may influence how well transgenic fish survive in nature, depending on food availability. Measurements of metabolic efficiency can be made by: (i) examining *in vitro* levels of cellular and plasma components (metabolites, enzymes, mRNAs, etc.) (Krasnov *et al.*, 1999; Hill *et al.*, 2000; Blier *et al.*, 2002; Rise *et al.*, 2006); (ii) performing detailed nutritional studies examining energy inputs, conversions and outputs (e.g. Fu *et al.*, 1998; Cook *et al.*, 2000a; Martinez *et al.*, 2000; Raven *et al.*, 2006); and (iii) examining respiratory demands with or without food consumption (e.g. Cook *et al.*, 2000b; Leggatt *et al.*, 2003). It is important to note that improved feed conversion efficiency is often closely correlated with enhanced appetite and growth, and may thus trigger other ecological effects.

Reproductive competition

Spawning competition is anticipated to be a crucial component in determining the impact of escaped transgenic fishes on wild relatives and other ecosystem members. Reproductive biology varies considerably among common aquaculture species. For example, Nile tilapia (*Oreochromis niloticus*) spawns in firm sand in water depths between 0.6 and 2 m, and the males aggressively defend territories, which are visited by females that brood the fertilized eggs in their mouth (Trewavas, 1983). In salmon, females compete for suitable spawning areas (well-oxygenated gravel beds) and males compete aggressively for access to females (Fleming, 1996). In contrast, another important aquaculture species, common carp (*Cyprinus carpio*), does not compete for breeding territory. Spawning females, which are usually followed by several males, lay their sticky eggs in shallow vegetation and leave them without paternal care (Balon, 1990). Thus, the extent of any reproductive competition will depend on the existence of limiting resources, such as spawning habitats and mates, which are sought jointly by transgenic and non-transgenic fish.

Transgenic fish introduced into the wild will need to effectively compete for mates and breeding resources (e.g. breeding sites, nest building materials) in order to establish themselves in the ecosystem. Such competition will not only potentially affect the breeding success of wild fish whose breeding needs overlap with those of transgenics, but also the likelihood of introgression of transgenes into wild populations (see Chapter 5, this volume). Because competitive ability in many fish mating systems is positively associated with body size (Forsgren *et al.*, 2002), GH-transgenic fish could be of particular concern. Studies of GH-transgenic fish suggest that they can mature at an earlier age, and, depending on the species, may mature at either the same body size as wild relatives (e.g. coho salmon (Devlin *et al.*, 1995a; Devlin *et al.*, 2004a)) or at a considerably larger size (e.g. trout and mud loach (Devlin *et al.*, 2001; Nam *et al.*, 2001)). Maturation at a younger age, for example, has the potential to accelerate introgression of transgenes (Muir and Howard, 1999; Devlin *et al.*, 2004a,b; see Chapter 5, this volume, for detailed discussion), thus increasing the proportion of transgenic fish in populations. In addition, shifts in maturation age might alter the reproductive characteristics of the transgenic fish relative to the wild type. Evaluations of reproductive success of transgenic fish in artificial streams and ponds (Dunham *et al.*,1992; Bessey *et al.*, 2004) examined spawning behaviour, mating success and subsequent offspring survival. Studies of gamete quantity and quality in transgenics (i.e. sperm competition) can be also undertaken, using either natural or artificial spawning methods (Bessey *et al.*, 2004).

Growth-enhanced strains of transgenic fish have shown different reproductive abilities relative to unmodified conspecifics. Assessing reproductive success of transgenic fish versus wild fish is feasible under laboratory conditions (i.e. for confirmation of breeding potential), but care must be taken to ensure that effects observed for respective strains accurately mimic those anticipated under natural rearing conditions. For GH-transgenic coho salmon, reproductive success of transgenic individuals was significantly reduced compared to wild non-transgenic fish. However, non-transgenic fish raised under the same culture conditions as the transgenic strain also showed reproductive impairment relative to wild-reared fish. This indicates that the rearing environment can influence the reproductive abilities of both genotypes, making predictions about the transgene's effect on reproductive behaviour difficult, due to the potential genotype–by–environment interaction (see Bessey *et al.*, 2004; and below). Experiments with laboratory-reared GH-transgenic medaka showed that they had a significant size advantage at maturity over non-transgenics, allowing them to physically dominate non-transgenic male fish and giving them an approximate fourfold mating advantage (Howard *et al.*, 2004). Transgenic males of GH-transgenic tilapia also appear to have a breeding advantage over non-transgenic counterparts (de la Fuente *et al.*, 1999), whereas transgenic channel catfish appear to have normal spawning success (Dunham *et al.*, 1992). Whether such effects can be extrapolated to fish reared in the wild remains unknown.

Domestication during the propagation of cultured transgenic lines could contribute to impaired competitive ability and reproductive performance in the wild, irrespective of any effects of the transgene (Fleming *et al.*, 2000). Little is known about the potential for transgenic fishes to disrupt wild breeding populations through competition for mates and breeding resources, destruction of nests or disruption of mating and parental care behaviours.

Distribution and migratory behaviour

Many species of salmonids, eels and other species undertake large migrations to utilize habitats that afford better survival rates and access to resources. In such cases, ecological effects of transgenic fish may occur over wide geographic and ecological spaces, complicating ecological risk assessment. Another potential means of dispersal is through involuntary transportation by birds or animals, capture fishing or other human activities that facilitate dispersal (e.g. through water ballast, by purposeful introduction of pet species or as escaped food species) (Rixon *et al.*, 2005; Leung *et al.*, 2006). In some cases, fish themselves can move across land (e.g. *Clarias* spp.). GH-transgenic salmon can express traits that make them more vulnerable to bird or mammal predation, such as remaining closer to the surface (Sundström *et al.*, 2003), and display a greater tendency to disperse (Sundström *et al.*, 2007a).

Unintended phenotypic effects on migration

Changes in growth patterns in GH-transgenic fish may alter migration behaviour. For example, development of the ability to regulate ions to allow survival in sea water (smoltification) is attained prematurely and at an inappropriate season in GH-transgenic salmon (Devlin *et al.*, 1995a, 2000; Saunders *et al.*, 1998). Times at which critical developmental stages are reached may influence a transgenic fish's decision to migrate from nursery to adult feeding grounds.

If premature migration occurs, fish may enter feeding grounds at an inappropriate time of year when conditions are suboptimal for survival. Alternatively, failure to migrate may cause transgenic fish to remain in nursery environments longer, placing a greater burden on food resources in these habitats. For growth-accelerated strains, there may not be enough suitable prey at certain times of the year (e.g. salmonids in streams in winter) to support their enhanced growth and avoid inducing semi-starvation physiological conditions.

Growth enhancement in transgenic fish appears to increase foraging behaviour (Abrahams and Sutterlin, 1999), which is expected to enhance the dispersal behaviour of fish under some circumstances (e.g. low food availability). In stream mesocosms, transgenic GH-coho salmon have shown different dispersal behaviours than control salmon, and these effects were influenced by food supply (Sundström *et al.*, 2005).

GH-transgenesis may also influence transgenic fish mobility because of morphological changes, such as cranial deformities (Devlin *et al.*, 1995b; Ostenfeld *et al.*, 1998), muscle fibre structure (Hill *et al.*, 2000; Pitkänen *et al.*, 2000) or gill surface area (Stevens and Sutterlin, 1999). Such alterations to muscles and organs in transgenic fish may affect swimming performance (Farrell *et al.*, 1997) and indirectly alter their ability to migrate or capture evasive prey.

Measuring movement behaviour

Species' dispersal can influence their rate of spread and ability to locate suitable microhabitats. These factors can play an important role in determining the invasiveness of species, particularly in fragmented, non-homogeneous habitats (With, 2004). Movement behaviour of transgenic fish can be assessed on a small scale using simple laboratory and semi-natural environments. Although it is relatively feasible to test transgenic fish movement within simulated small habitats (e.g. a pond or stream section), such experiments can only approximate the movement of a species that migrates over long distances and from one habitat to another (e.g. from a stream to the ocean). Tracking transgenic fish during confined tests of migratory behaviour can be done with electronic tag systems, or in mesocosms (confined tanks, ponds or artificial streams) designed to identify immigration to or emigration from a test area (as has been employed for GH-transgenic salmon in Tymchuk *et al.*, 2005). Ideally, such mesocosm studies would combine migratory potential and prey choice to test the ability of transgenic individuals to exploit alternate habitats with less nutritious foods, which may be capable of supporting transgenic but not non-transgenic fish (Raven *et al.*, 2006). For large-scale movements, radio telemetry (i.e. tags which emit acoustic or radio signals that can be detected in water) can be used to track transgenic fish or their surrogate models (see below).

Tolerance of abiotic factors

The extent to which transgenic fish affect the environment may also depend on how they tolerate or react to varying abiotic conditions. Major physical factors to investigate include water flow rates, temperature (preferred temperatures, and upper and lower temperature limits required to support viability and reproduction), dissolved gases (e.g. oxygen), hardness and osmolality, salinity and light (intensity, wavelength and diurnal and seasonal variation). More recently, organic and inorganic pollutants have also been found to influence fish phenotypes (Sumpter and Jobling, 1995). Simple physiological apparatuses (e.g. incubation chambers) can be used to measure tolerance of transgenic fish to these variables. For example, a strain of GH-transgenic Atlantic salmon required elevated oxygen relative to wild type (Cook *et al.*, 2000b), whereas a strain of GH-coho salmon required elevated oxygen only in underfed conditions (Leggatt *et al.*, 2003). Low levels of dissolved oxygen recently have been shown to differentially affect the survival of wild-type and GH-transgenic coho salmon (Sundt-Hansen *et al.*, 2007), which could have important fitness effects in the wild.

It is important to know what factors limit the distribution of the wild species in nature when designing empirical studies. For example, if temperature or pH limits a fish's spread to certain areas, it is important that transgenic individuals be studied beyond those restricting conditions (e.g. at a lower temperature than that tolerated by non-transgenic conspecifics). Intolerance to low water temperatures may prevent many tropical species from thriving (or even surviving long) in temperate climates, and vice versa. Transgenes conferring tolerance to cold water conditions (e.g. antifreeze protein genes (Shears *et al.*, 1991)) may enhance the fitness of transgenic fish in cold water climates, possibly extending their range into ecosystems where they would not normally survive.

Tolerance of biotic factors

Fish encounter many biotic factors in their environments (Fig. 6.3) in addition to competitors and predators as discussed above. By-products of metabolism (e.g. carbon dioxide, ammonia) and faecal waste must be tolerated, as well as products from decaying plants and animals. In some cases, organisms (primarily algae) can influence fish survival by secreting bioactive toxins into surrounding water or by causing physical irritation to gill membranes. Thus, transgenesis affecting gill structure (Stevens and Sutterlin, 1999) may influence susceptibility to such factors.

Although it is possible to develop transgenic fish specifically to resist certain biotic factors, it is also the case that pleiotropic actions of a transgene designed to affect another trait may alter their tolerance. In either case, subjecting transgenic and wild fish to the suspected factors should reveal whether their tolerance levels have been altered.

Additionally, transgenesis may affect the concentration of waste compounds in ecosystems. Transgenesis for enhanced growth, for example, has been shown to influence ammonia production (Krasnov *et al.*, 1999), which may affect the transgenic fish and other members of an ecosystem, particularly in smaller water bodies.

Disease resistance and pathogen carrier status

Pathogens can play an indirect but important role in resource acquisition through disease outbreaks, which alter numbers and types of species present. Transgenic fish strains with resistance to bacterial pathogens exist (Dunham *et al.*, 2002; Sarmasik *et al.*, 2002; Zhong *et al.*, 2002). A transgenic strain possessing enhanced disease resistance may increase the population size of the transformed species by reducing its natural mortality, thus increasing its resource acquisition at the expense of wild relatives and species with similar resource needs.

Disease-resistant transgenic fish may act as pathogen carriers, altering the potential for disease transmission to conspecifics and other species in the ecosystem. The magnitude of such ecological effects will likely depend on the specificity and intensity of the disease resistance, the epidemiology of pathogen exposure and the relative importance of other population-limiting factors (e.g. density-dependent effects on pathogen transfer or resource acquisition). It is also important to consider whether increased disease resistance confers costs to other fitness-related traits. Gene constructs which reduce disease resistance may result in enhanced pathogen carrier frequency in populations, or, depending on the severity of the impairment, may actually reduce pathogen loads in populations by eliminating infected individuals rapidly. Transgenic fish who are disease vectors may suffer from negative effects in many ways (e.g. swimming ability, food intake rate). Thus, ecological risk assessment of disease-resistant transgenic fish strains requires measuring both their mortality rates and their status as a carrier for various pathogens under different environmental conditions.

The carrier status of transgenic fish can be assessed by pathological assessments, molecular detection of pathogens and pathogen challenge trials with native fish. Disease resistance in transgenic fish can be assessed in quarantine facilities through disease challenges, where the fish are exposed to specific pathogens, either by direct administration or by passive horizontal transmission between carriers. Determining which pathogens to test should be based on known disease conditions for wild fish, as well as other pathogens known to exist in the ecosystem. It is well known that environmental variables influence immune physiology; hence, conducting trials under a wide range of environmental conditions is ideal. Such conditions should be selected so as to vary the physiology of the transgenic fish, as well as the survival and pathogenicity of the disease organism. The ability of transgenic fish to pass the disease to the offspring may also be altered by pathogen spread through vertical transmission. This can be assessed via quarantined tests using maternal broodstock known to transmit the pathogen, which could include a transgenic line whose disease resistance is compromised as a consequence of transgene expression.

Disease resistance may be altered in transgenic strains containing transgenes modifying non-disease-related traits. There is some evidence that rapid growth occurs at the cost of maintaining a healthy immune system (Arendt, 1997); indeed a growth-enhanced transgenic salmon strain has shown impaired resistance to bacterial pathogens (Jhingan *et al.*, 2003). Other kinds of genetic alterations, such as triploidy, may also influence disease resistance (Benfey, 1999; Jhingan *et al.*, 2003). It is therefore important to evaluate all transgenic strains for unknown or unanticipated pleiotropic actions that might alter their disease resistance.

Experimental Approaches for Assessing Transgenic Fish Phenotypes

At this time, the only way to study phenotypic effects of transgenes is within confined laboratories, using apparatuses to examine specific traits and involving ecologically simple habitats (see Table 6.3 and Devlin *et al.*, 2006). Molecular and cellular biology and biochemistry studies can also provide information on patterns of gene expression and transgene effects on cellular and organismal processes (e.g. microarray analysis of GH-transgenic coho salmon; Rise *et al.*, 2006). Laboratory tests comparing transgenic and non-transgenic fish under controlled environmental conditions can reveal differences in a variety of phenotypic traits (e.g. growth, disease resistance, swimming ability, predator avoidance, feeding motivation). Environmental conditions can be manipulated to systematically assess the importance of specific variables on traits (e.g. examining effects of different oxygen levels on growth and survival), providing important information on phenotypic plasticity and genotype-by-environment interactions (see below). Although confined laboratory experiments cannot completely mimic actual environmental conditions, they are critical for identifying phenotypic differences between transgenic and wild-type strains; they also provide a foundation for designing more complex experiments (Devlin *et al.*, 2006).

Application of semi-natural conditions in confined mesocosms and microcosms

Tests conducted in confined mesocosms or microcosms provide a better understanding of the phenotypic effects of a transgene under more complex environmental conditions. For example, it is possible to include key features of a natural ecosystem in artificial streams and ponds or tanks to simulate portions of rivers, lakes or marine habitats. Such facilities should mimic natural conditions as closely as possible and include three-dimensional habitats, other ecosystem species, live natural prey items, natural predators and opportunities for pathogen effects. Such conditions allow multiple factors to operate simultaneously in a spatial context more representative of nature, minimizing phenotypic effects resulting from artificial culture conditions (Sundström *et al.*, 2004c, 2005, 2007b). Observations are also more likely to reflect the range of stochastic outcomes from the myriad of species–ecosystem interactions which drive ecosystem dynamics on both small and large scales.

Added complexity can reduce experimental control, yielding stochasticity and observations not previously hypothesized (in some cases making it more difficult to interpret results). Thus, mesocosm experiments ideally should be

Table 6.3. Examples of laboratory methods examining physiological and behavioural phenotypes of transgenic fish.

Type of study	Example phenotypes	Example methodology	Example references
Molecular/cellular/ morphological studies	Hormone levels	RIA, ELISA	Devlin *et al.*, 1994, 2000
	Altered metabolic pathways	Enzymology, microarray analysis	Krasnov *et al.*, 1999; Blier *et al.*, 2002; Rise *et al.*, 2006
	Muscle and other organ or tissue structure	Histopathology, immunocytochemistry	Hill *et al.*, 2000; Pitkänen *et al.*, 2001
	Transgene structure	Southern blot, cloning, sequencing	Uh *et al.*, 2006; Yaskowiak *et al.*, 2006;
	Transgene expression	Quantitative PCR, Northern blot and microarray analysis	Mori *et al.*, 1999; Rise *et al.*, 2006
	Stress tolerance	Cortisol and hsp measurements	Jhingan *et al.*, 2003
Tank/laboratory studies	Growth and nutrition	Weight and length gain studies, condition factor	Most studies
	Feeding motivation	Competitive feeding ability and growth, assessment of prey utilization	Abrahams and Sutterlin, 1999; Devlin *et al.*, 1999, 2004b; Sundström *et al.*, 2003, 2004a
	Metabolism	Feed efficiency trials, assessment of feed ingredient requirements (and efficiencies and tolerances)	Cook *et al.*, 2000a; Martinez *et al.*, 2000; Raven *et al.*, 2006
	Toxicity	Examine the nutritive value of transgenic species as prey items; feeding trials with aquatic species followed by pathological assessments	Guillén *et al.*, 1999
	Disease resistance	Challenges with specific pathogens, haematology, cellular and humeral responses	Dunham *et al.*, 2002; Sarmasik *et al.*, 2002; Zhong *et al.*, 2002; Jhingan *et al.*, 2003

	Reproduction	Gamete quality and quantity: *in vitro* fertilization, gamete number and gonad development, maturation timing	Muir and Howard, 2001, 2002; Bessey *et al.*, 2004; Howard *et al.*, 2004
	Osmoregulation	Seawater challenges, plasma ion levels, mortality	Saunders *et al.*, 1998; Devlin *et al.*, 2000
	Dispersal and migration	Tracking movement over time	Sundström *et al.*, 2007a
	Social behaviour	Schooling/shoaling behaviour, aggressive interactions	Sundström *et al.*, 2007a
Performance testing	Swimming ability	Swim tunnel studies	Farrell *et al.*, 1997; Stevens *et al.*, 1998; McKenzie *et al.*, 2001; Lee *et al.*, 2003; Leggatt *et al.*, 2003
	Intraspecific interactions	Dominance hierarchy evaluations, habitat selection	
	Respiration	Closed respirometers, starvation trials	Stevens *et al.*, 1998; Cook *et al.*, 2000a; McKenzie *et al.*, 2001; Leggatt *et al.*, 2003
Semi-natural mesocosms	Interspecific interactions	Effects of food availability, habitat structure, predator presence on predation activity and sensitivity to predation, feeding competition trials	Dunham *et al.*, 1999; Sundström *et al.*, 2004b,c, 2005
	Migration	Immigration/emigration studies, schooling/ shoaling behaviour	Sundström *et al.*, 2005
	Development	Hatch timing and survival effects	Sundström *et al.*, 2005
	Spawning	Spawning arenas for assessment of reproductive behaviours, mate choice (intra and interspecific) and spawning success, nest building	Bessey *et al.*, 2004

replicated as many times as needed to achieve the statistical power needed to detect effects, particularly when experimental design involves varying parameters such as predator pressure, food abundance and water conditions (see Chapter 9, this volume, for discussion of statistical power considerations). Mesocosms have the greatest utility for providing the data necessary for assessing the ecological risks of transgenic species in relatively small and simple environments. For species with greater ranges over their full life cycle, mesocosms can provide useful insights about certain life-history stages (e.g. very early larval life of salmon). For example, transgenic coho salmon reared in mesocosms with naturalized stream conditions (versus standard tank culture conditions) profoundly affected their phenotypic differences compared to wild-type individuals (Sundström *et al.*, 2007b). Mesocosms can also be used to rear fish for other experiments, while minimizing culture effects arising from phenotypic plasticity.

Use of surrogate models in nature

In the absence of data from transgenic fish in nature, it may be possible to utilize information derived from surrogate models. Three main types of surrogate fish include: (i) specific selectively bred or domesticated strains that possess traits similar to those of the transgenic strain; (ii) sterile transgenic fish; and (iii) wild-type fish which have undergone non-genetic manipulations to induce a similar phenotypic change.

Domesticated and selectively bred strains

Fish developed through conventional breeding or domestication may be useful for simulating the effects of transgenic fish on environmental variables. Some transgenic fish currently under development have selectively bred conspecifics (e.g. genetically improved farmed tilapia (GIFT) – see Chapter 1, this volume); selectively bred salmonid strains currently used in aquaculture can show strong phenotypic changes in growth rates (Hershberger *et al.*, 1990; Gjedrem, 1998; Tymchuk and Devlin, 2005; Tymchuk *et al.*, 2006a) and phenotypes similar to those of transgenic strains (Devlin *et al.*, 2001). Domesticated strains may also show similar behavioural changes (Johnsson *et al.*, 1996, 2001; Einum and Fleming, 1997; Tymchuk *et al.*, 2006a) and endocrinological characteristics (Fleming *et al.*, 2002) as those of transgenic fish (Abrahams and Sutterlin, 1999; Devlin *et al.*, 1999, 2000; Sundström *et al.*, 2003, 2004b,c). Risk assessors can gain insights from studies on the effects of domesticated salmon entering natural ecosystems containing conspecifics (for reviews, see Weber and Fausch, 2003; Weir and Grant, 2005; Tymchuk *et al.*, 2006b). Some studies show impaired overall fitness of domesticated salmonids relative to wild salmon (Fleming *et al.*, 2000; McGinnity *et al.*, 2003; Biro *et al.*, 2004) and suggest that interbreeding between frequently escaping domesticated and wild individuals could depress survival and abundance of wild populations. Thus, frequent escapes of transgenic fish derived from highly domesticated strains could pose the same hazard. (See Chapter 5, this volume, for additional points to consider when assessing gene flow from transgenic fish derived from highly selected fish lines.)

A key difference between domesticated or selectively bred fish and transgenic fish with similar phenotypes is the genetic basis for their phenotypes. Directed selection and domestication involves the collection of many polygenic alleles from natural genomes (McClelland *et al.*, 2005; Tymchuk *et al.*, 2006b), from which an integrated phenotype is chosen. Many traits are simultaneously selected, such as growth rate and survival, thus maintaining significant physiological balance for the overall performance of the whole organism. In contrast, a novel transgenic construct may be less adapted to the entire genome of the transgenic fish, and may induce more diverse pleiotropic interactions and variable phenotypes than those generated by selective breeding. Hence, the different methods employed to produce selectively bred and transgenic fish may complicate the use of selectively bred individuals as surrogates for transgenics when examined over multiple generations (see also Chapter 5, this volume).

Sterile transgenic organisms

Introducing sterile transgenic strains of fish into nature could be a useful way to examine their single-generation ecological effects without risk of long-term consequences arising from their reproduction in nature. However, it is recognized that long-term effects may occur if sterile fish directly affect keystone species. This approach must be considered with extreme caution and preferably implemented only under conditions with minimal risk for emigration. Triploidy is a commonly employed technique for sterilizing fish, but in most cases it is not completely effective (Devlin and Donaldson, 1992; Benfey, 1999; NRC, 2004; Chapter 8, this volume). Thus, before placing any triploid transgenic fish into a natural ecosystem, it is absolutely essential to confirm the sterility of every individual (including identification of germline diploid and triploid mosaics). A further consideration is the degree to which the sterilization approach alters other traits, and hence ecological effects, of the transgenic strain. For example, triploidy has been observed to reduce growth rate and survival of GH-transgenic tilapia and salmon (Razak *et al.*, 1999; Jhingan *et al.*, 2003; Devlin *et al.*, 2004a). Nevertheless, the use of individually verified sterilized transgenic fish could provide useful risk assessment data in well-selected contexts.

Induced phenotype in wild-type fish

Another approach for gathering data about transgenic fish in nature is to use non-transgenic fish treated to have a similar phenotype. This approach would allow single-generation effects of phenotypically distinct fish to be studied in nature, providing data free from laboratory influences other than those of the treatment itself. For example, slow-release formulations of GH (Garber *et al.*, 1995; McLean *et al.*, 1997) can mimic the effects of growth-enhancing transgenes to some degree. Similarly, use of small pumps to deliver antibiotic peptides (Jia *et al.*, 2000) can mimic the effects of transgenes designed to confer disease resistance (Dunham *et al.*, 2002; Sarmasik *et al.*, 2002). Fish implanted with slow-release formulations of bovine GH (e.g. Posilac, Monsanto Corporation) have been used to assess the potential effects of GH-transgenesis on growth and survival of individuals in the wild (Johnsson *et al.*, 1999, 2000; Johnsson and Björnsson, 2001). Together, these studies suggest that growth-enhanced salmonid parr may compete successfully with wild conspecifics. A similar approach could assess other ecologically important changes in transgenic fish, such as altered movement or migration behaviours.

When using treated fish to gather data for risk assessments, one should ensure that the phenotype of the treated fish approximates that of the transgenic strain being evaluated. It is essential to account for differences between duration of the treatment effect and duration of transgene expression. In many cases, the surrogate treatment will be active for a shorter part of the life cycle than will the expression of the transgene. Moreover, effects of the treatment arising early in development may affect traits expressed later in life, even after the treatment is no longer active. Thus, ideally, a surrogate treatment should act throughout development (to most closely simulate the expression of a constitutive transgene) and should induce the same magnitudes of phenotypic changes as the transgene. Although fully mimicking a transgenic phenotype may be difficult to achieve by using a specific treatment, valuable information can nevertheless be gained from the use of surrogate models.

Introductions of non-indigenous species

It may be possible to compare traits of a transgenic fish line to an invasive fish species to help identify which traits influence successful or unsuccessful invasions of a particular aquatic ecosystem. This analysis could draw on approaches used previously to identify invasiveness traits in non-indigenous species (Lodge, 1993; Williamson, 1999; Kolar and Lodge, 2002; Zanden *et al.*, 2004). For instance, experiments comparing four closely related mosquitofish species (*Gambusia* spp.) suggest that species invasiveness is linked to dispersal behaviour and boldness (Rehage and Sih, 2004). Similarly, successful invasion by introduced tilapiines (particularly *O. niloticus*) in freshwater bodies can be attributed to rapid growth rates, mouth brooding behaviour, wide physiological tolerance, wide habitat preference and ability to feed on different components of the aquatic food chain (McKaye, 1984). Also, it may be possible to assess which ecosystem variables make a system more vulnerable to invasion and subsequent adverse ecological effects (e.g. Bohn and Amundsen, 1998; Marchetti, 1999; Drake, 2005).

There are important limitations to drawing on experience with aquatic invasive species to predict the ecological impacts of transgenic fish (Devlin *et al.*, 2006). It is essential to have data on establishment and effects of invasive species in the ecosystem under consideration; relying solely on invasive species data from very different ecosystems greatly increases uncertainty about the relevance of the data to the transgenic fish case under consideration (Hayes and Barry, in press). Using expert opinion to determine the degree to which a species may be a nuisance can also help identify the traits that are most likely associated with ecological effects; this approach proved useful in a retrospective assessment that reviewed many cases of fish invasions in North America's Great Lakes (Kolar and Lodge, 2002). This study also illuminates the limits to using invasive species data to inform ecological risk assessment of transgenic fish: a single trait (faster growth) was positively associated with establishment, negatively associated with spread, and not associated with nuisance status across different cases. Due to the vast complexity and variation involved in species–ecosystem interactions, it is unlikely that characteristics of invasive species and strains will provide definitive information regarding consequences of a transgenic fish in nature. However, because no data exist on the success and consequences of transgenic strains in the wild, information from invasive introductions may be one useful source of information to help prioritize the most relevant traits of the transgenic fish and characteristics of the receiving ecosystem to examine through appropriate studies.

Factors Contributing to Uncertainty in Phenotypic and Ecological Assessments

Several of the major sources of uncertainty frequently encountered in empirical assessments of transgenic fish phenotypes, and their impact on ecological risk assessment, are discussed below and by Devlin *et al*. (2006). Methods for dealing with uncertainty in risk assessments are discussed in Chapter 7.

Extrapolating results from confined tests to natural ecosystems

It is difficult to extrapolate laboratory observations of individual traits to quantitative consequences in nature. For example, a transgenic fish's disease resistance might be enhanced in a laboratory test (e.g. Jhingan *et al*., 2003), but it will be difficult to predict ecological effects from these data alone because pathogen loads and types vary over time and space in nature. Similarly, nutrition trials assessing feed conversion efficiency with formulated diets may allow identification of changes in a transgenic fish's metabolic efficiency (e.g. Raven *et al*., 2006), but unless the trials are conducted with natural water temperatures and prey species and availabilities, it will be difficult to determine the quantitative consequences of such metabolic effects in nature. Despite such limitations, laboratory data are critical for identifying potential effects of a transgene on specific traits, and they can be used to predict the direction of an effect (e.g. increased, decreased or unaffected) relative to non-transgenic fish. Such data are also important for designing more detailed experiments in complex semi-natural environments.

Pleiotropic effects and phenotypic trade-offs between traits

Ecological risk assessments of transgenic fish should include assessment of potential effects from pleiotropic actions of the gene product. The action of a specific gene product rarely, if ever, influences only a single phenotypic trait because of complex interactions between molecular and physiological pathways in organisms. Hence, phenotypic effects induced by the transgene may be correlated with other traits in different situations (Sih *et al*., 2004). However, it can be difficult to predict the effects of transgenes even when the physiological action of the transgene is known. For example, growth-enhancement is known to influence many processes in fish (Björnsson, 1997); indeed, GH-transgenes have shown a wide range of additional effects beyond growth. For example, some GH-transgenic salmonid strains show that over-stimulation of cartilage deposition can result in cranial deformities (Devlin *et al.*, 1995b; Ostenfeld *et al.*, 1998). Pleiotropic effects can also occur as a secondary consequence of altered phenotypic states. For example, appetite stimulation in GH-transgenic salmon can affect intestinal surface area independent of the action of a GH-transgene (Stevens and Devlin, 2005), arising in part as an indirect effect of enhanced food intake.

Trade-offs between transgene-influenced pleiotropic traits (e.g. foraging behaviour and predation risk; Lima and Dill, 1990) may limit growth rates. In GH-transgenic salmon, this trade-off has made transgenic fish more willing to incur predation risk because of enhanced appetite and reduced hiding behaviour (Abrahams and Sutterlin, 1999; Devlin *et al.*, 1999; Sundström *et al.*, 2003, 2004c). Such shifts in microdistribution may alter the kinds of prey utilized by transgenic fish, as well as make them more susceptible to predation. Predation risk may be further enhanced by altered colouration (as observed in emerging GH-transgenic coho salmon fry (Devlin *et al.*, 1995b)). Such effects on feeding and risk-taking behaviour need to be considered carefully as they may significantly alter the invasive potential and ecological risk of transgenic individuals.

Pleiotropic effects and trade-offs between different traits have direct relevance when designing and analysing laboratory assessments. Performing exploratory investigations of phenotypic changes at various levels (e.g. morphological, behavioural, physiological and molecular) is important for revealing potential pleiotropic effects from the transgene. Careful comparison of possible interactions among multiple traits is required, and in some cases the effects of changes in a single trait may be serious enough to offset the effect of pleiotropic changes in other traits.

Genotype-by-environment interactions (G × E effects)

Some phenotypic traits change in response to environmental variables, and are thus phenotypically plastic. When two genotypes differ in their phenotypic responses to multiple environmental conditions, these phenotypes are influenced by genotype-by-environment (G × E) interactions. Strong G × E effects have been noted for GH-transgenic coho salmon strains, particularly on phenotypes affecting growth, reproduction and survival (Bessey *et al.*, 2004; Devlin *et al.*, 2004b; Sundström *et al.*, 2004c, 2005, 2007b; Tymchuk *et al.*, 2005). Hence, ecological risk assessment-related experiments should be conducted in multiple environments under varied ecological parameters (e.g. as identified in Box 6.1, steps 3–6). Tests should also be conducted over multiple generations to identify maternal and paternal effects, which can influence phenotype primarily during early life-history stages. Because such complex studies are difficult using only confined experimental units, G × E effects will be an important source of incertitude in ecological risk assessment for transgenic fish.

Plastic responses to rearing conditions (e.g. feed type, exercise training, etc.) may also affect the phenotype of fish reared for risk assessment experiments; careful consideration of the rearing history of test fish is warranted. To understand G × E effects better, cultured non-transgenic fish should be compared with conspecific fish obtained from nature. A difference in phenotype between the two (e.g. Bessey *et al.*, 2004; Sundström *et al.*, 2007b) may indicate that laboratory culture conditions are influencing phenotype. However, lack of phenotypic differences between cultured and wild-reared non-transgenic fish does not provide definitive evidence that a transgenic fish reared in the same environment fully represents a hypothetical transgenic fish living its entire life in the wild.

Genetic stability: transgene structure and background genetic effects

Assessments of phenotype in transgenic fish are performed using defined strains with a particular transgene, and as such, factors which may alter the structure and hence expression characteristics of the transgene may change phenotypes from that determined under an original assessment (Devlin and Donaldson, 1992). Determining the structure and stability of an integrated transgene is important when determining whether genetic changes may occur that result in phenotypic changes within a transgenic strain (Devlin *et al.*, 2004a,b; Uh *et al.*, 2006; Yaskowiak *et al.*, 2006). In strains with stable transgenic constructs and expression, phenotypic assessments between transgenic and non-transgenic fish are usually performed in a limited diversity of genetic backgrounds because the fish are most often derived from a single laboratory or wild population. However, transgenic fish entering nature may introgress their transgenes into a greater diversity of genetic backgrounds, resulting from interbreeding with conspecifics of different strains and even other reproductively compatible species. Such changes in the transgene's genetic background would be expected to influence the expression of the transgene in some cases (Devlin and Donaldson, 1992), with potential effects on phenotype and ecological consequences. Indeed, a GH-transgene has been observed to have strong growth-stimulating effects in a wild (slow-growing) strain and reduced growth effects in a domesticated (fast-growing) strain (Devlin *et al.*, 2001). (See Chapter 5 for further discussion of this issue and suggestions for designing experiments to test for such effects of genetic background on transgenic fish.)

Chapter Summary

Testing for the potential ecological effects of transgenic fish prior to their entry to natural ecosystems is a major component of assessing their potential environmental risk. This chapter outlines approaches to divide the assessment of ecological effects into four phases that build upon each other: (i) characterize the relevant biotic and abiotic properties of the receiving environment(s); (ii) measure the intended and unintended changes in traits (phenotypic characteristics) of the transgenic fish strain; (iii) determine the interactions anticipated between transgenic fish and the ecosystem (ecosystem resources and services both used and provided by transgenic fish); and (iv) estimate the scale and likelihood of ecological effects resulting from each transgenic fish–ecosystem interaction.

Information for each phase of the ecological assessment is generated through an integration of expert and (where appropriate) stakeholder knowledge, baseline data about potential receiving ecosystems, results of empirical tests of transgenic fish in confined laboratories or mesocosms and relevant field data from surrogate fish exhibiting traits similar to the transgenic fish. Identification of key traits of the host fish species and controlling ecosystem variables forms the basis for proper design and conduct of experiments to assess whether and how these traits have been modified in the transgenic strain. Significant sources of uncertainty when conducting prospective ecological risk assessments of transgenic fish include: (i) limited ability to extrapolate results from confined laboratory and mesocosm tests to nature because of the lack of correlated field and lab data; (ii) the potential for the inserted transgene to alter multiple traits at different life stages (pleiotropic effects), which might be missed if confined tests do not cover the entire life cycle; (iii) genotype-by-environment interactions that alter expression of traits of transgenic fish reared in the laboratory, preventing direct extrapolation of effects seen in a confined testing environment to the natural environment; and (iv) potentially different expression of traits of a transgenic fish strain in nature due to effects of its genetic background (i.e. natural genetic variation) on the trait targeted by the transgene.

Assessing the ecological effects of a specific transgenic strain of fish prior to its entry into nature is a complex and daunting task requiring careful selection of assessment endpoints (see Chapter 1, this volume). Despite the limitations of laboratory-based studies, they do provide valuable information regarding the basic phenotypes of the transgenic strain and possible interactions with specific biotic and abiotic components of an ecosystem. They currently represent the only approach for obtaining empirical data specific to a transgenic fish strain and a certain kind of ecosystem without undertaking field trials in nature. It is critical for such laboratory studies of ecosystem consequences to be undertaken under adequate confinement because the recovery of escaped transgenic fish would be extremely difficult or impossible.

References

ABRAC (Agricultural Biotechnology Research Advisory Committee) (1995) *Performance Standards for Safely Conducting Research with Genetically Modified Fish and Shellfish*, Parts I and II, Document Nos. 95-04 and 95-05. United States Department of Agriculture, Office of Agricultural Biotechnology, Washington, DC. Available at: www.isb.vt.edu/perfstands/psmain.cfm

Abrahams, M.V. and Sutterlin, A. (1999) The foraging and anti-predator behaviour of growth-enhanced transgenic Atlantic salmon. *Animal Behaviour* 58, 933–942.

Andow, D. and Hilbeck, A. (2004) Science-based risk assessment for non-target effects of transgenic crops. *Bioscience* 54, 637–649.

Arendt, J.D. (1997) Adaptive intrinsic growth rates: an integration across taxa. *The Quarterly Review of Biology* 72, 149–177.

Balon, E.K. (1990) Epigenesis of an epigeneticist: the development of some alternative concepts on the early ontogeny and evolution of fishes. *Guelph Ichthyology Reviews* 1, 1–48.

Benfey, T.J. (1999) The physiology and behaviour of triploid fish. *Reviews in Fisheries Science* 7, 39–67.

Bessey, C., Devlin, R.H., Liley, N.R. and Biagi, C.A. (2004) Reproductive performance of growth-enhanced transgenic coho salmon (*Oncorhynchus kisutch*). *Transactions of the American Fisheries Society* 133, 1205–1220.

Birch, A.N.E., Wheatley, R., Anyango, B., Arpaia, S., Capalbo, D., Getu Degaga, E., Fontes, D., Kalama, P., Lelmen, E., Lovei, G., Melo, I., Muyekho, F., Ngi-Song, A., Ochieno, D., Ogwang, J., Pitelli, R., Schuler, T., Sétamou, M., Srinivasan, S., Smith, J., Van Son, N., Songa, J., Sujii, E., Tan, T., Wan, F.-H. and Hilbeck, A. (2004) Biodiversity and non-target impacts: a case study of Bt maize in Kenya. In: Hilbeck, A. and Andow, D.A. (eds) *Environmental Risk Assessment of Genetically Modified Organisms: A Case Study of Bt Maize in Kenya*. CAB International, Wallingford, UK, pp. 117–185.

Biro, P.A., Abrahams, M.V., Post, J.R. and Parkinson, E.A. (2004) Predators select against high growth rates and risk-taking behaviour in domestic trout populations. *Proceedings of the Royal Society of London, Series B* 271, 2233–2237.

Björnsson, B.T. (1997) The biology of salmon growth hormone: from daylight to dominance. *Fish Physiology and Biochemistry* 17, 9–24.

Blier, P., Lemieux, H. and Devlin, R.H. (2002) Is the growth rate of fish set by digestive enzymes or metabolic capacity of the tissues? Insight from transgenic coho salmon. *Aquaculture* 209, 379–384.

Bohlin, T., Sundström, L.F., Johnsson, J.I., Höjesjö, J. and Pettersson, J. (2002) Density-dependent growth in brown trout: effects of introducing wild and hatchery fish. *Journal of Animal Ecology* 71, 683–692.

Bohn, T. and Amundsen, P.A. (1998) Effects of invading vendace (*Coregonus albula* L.) on species composition and body size in two zooplankton communities of the Pasvik River System, northern Norway. *Journal of Plankton Research* 20, 243–256.

Coleman, F.C. and Williams, S.L. (2002) Overexploiting marine ecosystem engineers: potential consequences for biodiversity. *Trends in Ecology and Evolution* 17, 40–43.

Cook, J.T., McNiven, M.A., Richardson, G.F. and Sutterlin, A.M. (2000a) Growth rate, body composition and feed digestibility/conversion of growth-enhanced transgenic Atlantic salmon (*Salmo salar*). *Aquaculture* 188, 15–32.

Cook, J.T., McNiven, M.A. and Sutterlin, A.M. (2000b) Metabolic rate of pre-smolt growth-enhanced transgenic Atlantic salmon (*Salmo salar*). *Aquaculture* 188, 33–45.

de la Fuente, J., Guillen, I., Martinez, R. and Estrada, M.P. (1999) Growth regulation and enhancement in tilapia: basic research findings and their applications. *Genetic Analysis: Biomolecular Engineering* 15, 85–90.

Devlin, R.H. and Donaldson, E.M. (1992) Containment of genetically altered fish with emphasis on salmonids. In: Hew, C.L. and Fletcher, G.L. (eds) *Transgenic Fish*. World Scientific Press, Singapore, pp. 229–265.

Devlin, R.H., Yesaki, T.Y., Biagi, C.A., Donaldson, E.M., Swanson, P. and Chan, W.K. (1994) Extraordinary salmon growth. *Nature* 371, 209–210.

Devlin, R.H., Yesaki, T.Y., Donaldson, E.M., Du, S.J. and Hew, C.L. (1995a) Production of germline transgenic Pacific salmonids with dramatically increased growth performance. *Canadian Journal of Fisheries and Aquatic Sciences* 52, 1376–1384.

Devlin, R.H., Yesaki, T.Y., Donaldson, E.M. and Hew, C.L. (1995b) Transmission and phenotypic effects of an antifreeze/GH gene construct in coho salmon (*Oncorhynchus kisutch*). *Aquaculture* 137, 161–169.

Devlin, R.H., Johnsson, J.I., Smailus, D.E., Biagi, C.A., Jönsson, E. and Björnsson, B.T. (1999) Increased ability to compete for food by growth hormone-transgenic coho salmon *Oncorhynchus kisutch* (Walbaum). *Aquaculture Research* 30, 479–482.

Devlin, R.H., Swanson, P., Clarke, W.C., Plisetskaya, E., Dickhoff, W., Moriyama, S., Yesaki, T.Y. and Hew, C.L. (2000) Seawater adaptability and hormone levels in growth-enhanced transgenic coho salmon, *Oncorhynchus kisutch*. *Aquaculture* 191, 367–385.

Devlin, R.H., Biagi, C.A., Yesaki, T.Y., Smailus, D.E. and Byatt, J.C. (2001) Growth of domesticated transgenic fish. *Nature* 409, 781–782.

Devlin, R.H., Biagi, C.A. and Yesaki, T.Y. (2004a) Growth, viability and genetic characteristics of GH transgenic coho salmon strains. *Aquaculture* 236, 607–632.

Devlin, R.H., D'Andrade, M., Uh, M. and Biagi, C.A. (2004b) Population effects of GH transgenic salmon are dependant upon food availability and genotype by environment interactions. *Proceedings of the National Academy of Sciences USA* 101, 9303–9308.

Devlin, R.H., Sundstrom, L.F. and Muir, W.F. (2006) Interface of biotechnology and ecology for environmental risk assessments of transgenic fish. *Trends in Biotechnology* 24, 89–97.

Drake, J. (2005) Risk analysis for species introductions: forecasting population growth of Eurasian ruffe (*Gymnocephalus cernuus*). *Canadian Journal of Fisheries and Aquatic Sciences* 62, 1053–1059.

Dunham, R.D., Ramboux, A.C., Duncan, P.L., Hayat, M., Chen, T.T., Lin, C.M., Kight, K., Gonzalez-Villasenor, I. and Powers, D.A. (1992) Transfer, expression, and inheritance of salmonid growth hormone genes in channel catfish, *Ictalurus punctatus*, and effects on performance traits. *Molecular Marine Biology and Biotechnology* 1, 380–389.

Dunham, R.A., Chitmanat, C., Nichols, A., Argue, B., Powers, D.A. and Chen, T.T. (1999) Predator avoidance of transgenic channel catfish containing salmonid growth harmone genes. *Marine Biotechnology* 1, 545–551.

Dunham, R.A., Warr, G.W., Nichols, A., Duncan, P.L., Argue, B., Middleton, D. and Kucuktas, H. (2002) Enhanced bacterial disease resistance of transgenic channel catfish *Ictalurus punctatus* possessing cecropin genes. *Marine Biotechnology* 4, 338–344.

Einum, S. and Fleming, I.A. (1997) Genetic divergence and interactions in the wild among native, farmed and hybrid Atlantic salmon. *Journal of Fish Biology* 50, 634–651.

Elliott, J.M. (1994) *Quantitative Ecology and the Brown Trout*. Oxford University Press, Oxford, UK.

Farrell, A.P., Bennett, W. and Devlin, R.H. (1997) Growth-enhanced transgenic salmon can be inferior swimmers. *Canadian Journal of Zoology* 75, 335–337.

Fleming, I., Agustesson, T., Finstad, B., Johnsson, J. and Björnsson, B. (2002) Effects of domestication on growth physiology and endocrinology of Atlantic salmon (*Salmo salar*). *Canadian Journal of Fisheries and Aquatic Sciences* 59, 1323–1330.

Fleming, I.A. (1996) Reproductive strategies of Atlantic salmon: ecology and evolution. *Reviews in Fish Biology and Fisheries* 6, 379–416.

Fleming, I.A., Hindar, K., Mjoelneroed, I.B., Jönsson, B., Balstad, T. and Lamberg, A. (2000) Lifetime success and interactions of farm salmon invading a native population. *Proceedings of the Royal Society of London, Series B* 267, 1517–1523.

Forsgren, E., Reynolds, J.D. and Berglund, A. (2002) Behavioural ecology of reproduction in fish. In: Hart, P.J.B. and Reynolds, J.D. (eds) *Handbook of Fish Biology and Fisheries*. Blackwell, Oxford, UK, pp. 225–247.

Frank, K.T., Petrie, B., Choi, J.S. and Leggett, W.C. (2005) Trophic cascades in a formerly cod-dominated ecosystem. *Science* 308, 1621–1623.

Fu, C., Cui, Y., Hung, S.S.O. and Zhu, Z. (1998) Growth and feed utilization by F_4 human growth hormone transgenic carp fed diets with different protein levels. *Journal of Fish Biology* 53, 115–129.

Garber, M.J., Deyonge, K.G., Byatt, J.C., Lellis, W.A., Honeyfield, D.C., Bull, R.C., Schelling, G.T. and Roeder, R.A. (1995) Dose-response effects of recombinant bovine somatotropin

(Posilac) on growth performance and body composition of two-year-old rainbow trout (*Oncorhynchus mykiss*). *Journal of Animal Science* 73, 3216–3222.
Gjedrem, T. (1998) Selective breeding in aquaculture. *InfoFish International* 3, 44–48.
Goldschmidt, T., Witte, F. and Wanink, J. (1993) Cascading effects of the introduced Nile perch on the detrivorous/phytoplantivorous species in the sublittoral areas of Lake Victoria. *Conservation Biology* 7, 686–700.
Guillén, I., Berlanga, J., Valenzuela, C.M., Morales, A., Toledo, J., Estrada, M.P., Puentes, P., Hayes, O. and de la Fuente, J. (1999) Safety evaluation of transgenic tilapia with accelerated growth. *Marine Biotechnology* 1, 2–14.
Hansson, L.A., Annadotter, H., Bergman, E., Hamrin, S.F., Jeppesen, E., Kairesalo, T., Luokkanen, E., Nilsson, P.A., Sondergaard, M. and Strand, J. (1998) Biomanipulation as an application of food-chain theory: constraints, synthesis, and recommendations for temperate lakes. *Ecosystems* 1, 558–574.
Hayes, K.R. and Barry, S.C. (in press) Are there any consistent predictors of invasion success? *Biological Invasions.*
Hershberger, W.K., Myers, J.M., Iwamoto, R.N., Mcauley, W.C. and Saxton, A.M. (1990) Genetic changes in the growth of coho salmon (*Oncorhynchus kisutch*) in marine net-pens produced by ten years of selection. *Aquaculture* 85, 187–197.
Hill, J.A., Kiessling, A. and Devlin, R.H. (2000) Coho salmon (*Oncorhynchus kisutch*) transgenic for a growth hormone gene construct exhibit increased rates of muscle hyperplasia and detectable levels of differential gene expression. *Canadian Journal of Fisheries and Aquatic Sciences* 57, 939–950.
Howard, R.D., DeWoody, J.A. and Muir, W.M. (2004) Transgenic male mating advantage provides opportunity for Trojan gene effect in a fish. *Proceedings of the National Academy of Sciences USA* 101, 2934–2938.
Hurlbert, S.H. (1978) The measurement of niche overlap and some relatives. *Ecology* 59, 67–77.
Jhingan, E., Devlin, R.H. and Iwama, G.K. (2003) Disease resistance, stress response and effects of triploidy in growth hormone transgenic coho salmon. *Journal of Fish Biology* 63, 806–823.
Jia, X., Patrzykat, A., Devlin, R.H., Ackerman, P.A., Iwama, G.K. and Hancock, R.E.W. (2000) Antimicrobial peptides protect coho salmon from *Vibrio anguillarum* infections. *Applied and Environmental Microbiology* 66, 1928–1932.
Johnsson, J. and Björnsson, B. (2001) Growth-enhanced fish can be competitive in the wild. *Functional Ecology* 15, 654–659.
Johnsson, J.I., Jönsson, E. and Björnsson, B.T. (1996) Dominance, nutritional state, and growth hormone levels in rainbow trout (*Oncorhynchus mykiss*). *Hormones and Behaviour* 30, 13–21.
Johnsson, J.I., Petersson, E., Jönsson, E., Järvi, T. and Björnsson, B.T. (1999) Growth hormone-induced effects on mortality, energy status and growth: a field study on brown trout (*Salmo trutta*). *Functional Ecology* 13, 514–522.
Johnsson, J.I., Jönsson, E., Petersson, E., Järvi, T. and Björnsson, B.T. (2000) Fitness-related effects of growth investment in brown trout under field and hatchery conditions. *Journal of Fish Biology* 57, 326–336.
Johnsson, J.I., Höjesjö, J. and Fleming, I.A. (2001) Behavioural and heart rate responses to predation risk in wild and domesticated Atlantic salmon. *Canadian Journal of Fisheries and Aquatic Sciences* 58, 788–794.
Kapuscinski, A.R. and Hallerman, E.M. (1990) Transgenic fish and public policy: anticipating environmental impacts of transgenic fish. *Fisheries* 15, 2–11.
Kolar, C. and Lodge, D. (2002) Ecological predictions and risk assessment for alien fishes in North America. *Science* 298, 1233–1236.
Krasnov, A., Aegren, J.J., Pitkänen, T.I. and Moelsae, H. (1999) Transfer of growth hormone (GH) transgenes into Arctic charr (*Salvelinus alpinus* L.) II. Nutrient partitioning in rapidly growing fish. *Genetic Analysis Biomolecular Engineering* 15, 99–105.

Lee, C.G., Devlin, R.H. and Farrell, A.P. (2003) Swimming performance, oxygen consumption and excess post-exercise oxygen consumption in adult transgenic and ocean-ranched coho salmon. *Journal of Fish Biology* 62, 753–766.

Leggatt, R.A., Devlin, R.H., Farrell, A.P. and Randall, D.J. (2003) Oxygen uptake of growth hormone transgenic coho salmon during starvation and feeding. *Journal of Fish Biology* 62, 1053–1066.

Leung, B., Bossenbroek, J. and Lodge, D. (2006) Boats, pathways, and aquatic biological invasions: Estimating dispersal potential with gravity models. *Biological Invasions* 8, 241–254.

Leveque, C. (1997) *Biodiversity Dynamics and Conservation, the Freshwater Fish of Tropical Africa*. Cambridge University Press, Cambridge.

Lima, S.L. and Dill, L.M. (1990) Behavioral decisions made under the risk of predation: a review and prospectus. *Canadian Journal of Zoology* 68, 619–640.

Lodge, D.M. (1993) Biological invasions: lessons for ecology. *Trends in Ecology and Evolution* 8, 133–137.

Marchetti, M. (1999) An experimental study of competition between the native Sacramento perch (*Archoplites interruptus*) and introduced bluegill (*Lepomis macrochirus*). *Biological Invasions* 1, 55–65.

Martinez, R., Juncal, J., Zaldivar, C., Arenal, A., Guillen, I., Morera, V., Carrillo, O., Estrada, M., Morales, A. and Estrada, M.P. (2000) Growth efficiency in transgenic tilapia (*Oreochromis* sp.) carrying a single copy of an homologous cDNA growth hormone. *Biochemical and Biophysical Research Communications* 267, 466–472.

McClelland, E.K., Myers, J.M., Hard, J.J., Park, L.K. and Naish, K.A. (2005) Two generations of outbreeding in coho salmon (*Oncorhynchus kisutch*): effects on size and growth. *Canadian Journal of Fisheries and Aquatic Science* 62, 2538–2547.

McGinnity, P., Prodöhl, P., Ferguson, A., Hynes, R., Ó Maoiléidigh, N., Baker, N., Cotter, D., O'Hea, B., Cooke, D., Rogan, G., Taggart, J. and Cross, T. (2003) Fitness reduction and potential extinction of wild populations of Atlantic salmon, *Salmo salar*, as a result of interactions with escaped farm salmon. *Proceedings of the Royal Society of London, Series B* 270, 2443–2450.

McKaye, K.R. (1984) Behavioral aspect of cichlids reproductive strategies. Patterns of territoriality and brood defence in Central American substratum-spawners and African mouth brooders. In: Wootton, R.J. and Potts, G.W. (eds) *Fish Reproduction: Strategies and Tactics*. Academic Press, London, pp. 245–273.

McKenzie, D.J., Martinez, R., Morales, A., Acosta, J., Taylor, E.W., Steffensen, J.F. and Estrada, M.P. (2001) Growth hormone transgenic tilapia (*Oreochromis* sp.) compensate for increased metabolic rate to preserve exercise performance and hypoxia tolerance. *Journal of Fish Physiology* 531P, 217P.

McLean, E., Devlin, R.H., Byatt, J.C., Clarke, W.C. and Donaldson, E.M. (1997) Impact of a controlled release formulation of recombinant bovine growth hormone upon growth and seawater adaptation in coho (*Oncorhynchus kisutch*) and chinook (*Oncorhynchus tshawytscha*) salmon. *Aquaculture* 156, 113–128.

Milinski, M. and Parker, G.A. (1991) Competition for resources. In: Krebs, J.R. and Davies, N. B. (eds) *Behavioural Ecology: An Evolutionary Approach*. Blackwell Scientific, Oxford, UK, pp. 137–168.

Mori, T. and Devlin, R.H. (1999) Transgene and host growth hormone gene expression in pituitary and nonpituitary tissues of normal and growth hormone transgenic salmon. *Molecular and Cellular Endocrinology* 149, 129–139.

Muir, W.M. and Howard, R.D. (1999) Possible ecological risks of transgenic organism release when transgenes affect mating success: Sexual selection and the trojan gene hypothesis. *Proceeding of the National Academy of Sciences USA* 96, 13853–13856.

Muir, W.M. and Howard, R.D. (2001) Fitness components and ecological risk of transgenic release: a model using Japanese medaka (*Oryzias latipes*). *American Naturalist* 158, 1–16.

Muir, W.M. and Howard, R.D. (2002) Assessment of possible ecological risks and hazards of transgenic fish with implications for other sexually reproducing organisms. *Transgenic Research* 11, 101–114.

Nam, Y.K., Noh, J.K., Cho, Y.S., Cho, H.J., Cho, K.N., Kim, C.G. and Kim, D.S. (2001) Dramatically accelerated growth and extraordinary gigantism of transgenic mud loach *Misgurnus mizolepis*. *Transgenic Research* 10, 353–362.

Neat, F.C. and Mayer, I. (1999) Plasma concentrations of sex steroids and fighting in male *Tilapia zillii*. *Journal of Fish Biology* 54, 695–697.

NRC (2004) *Biological Confinement of Genetically Engineered Organisms*. National Academy Press, Washington, DC.

Ostenfeld, T.H., McLean, E. and Devlin, R.H. (1998) Transgenesis changes body and head shape in Pacific salmon. *Journal of Fish Biology* 52, 850–854.

Parker, I.M., Simberloff, D., Lonsdale, W.M., Goodell, K., Wonham, M., Williamson, M.H., von Holle, B., Moyle, P.B., Byers, J.E. and Goldwasser, L. (1999) Impact: toward a framework for understanding the ecological effects of invaders. *Biological Invasions* 1, 3–19.

Persson, L. (2002) Community ecology of freshwater fishes. In: Hart, P.J.B. and Reynolds, J.D. (eds) *Handbook of Fish Biology and Fisheries*. Blackwell, Oxford, UK, pp. 97–122.

Pitkänen, T.I., Krasnov, A., Teerijoki, H., Xie, S.Q., Stickland, N.C. and Mölsä, H. (2000) Growth, metabolism and tissue cellularity in fast-growing, genetically modified salmonid, Arctic charr. *Comparative Biochemistry and Physiology – Part A: Molecular and Integrative Physiology* 126, S120.

Polis, A.G. and Holt, R.D. (1992) Intraguild predation: the dynamics of complex trophic interactions. *Trends in Ecology and Evolution* 7, 151–154.

Polunin, N.V.C. and Pinnegar, J.K. (2002) Trophic ecology and the structure of marine food webs. In: Hart, P.J.B. and Reynolds, J.D. (eds) *Handbook of Fish Biology and Fisheries* Blackwell, Oxford, UK, pp. 299–320.

Paine, R.T. (1969) A note on trophic complexity and community stability. *American Naturalist* 100, 65–75.

Pitkänen, T.I., Xie, S.Q., Krasnor, A., Mason, P.S., Mölsä, H. and Strickland, N.C. (2001) Changes in tissue cellularity are associated with growth enhancement in genetically modified Arctic charr (salvelinus alpinus L.) carrying recombinant growth harmone gene. *Marine Biotechnology* 3, 188–197.

Raven, P.A., Devlin, R.H. and Higgs, D.A. (2006) Influence of dietary digestible energy content on growth, protein and energy utilization and body composition of growth hormone transgenic and non-transgenic coho salmon (*Oncorhynchus kisutch*). *Aquaculture* 254, 730–747.

Razak, S.A., Hwang, G.L., Rahman, M.A. and Maclean, N. (1999) Growth performance and gonadal development of growth enhanced transgenic tilapia *Oreochromis niloticus* (L.) following heat-shock-induced triploidy. *Marine Biotechnology* 1, 533–544.

Rehage, J.S. and Sih, A. (2004) Dispersal behavior, boldness, and the link to invasiveness: a comparison of four *Gambusia* species. *Biological Invasions* 6, 379–391.

Rise, M., Douglas, S., Sakhrani, D., Williams, J., Ewart, K.V., Rise, M., Davidson, W., Koop, B. and Devlin, R.H. (2006) Multiple microarray platforms utilized for hepatic gene expression profiling of GH transgenic coho salmon with and without ration restriction. *Journal of Molecular Endocrinology* 37, 259–282.

Rixon, C.A.M., Duggan, I.C., Bergeron, N.M.N., Ricciardi, A. and Macisaac, H.J. (2005) Invasion risks posed by the aquarium trade and livefish markets on the Laurentian Great lakes. *Biodiversity and Conservation* 14, 1365–1381.

Sarmasik, A., Warr, G. and Chen, T.T. (2002) Production of transgenic medaka with increased resistance to bacterial pathogens. *Marine Biotechnology* 4, 310–322.

Saunders, R.L., Fletcher, G.L. and Hew, C.L. (1998) Smolt development in growth hormone transgenic Atlantic salmon. *Aquaculture* 168, 177–193.

Scientists' Working Group on Biosafety. (1998) *Manual for Assessing Ecological and Human Health Effects of Genetically Engineered Organisms. Part One: Introductory Text and Supporting Text for Flowcharts*. Part Two: *Flowcharts and Worksheets*. The Edmonds Institute. Available at: www.edmonds-institute.org.manual.html

Shears, M.A., Fletcher, G.L., Hew, C.L., Gauthier, S. and Davies, P.L. (1991) Transfer, expression, and stable inheritance of antifreeze protein genes in Atlantic salmon (*Salmo salar*). *Molecular Marine Biology and Biotechnology* 1, 58–63.

Sih, A., Bell, A. and Johnson, J.C. (2004) Behavioral syndromes: an ecological and evolutionary overview. *Trends in Ecology and Evolution* 19, 372–378.

Stevens, E.D. and Sutterlin, A. (1999) Gill morphometry in growth hormone transgenic Atlantic salmon. *Environmental Biology of Fishes* 54, 405–411.

Stevens, E.D., Sutterlin, A. and Cook, T. (1998) Respiratory metabolism and swimming performance in growth hormone transgenic Atlantic salmon. *Canadian Journal of Fisheries and Aquatic Sciences* 55, 2028–2035.

Stevens, E.D. and Devlin, R.H. (2005) Is enhancement of digestive capacity a direct effect of GH transgenesis or an indirect effect of enhanced appetite? *Journal of Fish Biology* 66, 1–16.

Sumpter, J.P. and Jobling, S. (1995) Vitellogenesis as a biomarker for estrogenic contamination of the aquatic environment. *Environmental Health Perspectives* 103, 173–178.

Sundström, L.F., Devlin, R.H., Johnsson, J.I. and Biagi, C.A. (2003) Vertical position reflects increased feeding motivation in growth hormone transgenic coho salmon (*Oncorhynchus kisutch*). *Ethology* 109, 701–712.

Sundström, L.F., Bohlin, T. and Johnsson, J.I. (2004a) Density-dependent growth in hatchery-reared brown trout released into a natural stream. *Journal of Fish Biology* 65, 1385–1391.

Sundström, L.F., Lõhmus, M., Devlin, R.H., Johnsson, J.I., Biagi, C.A. and Bohlin, T. (2004b) Feeding on profitable and unprofitable prey: comparing behaviour of growth-enhanced transgenic and normal coho salmon (*Oncorhynchus kisutch*). *Ethology* 110, 381–396.

Sundström, L.F., Lõhmus, M., Johnsson, J.I. and Devlin, R.H. (2004c) Growth hormone transgenic salmon pay for growth potential with increased predation mortality. *Proceedings of the Royal Society of London, Series B* 271, S350–S352.

Sundström, L.F., Lõhmus, M. and Devlin, R.H. (2005) Selection on increased intrinsic growth rates in coho salmon, *Oncorhynchus kisutch. Evolution* 59, 1560–1569.

Sundström, L.F., Lõhmus, M., Johnsson, J.I. and Devlin, R.H. (2007a) Dispersal potential is affected by growth-hormone transgenesis in coho salmon (*Oncorhynchus kisutch*). *Ethology* 113, 403–410.

Sundström, L.F., Lõhmus, M., Tymchuk, W.E. and Devlin, R.H. (2007b) Gene–environment interactions influence ecological consequences of transgenic animals. *Proceedings of the National Academy of Sciences USA* 104, 3889–3894.

Sundt-Hansen, L., Sundström, L.F., Einum, S., Hindar, K., Fleming, I.A., and Devlin, R.H. (2007) Genetically enhanced growth causes increased mortality in hypoxic environments. *Biology Letters* 3, 165–168.

Tiedje, J.M., Colwell, R.K., Grossman, Y.L., Hodson, R.E., Lenski, R.E., Mack, R.N. and Regal, P.J. (1989) The planned introduction of genetically engineered organisms: ecological considerations and recommendations. *Ecology* 70, 298–315.

Trewavas, E. (1983) *Tilapiine Fishes of the Genera* Sarotherodon, Oreochromis, Danakilia. British Museum (Natural History), London.

Tymchuk, W.E. and Devlin, R.H. (2005) Growth differences among first and second generation hybrids of domesticated and wild rainbow trout (*Oncorhynchus mykiss*). *Aquaculture* 245, 295–300.
Tymchuk, W.E., Abrahams, M.V. and Devlin, R.H. (2005) Competitive ability and mortality of growth-enhanced transgenic coho salmon fry and parr when foraging for food. *Transactions of the American Fisheries Society* 134, 381–389.
Tymchuk, W.E., Biagi, C., Withler, R.E. and Devlin, R.H. (2006a) Impact of domestication on growth and behaviour of coho salmon (*Oncorhynchus kisutch*) and rainbow trout (*Oncorhynchus mykiss*). *Transactions of the American Fisheries Society* 135, 442–455.
Tymchuk, W.E., Devlin, R.H. and Withler, R.E. (2006b) The role of genotype and environment in phenotypic differentiation among wild and domesticated salmonids. *Canadian Technical Report of Fisheries and Aquatic Sciences* 2450, 1–63.
Uh, M., Khattra, J. and Devlin, R.H. (2006) Transgene constructs in coho salmon (*Oncorhynchus kisutch*) are repeated in a head-to-tail fashion and can be integrated adjacent to horizontally transmitted parasite DNA. *Transgenic Research* 15, 711–727.
Weber, E.D. and Fausch, K.D. (2003) Interactions between hatchery and wild salmonids in streams: differences in biology and evidence for competition. *Canadian Journal of Fisheries and Aquatic Sciences* 60, 1018–1036.
Weir, L.K. and Grant, J.W.A. (2004) The causes of resource monopolisation: interaction between resource dispersion and mode of competition. *Ethology* 110, 63–74.
Weir, L.K. and Grant, J.W.A. (2005) Effects of aquaculture on wild fish populations: a synthesis of data. *Environmental Reviews* 13, 145–168.
Williamson, M. (1999) *Biological Invasions*. Chapman & Hall, London.
With, K.A. (2004) Assessing the risk of invasive spread in fragmented landscapes. *Risk Analysis* 24, 803–815.
Yaskowiak, E., Shears, M., Agarwal-Mawal, A. and Fletcher, G. (2006) Characterization and multi-generational stability of the growth hormone transgene (*EO-1α*) responsible for enhanced growth rates in Atlantic salmon. *Transgenic Research* 15, 465–480.
Zanden, V.J.M., Olden, J.D., Thorne, J.H. and Mandrak, N. (2004) Predicting occurrences and impacts of smallmouth bass introductions in north temperate lakes. *Ecological Applications* 14, 132–148.
Zhong, J., Wang, Y. and Zhu, Z. (2002) Introduction of the human lactoferrin gene into grass carp (*Ctenopharyngodon idellus*) to increase resistance against GCH virus. *Aquaculture* 214, 93–101.

7 Introduction to the Concepts and Methods of Uncertainty Analysis

K.R. Hayes, H.M. Regan and M.A. Burgman

Introduction

Uncertainty is present in virtually all parts of risk assessment; risk analysts encounter it when identifying potential adverse effects, when estimating parameters that predict the probability of adverse effects and when interpreting terms such as 'adverse effects'. The key to any reliable risk assessment is to recognize and treat (i.e. analyse, eliminate or propagate through the risk assessment) the various sources of uncertainty. Burgman (2005) defines an 'honest' risk assessment as one that: (i) is faithful to assumptions and the kinds of uncertainty embedded in it; (ii) carries these uncertainties through chains of calculations and judgements; and (iii) represents and communicates them in a reliable and transparent manner. Thus, uncertainty analysis is crucial to the scientific credibility, accuracy and 'honesty' of a risk assessment.

This chapter provides a non-technical introduction to uncertainty and uncertainty analysis. It aims to help risk analysts complete an 'honest' risk assessment and policy advisers to correctly interpret the results of a risk assessment. The chapter discusses the three main types of uncertainty commonly encountered in environmental risk assessment, and identifies techniques to address them. It also briefly highlights a number of practical analysis issues. Since the theory and methodology of uncertainty analysis are broad and continue to develop, this chapter highlights only the most relevant methods for risk assessment of transgenic fish, and it directs readers to other literature sources for more detailed information.

Types of Uncertainty

The different kinds of uncertainty in environmental risk assessment, and methods for their treatment, are summarized in a variety of ways (Baybutt, 1989;

Morgan and Henrion, 1990; Haimes, 1998; Cullen and Frey, 1999; Regan *et al.*, 2002a, 2003). This chapter distinguishes between three main types of uncertainty: linguistic uncertainty, variability and incertitude. These distinctions are important because each type arises from very different mechanisms, and risk analysts must use different approaches to represent, propagate and communicate these uncertainties (Ferson, 1996; Ferson and Ginzburg, 1996; Regan *et al.*, 2002a). The most important reason for this distinction, however, is that some sources of uncertainty can be eliminated through appropriate treatment (e.g. some sources of linguistic uncertainty), others can be reduced by further data collection (e.g. incertitude), while still others can neither be reduced nor eliminated and can only be better represented and understood (e.g. variability).

Linguistic uncertainty

Linguistic uncertainty occurs in environmental risk assessment because the language used to describe events and processes is sometimes ambiguous, context-dependent, underspecified and vague. Linguistic uncertainty is present in all types of risk assessment, but it is particularly prevalent in qualitative risk assessment (Chapter 1, this volume). It also occurs in policy advice, management decisions and stakeholder deliberations because interpretations of institutional directives and broadly defined scientific terms can influence decisions about what to assess and how to interpret the results of assessments. It is therefore important to identify the relevant sources of linguistic uncertainty at the outset of any risk assessment, to apply appropriate treatments and to be aware of the possible consequences of linguistic uncertainty when it cannot be eliminated or treated.

Ambiguity

Ambiguity arises when words have more than one meaning, and it is not clear which one is meant. For example, the term 'genetically modified' could encompass organisms modified by traditional breeding methods. However, it is more generally used to refer to organisms modified in ways that do not occur naturally from mating and natural recombination (EU, 2001), and the term reflects important differences associated with modern genetic techniques relevant for a risk assessment (Regal, 1994). This source of linguistic uncertainty is relatively simple to address: clearly define the term to remove its ambiguity. Although simple to remove, ambiguity can often be a resilient source of uncertainty if there is strong disagreement about the working definition of terms.[1]

Context dependence

Context dependence is uncertainty caused by a failure to specify the context in which a term is to be understood. A clear way to deal with context dependence

[1] The PFOA process (Chapter 2, this volume) provides a mechanism to identify and resolve these types of disagreements.

is to ensure that the context is stated explicitly. Issues of context dependence can occur in risk assessments of transgenic fish. For example, 'large-scale' escapes of transgenic fish (Chapter 5, this volume) can mean many different things depending on the size and intensity of the production facility.

Underspecificity

Underspecificity occurs when there is unwanted generality; the statement in question does not provide the degree of specificity required in order to proceed with an assessment or decision. Statements relevant for risk assessment can be underspecified with respect to location, time, species, methods, etc. For example, the following statement is underspecified with respect to species and method: '... in a small percentage (generally <10%) of founder fish, foreign DNA is integrated into the host genome and thus permanently retained in the transgenic fish ...' (Chapter 4, this volume). In reality, the success rate of transgenic methods for integrating foreign (exogenous) DNA into the genome of a fish varies markedly and depends on the method, the fish species and the skill of the technician involved (see Chapter 3, this volume). Underspecificity is minimized by providing all relevant contextual data, thereby ensuring the narrowest possible bounds on the statement in question.

Vagueness

Vagueness arises because terms used in environmental risk assessment, such as categorical descriptions (high, medium and low) of consequence and likelihood, sometimes allow borderline cases. Vagueness is particularly prevalent in qualitative risk assessments. This is because such risk assessments often differentiate impacts spatially and temporally, assigning, for example, the same level of risk to 'local, long-term' and 'widespread, short-term' impacts. Words such as 'local', 'widespread', 'short-term' and 'long-term' are vague because some impacts may be borderline cases, i.e. neither local nor widespread, and neither short- nor long-term. It is important to note that these terms are also context-dependent, but the vagueness still persists after the context is defined. A common treatment for vagueness is to substitute potentially vague terms with precise definitions. For example, the term 'local' might be defined as less than 10 km from an initial release site, 'widespread' defined as 'greater than 10 km', 'short-term' defined as 'less than 5 years' and 'long-term' defined as 'greater than 5 years'.

It is important to realize that precise definitions may not always honour the spirit in which the words are intended to be used. For instance, in the previous example, a distribution of 11 km would be considered widespread, whereas a distribution of 9 km would not. Vague terms operate along a continuum where there is a degree to which a statement is satisfied, and as such they are a source of uncertainty that is not always desirable to eliminate with precise definitions. Perhaps the most widely applied treatment of vagueness in scientific applications is fuzzy sets (e.g. Fig. 7.1) and fuzzy logic (Zadeh, 1978). Other treatments include supervaluations, rough sets, three-valued logic and other logic systems (see Regan *et al.*, 2002a, and references therein).

Conventional statements relevant to the risk assessment of transgenic organisms which are subject to considerable linguistic uncertainty include terms

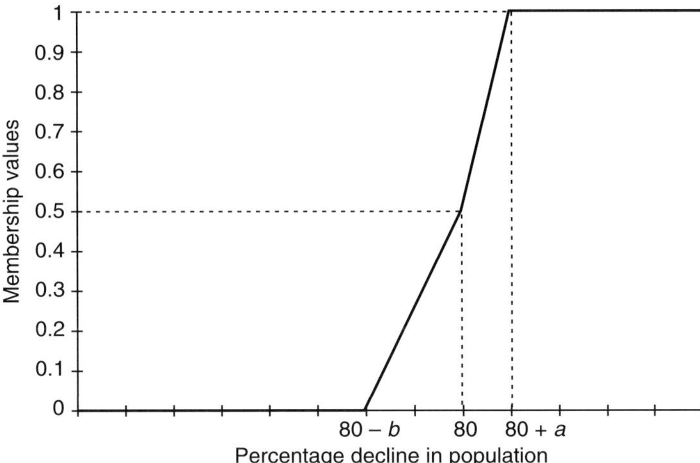

Fig. 7.1. The membership function for a fuzzy set that maps the percentage decline in a species population (*x*-axis) to the probability that it belongs to the set 'critically endangered' (*y*-axis). In this example, *a* and *b* are positive real numbers used to define the upper and lower bounds of the set, i.e. the percentage decline in a species population such that it is definitely endangered and definitely not endangered. Fuzzy sets are commonly used to eliminate vagueness from terms such as 'critically endangered'. (Reprinted from Regan *et al.*, 2000, with permission from Elsevier.)

like 'familiarity' (Van Dommelen, 1998; OECD, 2000). Consider the following statement found in a proposed risk assessment framework for genetically modified organisms:

> When the familiarity standard for a plant or micro-organism has been satisfied such that reasonable assurance exists that the organism and the other conditions of an introduction are essentially similar to known introductions ... the introduction is assumed suitable for field testing.
>
> (National Research Council, 1989)

Terms such as 'familiarity', 'essentially similar' and 'reasonable assurance' are vague ('reasonable assurance' permits borderline cases), context-dependent ('conditions of an introduction' vary dramatically between species, locations and over time) and underspecific ('other conditions' do not specify how many and what type of conditions must be considered). Hence, statements can be subject to multiple sources of linguistic uncertainty. These sources of linguistic uncertainty may be further compounded by variability and incertitude, as discussed below.

Variability

Variability is the uncertainty caused by fluctuations or differences in a quantity or process. Variability can occur because a parameter naturally fluctuates over time (e.g. water temperature in a given location), with location (e.g. average rainfall at different locations in April) or within a group (e.g. survival rates within a metapopulation). Parameters which vary through time and space in this manner are common in environmental risk assessment. Risk assessment parameters may also depend on other variables in ways that are difficult to quantify. Variation is therefore pervasive in all environmental risk assessments, and the term 'risk' is sometimes invoked to reflect this fact (Regan *et al.*, 2003).

Variability cannot be reduced by gathering additional data, but it can often be represented more accurately and communicated better with additional data. For example, structured population dynamics are essential elements of most models that predict fish population changes. Devlin *et al.* (Chapter 6, this volume) identify abundance, age structure and survival as key determinants of population behaviour. However, age- or stage-specific survivorship and fecundity are buffeted by unpredictable changes in environmental conditions, mitigated by temperature, currents, nutrients and trophic chain dynamics, all of which are inherently variable. Fortunately, there are a variety of mathematical methods available to characterize variability and propagate it through models (Box 7.1). Monte Carlo methods are the most common treatments of variability used in environmental risk assessment. These methods involve assigning probability distributions to parameters, then propagating the variability due to these multiple

Box 7.1. Mathematical methods for variability analysis.

First-order moment propagation

First-order moment propagation uses the rules of probability to estimate the means and variances of sums, products, differences and quotients based on the means and variances of the input variables. This method is useful when the means and variances of the variable parameters of a risk assessment model are known (or can be estimated) but their statistical distribution is not. The mean and variance of an uncertain parameter can be estimated from data or elicited based on subjective expert judgement. Although this approach has been widely used in conservation biology and traditional fisheries science (it is sometimes called the 'delta method': Seber, 1973), examples of first-order moment propagation in environmental risk assessment are rare. The authors are unaware of any examples relevant to transgenic fish.

Monte Carlo simulation

Monte Carlo simulation takes repeated random samples from statistical distributions specified for each variable parameter in the risk assessment, evaluates the risk algorithm many times and builds risk curves from the results (Fig. 7.2; Cullen and Frey, 1999). This method is useful when variable parameters in a risk assessment model can be represented by a statistical distribution. These statistical distributions may be based either on the probabilistic properties of the process being modelled, the empirical distribution of data, expert judgement or any information on the expected distribution of system behaviour (Gardner and O'Neill, 1983; Burgman, 2005). There are some ecological risk assessments relevant to transgenic fish that use or recommend Monte Carlo simulation to represent variability. Notable references include Murray (2002), Vose (2000) and Bartell and Nair (2003). It is important to note that Monte Carlo simulation analysis usually assumes that input parameters are independent or, at best, linearly dependent in a known manner. These assumptions, however, are rarely explored or supported by empirical data.

parameters through mathematical equations using random or stratified sampling techniques (Fig. 7.2). One or more of these methods could be applied, for example, to the variability associated with age- or stage-specific survivorship, growth rate, longevity, length or most traits identified in Chapter 6 that influence potential ecological impacts of transgenic fish. However, it is important to note that risk assessments based on Monte Carlo simulations typically make a

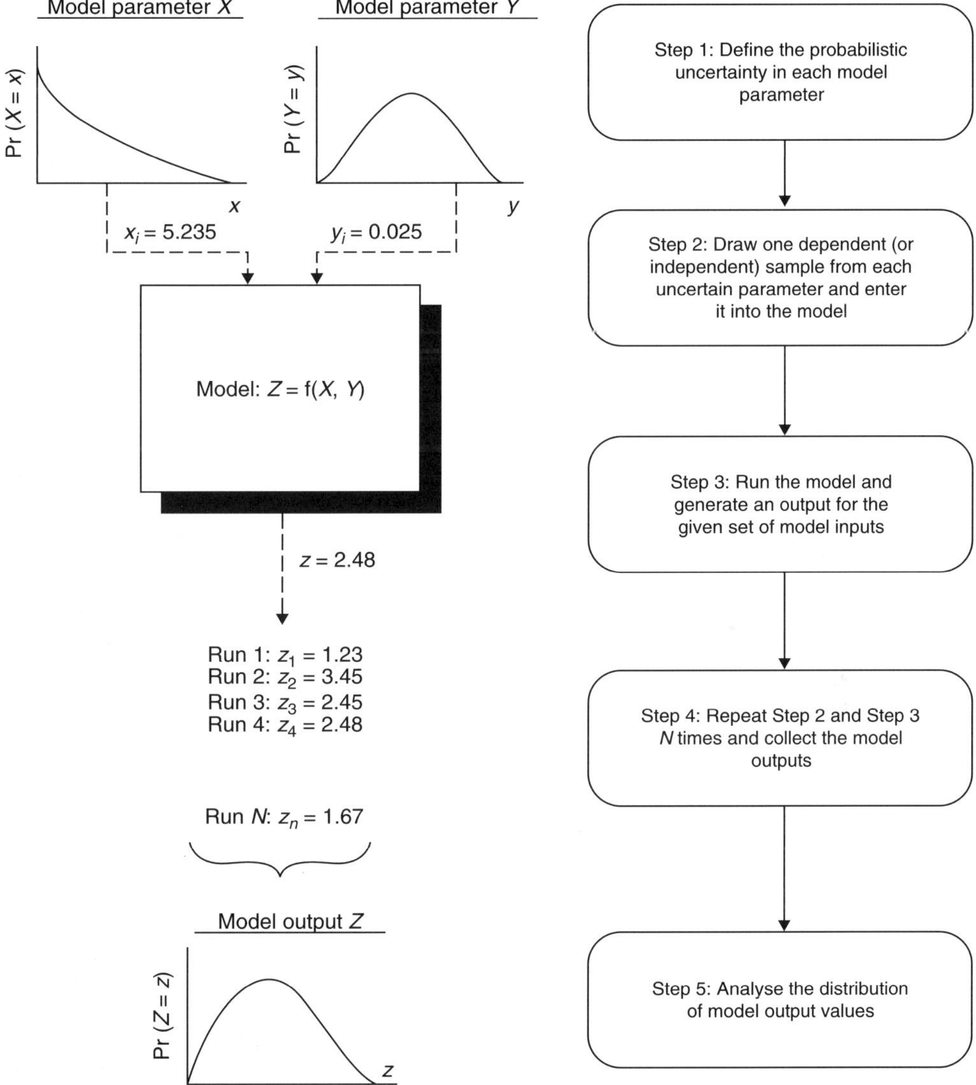

Fig. 7.2. Schematic and flowchart of a first-order Monte Carlo simulation. Monte Carlo simulation methods are commonly used to characterize the variability in model inputs parameters (*X* and *Y*) and explore the effect of this variability on the model output (*Z*) (Reprinted from Cullen and Frey, 1999, with permission of Springer Science and Business Media.)

number of often unwarranted assumptions that must be treated carefully in order to avoid overly optimistic results (see section on Dependence Between Random Variables, this chapter).

Incertitude

Incertitude is uncertainty caused by measurement error, systematic bias, missing data, censoring, use of surrogates, incomplete descriptions of a mechanism or process and other limitations of scientific knowledge. It is sometimes described in the literature as epistemic, subjective or Type I uncertainty, as well as knowledge uncertainty. The large number of sources of incertitude is partially responsible for its broad taxonomy. Its key defining criterion, however, is that it can be reduced by gathering additional data.

Measurement error

Measurement error occurs because observers and their equipment are imperfect. Measurement error can appear as random variation around a presumed true value and as systematic bias from this value. Random variation can be represented through simple intervals, confidence intervals or probability distributions (Edwards, 1996; Ferson, 2002). Systematic bias occurs through a variety of mechanisms: deliberate or accidental exclusion of certain data (censoring), poor calibration of measuring devices, non-random or poorly stratified field surveys, reference class problems[2] and extrapolation of surrogate information gathered at one site, or for one species, to other sites or species (Regan *et al.*, 2002a). Systematic bias is notoriously difficult to identify a priori and is best eliminated or minimized by careful study design and by validating predictions with independent data sets.

Model uncertainty

In a risk assessment, our understanding of ecological systems can be expressed by qualitative, statistical or mechanistic models of the system in question (Levins, 1974). Incomplete understanding of these systems is reflected in the assessment as 'model uncertainty', and this is a particularly important source of incertitude in quantitative environmental risk assessment. Model uncertainty occurs where variables and processes are omitted (via abstraction and simplification of complex realities) and because of the variety of ways mathematical equations can be used to represent ecological processes. Model results are also subject to compounding effects of uncertainty and variability in parameter estimates. Although complex models may be more realistic and relevant to the system under investigation (Bartell *et al.*, 2003), they usually involve many uncertain parameters. Hence, minimization of uncertainty in model results will often involve a trade-off between model complexity and parameter uncertainty. Simpler models will usually be subject to greater model uncertainty but contain

[2] Reference class problems occur because there is often no 'natural' or unique class of events that underlie a frequency probability. The event class is more usually the subjective choice of the analyst. For example, there is often no obvious way to group data into discrete categories when constructing a histogram.

few uncertain parameters, while more complex models will usually contain more realism but rely on a greater number of uncertain parameters.

Some subjective judgement is usually required when an analyst chooses which model to use in a risk assessment. The literature provides guidance on how to measure important ecological processes (see Chapters 6 and 9, this volume, for some examples), represent these processes mathematically (Crawford-Brown, 2001) and choose optimally complex models (Jackson *et al*., 2000; Pastorok *et al*., 2002; Bartell *et al*., 2003). Furthermore, there are a variety of techniques that can help analysts make more objective model choices. These techniques help the analyst explore alternative model specifications, identify important interactions in complex systems, distinguish genuine incertitude from variability, choose the most parsimonious model from a range of possible models and identify at what point uncertainty may change a risk-based decision (Box 7.2).

Box 7.2. Mathematical methods for incertitude analysis.

Sensitivity analysis

Sensitivity analysis explores the effect of different modelling decisions and assumptions on the results of the risk assessment (e.g. different parameter values, alternative model structures, statistical models or dependence between parameters). This method is useful because all risk assessment models include assumptions and inherent modelling decisions. There are usually many possible combinations of parameter values, model structures and dependence relations that could be explored. In practice, analysts usually select only two or three, representing worst, anticipated and best case scenarios (if it is possible to discern them all). Figure 7.3, for example, shows the results of a sensitivity analysis applied to two out of six parameters of a net fitness model for transgenic fish (Muir and Howard, 2002). There are a variety of ways to perform sensitivity analyses, depending on the model and decision context (Morgan and Henrion, 1990; Helton and Davis, 2002; Burgman, 2005).

Interval analysis

Interval analysis (Dwyer, 1951; Moore, 1966; Alefield and Herzberger, 1983; Neumaier, 1990) is one of the simplest ways to represent and propagate incertitude in quantitative and semi-quantitative risk assessment. This method is useful in data-poor situations when the bounds (max/min or best estimate ± some error) of uncertain model parameters can be expressed as an interval. Arithmetic operations on intervals are defined so that the results enclose the true value with certainty, given the input intervals (Ferson, 2002). The interval for a parameter may be calculated from data, estimated based on expert judgement, or simply reflect optimistic and pessimistic model assumptions. There are a few examples that employ interval analysis in ways relevant to risk assessment of transgenic fish, including Brown and Patil (1986), Hayes *et al.* (2005) and Nyberg and Wallentinus (2005). While interval analysis is most often used to deal with incertitude alone, it can also be applied to compound variability and incertitude (see Box 7.3).

Qualitative modelling

Qualitative modelling is an extension of loop analysis (Puccia and Levins, 1985) that can address incertitude associated with the structure of the 'community matrix' (see the right-hand side of Fig. 7.4). It is a useful way to explore the effect of different

Continued

Box 7.2. *Continued*

models of the community's structure, or different interactions within a structure, on the community's response to sustained stress (Fig. 7.4). Qualitative modelling focuses on the sign (positive or negative) of the interaction between species and their physical resources within a community matrix, but it ignores the strength (magnitude) of this interaction. By ignoring the strength of the interaction, analysts can focus on the structural uncertainty in the community matrix and quickly explore the direction of change (increase, decrease or no change) of any particular component of any community structure (Dambacher *et al.*, 2002, 2003; Ramsey and Veltman, 2005). Importantly, qualitative modelling can be extended to examine the incertitude found in any structured set of interactions so long as the system in question is in, or close to, dynamic equilibrium (Dambacher *et al.*, 2007, see also Chapter 6, this volume).

Aikake Information Criteria (AIC)
The Aikake Information Criteria (AIC) provides an objective way of determining the most prudent model from a range of possible models. It measures the parsimony (conciseness) of the model and is therefore useful when the analyst must choose from among a selection of plausible risk assessment models. The AIC combines a measure of the model likelihood (given the data) with a penalty for number of model parameters that must be estimated. By itself, the AIC for a given data set has no meaning. It is only useful when the AIC of a series of specified models are compared. The model with the lowest AIC is regarded as the 'best' of those considered. The AIC does not, however, reduce the uncertainty of any particular model.

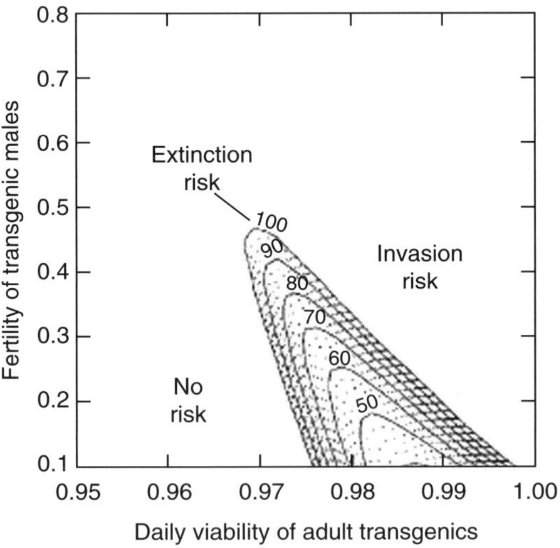

Fig. 7.3. Results of a sensitivity analysis highlighting the effect of varying the values of two model parameters (daily viability of adult transgenic fish and fertility of male transgenic fish) of a net fitness model for transgenic fish. (Reprinted from Muir and Howard, 2002, with permission from Springer Science and Business Media.) Refer to Chapter 5 for further discussion of this model.

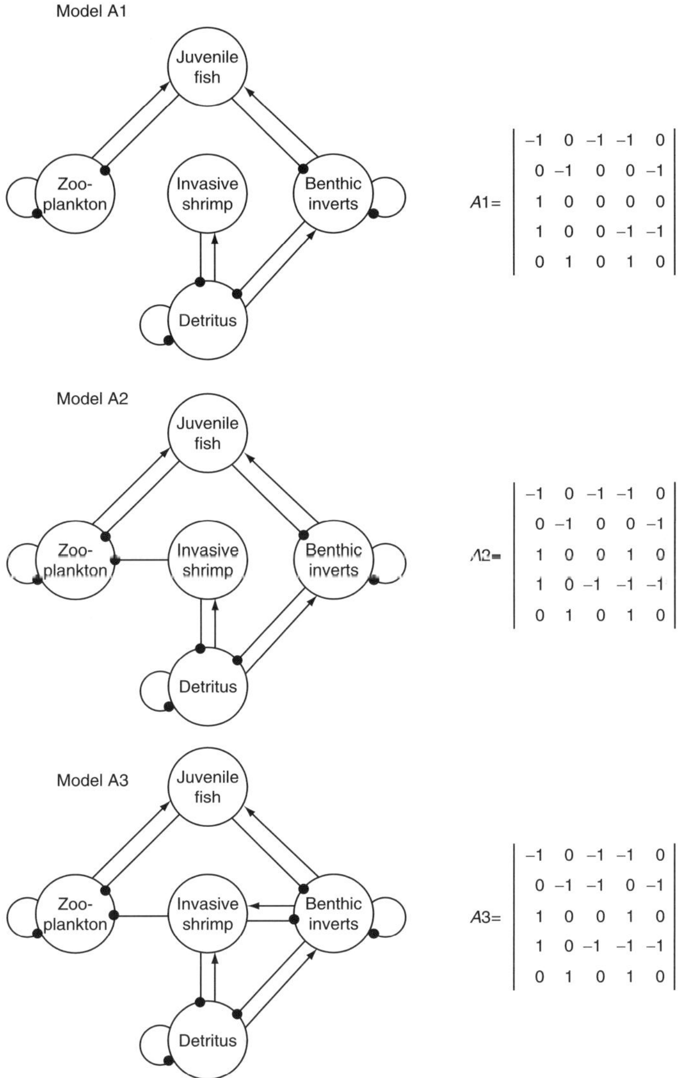

Fig. 7.4. Three hypothetical qualitative models (left) and the equivalent community matrix (right) showing possible interactions between a non-native, invasive shrimp and four components of the invaded ecosystem: detritus, zooplankton, benthic invertebrates and juvenile fish. Lines with arrows indicate positive effects, and those with filled circles denote negative effects. Self-effects are shown by lines that start and end at the same node (see Dambacher *et al.* (2003) for guidance on computing the community matrix). The different models are used to explore three different hypotheses: Model A1 – the shrimp feeds only on detritus; Model A2 – the shrimp feeds on detritus and competitively interferes with zooplankton; and Model A3 – the shrimp feeds on detritus, benthic invertebrates and competitively interferes with zooplankton. By analysing the sign of the interaction terms in the community matrices the analyst can predict the (in this case indirect) effect of the shrimp on juvenile fish. (From J. Dambacher, CSIRO, Australia, 2006, personal communication.)

Incertitude arises in transgenic fish risk assessment, for example, through the extrapolation of data from unmodified surrogates to genetically modified fish and through problems associated with genotype-by-environment (G × E) interactions (see Chapter 6, this volume). The presence of G × E interactions implies that the statistical characteristics of important phenotypic properties of transgenic fish in the wild (such as feeding rates, growth rates, predator avoidance, fecundity, etc.) may be substantially different from those measured in the laboratory. Therefore, there is both variability in, and incertitude about, these phenotypic properties.

Treatments for variability and incertitude

In practice, the distinction between variability and incertitude is not always clear and may be context-dependent. Moreover, most risk assessment problems must deal with both types of uncertainty. This is particularly true for transgenic fish because there is limited empirical information available to inform critical components of a risk assessment. Fortunately, there are a variety of mathematical tools that can accommodate both types of uncertainty, and there are a few that are specifically designed to handle both simultaneously (Box 7.3). Probability bounds analysis (Fig. 7.6; Ferson, 2002) and interval analysis (Moore, 1966) entail the least number of assumptions, and for this reason they are the simplest mathematical approaches. Probability bounds analysis and information gap theory are relatively new methods, and although there are some ecological applications available, the authors are unaware of any pertaining to transgenic fish.

Box 7.3. Mathematical methods that simultaneously treat variability and incertitude.

Two-dimensional Monte Carlo simulation

Two-dimensional (or second-order) Monte Carlo simulation nests one Monte Carlo simulation within another. This method is useful to simultaneously explore the effects of uncertainty and variability in a risk assessment model, while retaining a separate measure of the effect of each source of uncertainty on the risk assessment results. The inner simulation typically represents the natural variability in the physical or biological parameters of a risk assessment model. The outer simulation typically represents the analyst's incertitude about the parameters used to specify inputs to the inner simulation (Fig. 7.5). Each replication of the outer layer entails an entire Monte Carlo simulation. Two-dimensional Monte Carlo methods have been championed for use in risk analysis by Hoffman and Hammonds (1994), among others. Cullen and Frey (1999) give a good introduction to the technique. Wu and Tsang (2004) use this technique to explore the effects of incertitude in gravel-bed characteristics, and variability in the amount of sand accumulated in the bed surface, on the survival rate of salmonid embryos.

Continued

Box 7.3. *Continued*

Probability bounds analysis

Probability bounds analysis computes the results of arithmetic operations using only the bounds of statistical distributions used to represent variable input parameters (Frank *et al.*, 1987; Williamson and Downs, 1990; Ferson and Long, 1995; Berleant and Goodman-Strauss, 1998; Ferson, 2002). This method is useful because it can represent variability and incertitude in both data-poor and data-rich situations. If there are enough empirical data to estimate the statistical distribution of uncertain model parameters and the dependence between them, the bounds on the resulting arithmetic operation will approximate the distribution resulting from a Monte Carlo simulation. When there is very little information about these distributions and the dependence between them, the resulting bounds tend to be much wider, representing weaker confidence about the results of the arithmetic operation (Fig. 7.6). Probability bounds analysis reliably propagates uncertainty about dependence between random variables through a risk assessment. It is more comprehensive than sensitivity analysis and computationally less demanding and often easier to interpret than two-dimensional Monte Carlo methods. See Regan *et al.* (2002b,c) for examples of probability bounds analysis applied to environmental risk assessment.

Information gap theory

Information gap theory asks how wrong a model and its parameters can be before jeopardizing the quality of decisions made on the basis of this model (Ben-Haim, 2001, 2005; Takewaki and Ben-Haim, 2005). Information gap theory does not use probability theory to represent uncertainty and variability, and it is therefore a useful way to address the 'robustness' of decision making in situations where it is difficult to apply probability theory, such as extremely data-poor situations. Information gap theory requires: (i) a model of the system in question; (ii) a non-probabilistic description of parametric and model structure uncertainty; and (iii) decision criteria. Information gap theory helps the analyst to explore the effects of unbounded, non-probabilistic measures of parameter or model uncertainty on the results of the model and the resulting decisions. For example, the analyst may increase the fractional error of important model parameters, recording the point at which a decision threshold is crossed. Examples relevant to environmental risk assessment include Regan *et al.* (2005) and Fox *et al.* (2007).

Fuzzy sets and arithmetic

Fuzzy sets simultaneously specify the range of an uncertain variable and the plausibility or possibility of intermediate values. The level of 'presumption' for any number of values on the range describes the level of possibility of these values being between 0 and 1 (Fig. 7.1; Kaufmann and Gupta, 1985). Often, fuzzy sets are triangular or trapezoidal in shape. More complicated forms can be constructed by stacking a series of interval estimates or specifying three or more intervening values (and their associated level of presumption) on the interval range. Fuzzy arithmetic is simply arithmetic (e.g. risk calculations) with fuzzy sets. Fuzzy sets are useful because they can help to eliminate vagueness from risk assessment terms such as 'high', 'medium' or 'low'. Fuzzy arithmetic is useful because it can simultaneously yield 'worst-case' and 'best-estimate' risk assessment results in data-poor situations (Ferson, 2002). However, fuzzy arithmetic becomes cumbersome, with

Continued

Box 7.3. *Continued*

repeated variables, and it cannot use knowledge of correlations to tighten the risk bounds (Ferson *et al.*, 2001).

Hierarchical Bayesian analysis

In its simplest form Hierarchical Bayesian analysis is a Bayesian version of two-dimensional Monte Carlo analysis wherein the moments (e.g. mean and variance) of variable input distributions are themselves allowed to vary in a parametric fashion. It is a very powerful (but computationally intensive) method that is useful when variable and uncertain parameters in a risk assessment model can be represented by a statistical distribution. The computations associated with hierarchical Bayesian analysis were, until recently, prohibitively complex. However, recent numerical computation advances make the analysis much more tractable. Link *et al.* (2002) provide a general introduction to this subject matter. Clark (2007) provides a more comprehensive description of this technique and its application to environmental data sets.

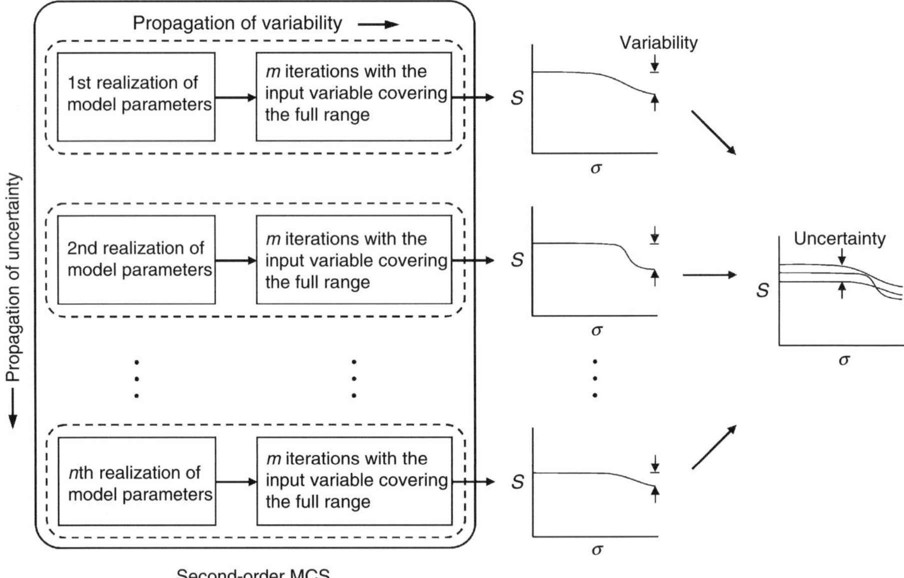

Fig. 7.5. Schematic showing the procedure and output of a second-order Monte Carlo simulation. In this example the salmonid embryo survival rate (S) is modelled as a function of the uncertain characteristics of the salmonid spawning nest and the variable sand composition (σ) of the stream bed. The m iterations of variability represent a first-order Monte Carlo simulation (refer to Fig. 7.2) in which S is calculated for a single value of each uncertain habitat parameter (drawn from triangular distributions) over the entire range of plausible values for σ. This procedure is repeated n times, drawing different values from the uncertain habitat characteristics each time. (Reprinted from Wu and Tsang, 2004, with permission from Elsevier.)

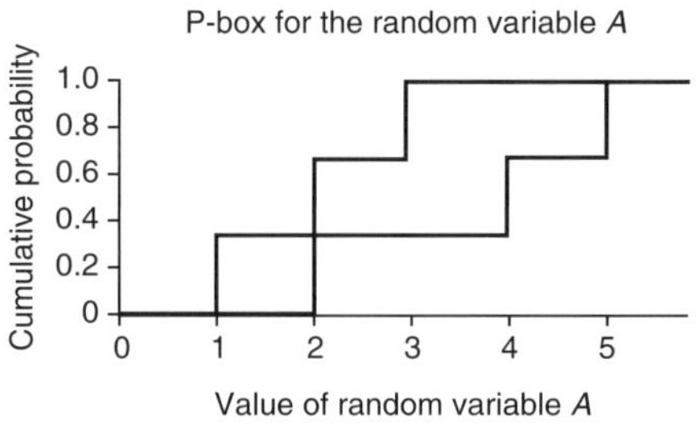

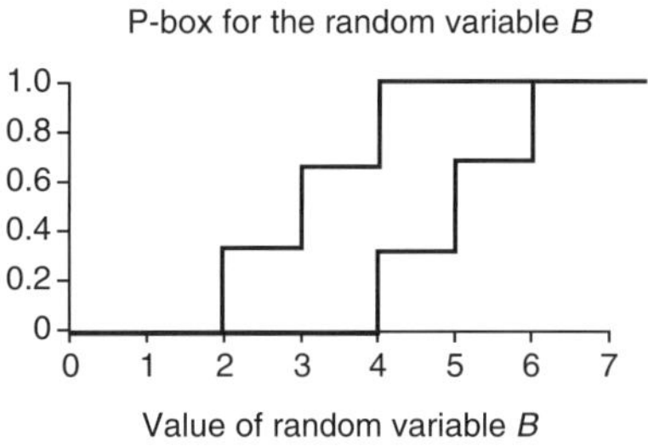

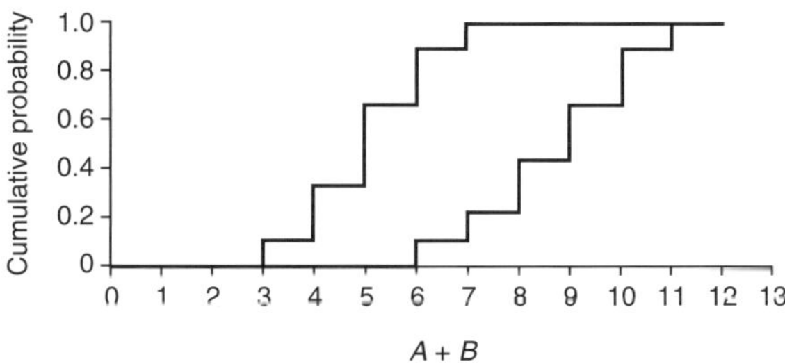

Fig. 7.6. Probability bounds analysis showing the result of the arithmetic operation $A + B$ when the inputs A and B are variable and uncertain (and assumed to be independent). Variability in A, for example, is reflected by the cumulative probability of it taking any particular value on the range [0,5]. Uncertainty in A is reflected in the upper and lower bounds of the cumulative probability. Note that a different result for $A + B$ would occur if no assumption were made about the dependence between these variables. (Reprinted from Burgman, 2005, with permission from Scott Ferson.)

Practical Issues when Dealing with Uncertainty

Data types and requirements

There are two general categories of data typically used in environmental risk assessment: (i) empirical observations of relevant parameters; and (ii) expert and stakeholder opinion. Empirical observations of relevant parameters are often considered to be 'objective' sources of data, although in reality experimental observations and measurements often incorporate important subjective decisions (Berger and Berry, 1988). Expert and stakeholder opinion may be based on their (sometimes different) interpretation of empirical measurements or on their accumulated experience and beliefs. As such, they are often labelled as 'subjective' sources of data.

The most appropriate representation of empirical variability depends on the quantity of observations available. Sample or empirical distribution functions (Gardner and O'Neill, 1983) are simple, do not require large amounts of data (e.g. 10 or more observations) and let the data speak for themselves. These functions assume that the data are collected randomly and are representative of the variable in question. Cullen and Frey (1999) reviewed strategies for computing empirical distributions from data. It has been found that kernel density estimates (Epanechnikov, 1969; Silverman, 1986) can be applied to small (e.g. at least 20 observations) data sets, but they require further assumptions, most notably about the maximum and minimum values that the variable can take. Parametric approaches based on discrete or continuous population distributions require larger data sets (e.g. at least 30 observations), and they must fit the most likely of a range of possible distributions using a variety of techniques and statistical tests (Palisade Corporation, 1996; Ferson *et al.*, 2001). Moore (1996) discusses some of the problems associated with parametric distribution choice in Monte Carlo simulations and makes a number of useful recommendations. Probability bounds analysis deals with uncertainty in a range of assumptions, including distribution shape, values of the mean, standard deviation and specified percentiles.

Subjective estimates of variation can be informally or formally elicited from experts using a variety of techniques (Morgan and Henrion, 1990). Informal probability estimates should be avoided wherever possible because of the 'psychological frailties' (such as overconfidence and insensitivity to sample size) discussed in Chapter 1. Burgman (2005), Vose (2000) and Cullen and Frey (1999) provide general summaries of formal elicitation and aggregation techniques, together with their advantages and potential pitfalls. However, these formal procedures are time consuming and are rarely employed in environmental risk assessment. To date, the authors are unaware of any examples relevant to transgenic fish.

Dependence between random variables

The preceding sections of this chapter emphasize the uncertainty associated with environmental risk assessment models and input variables. Risk assessments that acknowledge uncertain input parameters will inevitably entail arithmetic operations performed on random variables. It is important to recognize that the results of these operations, and hence the results of the risk assessment, are sensitive to the dependence (if any) between variables (see, e.g. Ferson and Burgman, 1995). Most probabilistic risk assessments assume independence among random variables, but this is not always a reasonable assumption. For instance, it is highly implausible that fish body size at maturity and fecundity are independent of one another (Jobling, 1995). The existence of dependent input variables can be checked by measuring their correlation and covariance using standard statistical techniques. However, it is important to note that correlation coefficients do not fully specify the dependence structure of two random variables, and zero correlations do not generally guarantee independence (Ferson *et al.*, 2004).

Dependence problems can be avoided by restructuring a risk assessment model to include only independent variables or by stratifying the entire assessment for subgroups of relatively homogenous input variables. The first approach re-specifies dependent input functions as functions of other related input variables. For example, a variable representing fish fecundity could be replaced with an allometric relationship using fish body mass that includes an independent (by construction) random error term reflecting the residual uncertainty in fecundity remaining after body mass is accounted for. Unfortunately, predictable relationships such as those between fish body mass and fecundity are relatively rare. The second approach stratifies populations into relevant subgroups (e.g. by age, stage or gender), making assumptions about independence between variables more plausible. This approach, however, makes the risk assessment more complex and cumbersome because it has to be repeated for each subgroup.

Monte Carlo simulation can account for linear dependence between input variables by using correlation coefficients (see, e.g. Vose, 2000), but this approach can only model one out of the many possible dependence structures. If this approach is adopted, the analyst must ensure that the matrix of adopted correlation coefficients is feasible; for example, one variable cannot be strongly positively correlated with each of two variables that are themselves strongly negatively correlated. This assurance can be achieved by checking the mathematical properties of the correlation matrix (Iman and Davenport, 1982). A more comprehensive simulation approach based on copulas is available (Haas, 1999; Clemen and Reilly, 1999). Copulas provide a way to study and measure dependence between random variables (Nelsen, 1999). In practice, linear correlation is often used to measure dependence. Linear correlation, however, can be misleading and should not be taken as a general measure of dependence (see Ferson *et al.*, 2004). The advantage of copulas is that they provide a formal way of dealing with the full spectrum of possible dependence between variables. The disadvantage of copulas is that they can be difficult to construct if there is little or no empirical information on the dependence in question. By contrast, probability bounds analysis (Box 7.3; Fig. 7.6) does not require *a priori* information on the strength of dependence between input variables, and it provides reliable results for all types of dependence. Interested readers should refer to Ferson and Long (1995) for further discussion about the problems associated with dependence between input variables, Monte Carlo analysis and the advantages of probability bounds analysis.

Resources for Capacity Building in Uncertainty Analysis

There are a variety of mathematical methods that, when carefully implemented, provide a comprehensive and honest account of uncertainty within quantitative risk assessment frameworks. These approaches range from relatively simple to very complex, and they all require some combination of training and practice to master. The Society for Risk Analysis (SRA) (available at: http://www.sra.org/) and the Society for Environmental Toxicology and Chemistry (SETAC) (available at: http://www.setac.org/) occasionally run uncertainty analysis seminars and workshops, and there is a small (but growing) number of universities offering courses on environmental risk assessment.

There are many free and commercially available software packages for dealing with uncertainty in calculations (Table 7.1). Ferson (2002) provides a

Table 7.1. Web sites for uncertainty analysis software.

Method	Software	Web site
Qualitative modelling	MAPLE	http://www.maplesoft.com/
Statistical modelling	R	http://www.r-project.org/
Bayesian modelling	WINBUGS	http://www.mrc-bsu.cam.ac.uk/bugs
Monte Carlo simulation	@RISK	http://www.palisade.com.au/risk/default.asp
Monte Carlo simulation	CRYSTAL BALL	http://www.decisioneering.com/
Interval and probability bounds analysis	RISK CALC	http://www.ramas.com/riskcalc.htm
Probability bounds analysis	STATOOL	http://class.ee.iastate.edu/berleant/home/Research/Pdfs/versions/statool/distribution/index.htm

comprehensive package that can perform risk assessment calculations using intervals and probability bounds. STATOOL also performs dependence bounds convolutions and probability bounds analysis (Berleant *et al.*, 2003). CRYSTAL BALL and @RISK (Palisade Corporation, 1996; Vose, 2000) enable first- and second-order Monte Carlo simulations within a Microsoft Excel spreadsheet environment. Qualitative modelling can be done using MAPLE. Information gap analysis, while conceptually quite complex, can be performed analytically and numerically using a variety of standard mathematical and statistical software packages, including spreadsheet packages. R, for example, is an excellent, and freely available, package capable of performing a variety of uncertainty analyses, including information gap theory, sensitivity analysis, Aikake Information Criteria (AIC) calculations and hierarchical Bayesian analysis. WINBUGS is similar to R, but it is specifically designed for Bayesian analysis.

Chapter Summary

Uncertainty is a pervasive phenomenon in biology that should be addressed in any environmental risk assessment. This chapter identifies three main types of uncertainty: linguistic uncertainty, variability and incertitude. Qualitative risk assessment is fraught with linguistic uncertainty and does not adequately deal with incertitude and variability. Although quantitative risk assessments are not immune from linguistic uncertainty, they are able to provide a more 'honest' account of variability and incertitude through a variety of mathematical methods. This chapter provides an introduction to such methods (Table 7.2), as well as resources and citations for further study.

Wherever possible, this chapter identifies examples of the application of quantitative uncertainty analysis techniques to environmental risks. There are, however, very few examples specific to transgenic fish from which practitioners can draw guidance. This statement, however, is true for virtually all forms of quantitative risk assessment for genetically modified organisms. Future efforts to provide tools to assess and manage biosafety of transgenic fish should not

Table 7.2. A summary of the various sources of uncertainty in environmental risk assessment, together with treatments identified in this chapter.

Source of uncertainty	Treatments identified in this chapter
Linguistic uncertainty	
Ambiguity	Clarify and agree on meaning (via PFOA, Chapter 2, this volume)
Context dependence	Clearly specify context
Underspecificity	Specify all available contextual data and provide narrowest possible bounds
Vagueness (non-numerical)	Construct numerical measures, then treat as numerical vagueness
Vagueness (numerical)	Fuzzy sets, supervaluations, rough sets, three-valued logic
Variability	
	First-order moment propagation, first-order Monte Carlo simulations
Incertitude	
Measurement error (random)	Standard statistical techniques, interval analysis
Measurement error (systematic)	Careful study design and validation of predictions with independent data
Model uncertainty	Sensitivity analysis, model validation, Aikake Information Criteria, qualitative modelling
Variability and incertitude	
	Second-order Monte Carlo simulation, probability bounds analysis, information gap theory, fuzzy sets and arithmetic, hierarchical Bayesian analysis.

only include developing strategies to identify and treat key sources of uncertainty, but also to build the capacity within regulatory programmes to address uncertainty. Moreover, future risk assessment research should aim to reduce these uncertainties or, at the very least, to increase the understanding of the consequences of uncertainty for decision making.

References

Alefield, G. and Herzberger, J. (1983) *Introduction to Interval Computations*. Academic Press, New York.

Bartell, S.M. and Nair, S.K. (2003) Establishment risks for invasive species. *Risk Analysis* 24, 833–845.

Bartell, S.M., Pastorok, R.A., Akcakaya, H.R., Regan, H., Ferson, S. and Mackay, C. (2003) Realism and relevance of ecological models used in chemical risk assessment. *Human and Ecological Risk Assessment* 9, 907–938.

Baybutt, P. (1989) Uncertainty in risk analysis. In: Cox, R.A. (ed.) *Mathematics in Major Accident Risk Assessment*. Clarendon Press, Oxford, UK, pp. 247–261.

Ben-Haim, Y. (2001) *Information Gap Decision Theory: Decisions Under Severe Uncertainty*. Academic Press, San Diego, California.

Ben-Haim, Y. (2005) Value at risk with info-gap uncertainty. *Journal of Risk Finance* 6, 388–403.

Berger, J.O. and Berry, D.A. (1988) Statistical analysis and the illusion of objectivity. *American Scientist* 76, 159–165.

Berleant, D. and Goodman-Strauss, C. (1998) Bounding the results of arithmetic operations on random variables of unknown dependency using intervals. *Reliable Computing* 4, 147–165.

Berleant, D., Xie, L. and Zhang, J. (2003) Statool: a tool for Distribution Envelope Determination (DEnv), an interval-based algorithm for arithmetic on random variables. *Reliable Computing* 9, 91–108.

Brown, B.E. and Patil, G.P. (1986) Risk analysis in the Georges Bank haddock fishery – a pragmatic example of dealing with uncertainty. *North American Journal of Fisheries Management* 6, 183–191.

Burgman, M.A. (2005) *Risks and Decisions for Conservation and Environmental Management*. Cambridge University Press, Cambridge.

Clark, J.S. (2007) *Models for Environmental Data: An Introduction*. Princeton University Press, USA.

Clemen, R.T. and Reilly, T. (1999) Correlations and copulas for decision and risk analysis. *Management Science* 45, 208–224.

Crawford-Brown, D.J. (2001) *Mathematical Methods of Environmental Risk Modeling*. Kluwer Academic, London.

Cullen, A.C. and Frey, H.C. (1999) *Probabilistic Techniques in Exposure Assessment: A Handbook for Dealing with Variability and Uncertainty in Models and Inputs*. Plenum Press, New York.

Dambacher, J.M., Hiram, W.L. and Rossignol, P.A. (2002) Relevance of community structure in assessing indeterminacy of ecological structure. *Ecology* 83, 1372–1385.

Dambacher, J.M., Hiram, W.L. and Rossignol, P.A. (2003) Qualitative predictions in model ecosystems. *Ecological Modelling* 161, 79–93.

Dambacher, J.M., Brewer, D.T, Dennis, D.M., Macintyre, M. and Foale, S. (2007) Qualitative modelling of gold mining impacts on Lihir Island's socio economic system and reef-edge fish community. *Environmental Science and Technology* 41, 555–562.

Dwyer, P. (1951) *Linear Computations*. Wiley, New York.

Edwards, D. (1996) The first data analysis should be journalistic. *Ecological Applications* 6, 1090–1094.

Epanechnikov, V.A. (1969) Non-parametric estimation of multivariate probability density. *Theory of Applied Probability and Its Applications* 14, 153–158.

European Union (EU) (2001) Directive 2001/18/EC of the European Parliament and of the Council of 12 March 2001 on the deliberate release into the environment of genetically modified organisms and repealing Council Directive 90/220/EEC. *Official Journal of the European Communities L106*, 1–38.

Ferson, S. (1996) What Monte Carlo methods cannot do. *Human and Ecological Risk Assessment* 2, 990–1007.

Ferson, S. (2002) *RAMAS Risk Calc: Risk Assessment with Uncertain Numbers*. Lewis, Boca Raton, Florida.

Ferson, S. and Burgman, M.A. (1995) Correlations, dependency bounds and extinction risks. *Biological Conservation* 73, 101–105.

Ferson, S. and Ginzburg, L.R. (1996) Different methods are needed to propagate ignorance and variability. *Reliability Engineering and System Safety* 54, 133–144.

Ferson, S. and Long, T.F. (1995) Conservative uncertainty propagation in environmental risk assessment. In: Hughes, J.S., Biddinger, G.R. and Mones, E. (eds) *Environmental Toxicology and Risk Assessment – Third Volume*. American Society for Testing and Materials, Philadelphia, pp. 97–110.

Ferson, S., Cooper, J.A. and Myers, D. (2001) *Beyond Point Estimates: Risk Assessment Using Interval, Fuzzy and Probabilistic Arithmetic*. Applied Biomathematics, Setauket, New York.

Ferson, S., Nelsen, R., Hajagos, J., Berleant, D., Zhang, J., Tucker, W.T., Ginzburg, L. and Oberkampf, W.L. (2004) *Dependence in Probabilistic Modeling, Dempster-Shafer Theory, and Probability Bounds Analysis*. Sandia National Laboratories, SAND2004-3072, Albuquerque, New Mexico.

Fox, D., Ben-Haim, Y., Hayes, K.R., McCarthy, M., Wintle, B. and Dunstan, P. (2007) An info-gap approach to power and sample size calculations. *Envirometrics* 18, 189–203.

Frank, M.J., Nelsen, R.B. and Schweizer, B. (1987) Best possible bounds for the distribution of a sum – a problem of Kolmogorov. *Probability Theory and Related Fields* 74, 199–211.

Gardner, R.H. and O'Neill, R.V. (1983) Parameter uncertainty and model predictions: a review of Monte Carlo results. In: Beck, M.B. and van Straten, G. (eds) *Uncertainty and Forecasting of Water Quality*. Springer, Berlin, Germany, pp. 245–257.

Haas, C.N. (1999) On modeling correlated random variables in risk assessment. *Risk Analysis* 19, 1205–1214.

Haimes, Y.Y. (1998) *Risk Modeling, Assessment, and Management*. Wiley, New York.

Hayes, K.R., Sliwa, C., Migus, S., McEnnulty, F. and Dunstan, P. (2005) *National Priority Pests – Part II: Ranking of Australian Marine Pests*. Final report for the Australian Government Department of Environment and Heritage, CSIRO Division of Marine Research, Hobart, Australia.

Helton, J.C. and Davis, F.J. (2002) Illustration of sample-based method for uncertainty and sensitivity analysis. *Risk Analysis* 22, 591–622.

Hoffman, F.O. and Hammonds, J.S. (1994) Propagation of uncertainty in risk assessments: the need to distinguish between uncertainty due to lack of knowledge and uncertainty due to variability. *Risk Analysis* 14, 707–712.

Iman, R.L. and Davenport, J.M. (1982) An interactive algorithm to produce a positive-definite correlation matrix from an approximate correlation matrix (with a program users' guide). SAND81-1376, Sandia National Laboratories, Albuquerque, New Mexico.

Jackson, L.L., Trebitz, A.S. and Cottingham, K.L. (2000) An introduction to the practice of ecological modeling. *BioScience* 50, 694–706.

Jobling, M. (1995) *Environmental Biology of Fishes*. Fish and Fisheries Series 16, Chapman & Hall, London.

Kaufmann, A. and Gupta, M.M. (1985) *Fuzzy Mathematical Models in Engineering and Management Science*. Van Nostrand Reinhold, New York.

Levins, R. (1974) The qualitative analysis of partially specified systems. *Annals of the New York Academy of Sciences* 231, 123–138.

Link, W.A., Cam, E., Nichols, J.D. and Cooch, E.G. (2002) Of BUGS and birds: Markov Chain Monte Carlo for hierarchical modeling in wildlife research. *Journal of Wildlife Management* 66, 277–291.

Moore, D.R.J. (1996) Using Monte Carlo analysis to quantify uncertainty in ecological risk assessment: are we gilding the lily or bronzing the dandelion? *Human and Ecological Risk Assessment* 2, 628–633.

Moore, R.E. (1966) *Interval Analysis*. Prentice-Hall, New Jersey.

Morgan, M.G. and Henrion, M. (1990) *Uncertainty: A Guide to Dealing with Uncertainty in Quantitative Risk and Policy Analysis*. Cambridge University Press, Cambridge.

Muir, W.M. and Howard, R.D. (2002) Assessment of possible ecological risks and hazards of transgenic fish with implications for other sexually reproducing organisms. *Transgenic Research* 11, 101–114.

Murray, N. (2002) Import risk analysis: animals and animal products. *New Zealand Ministry of Agriculture and Forestry*. Wellington, New Zealand.

National Research Council (1989) *Field Testing Genetically Modified Organisms: Framework for Decisions*. National Academy Press, Washington, DC.

Nelsen, R.B. (1999) *An Introduction to Copulas*. Lecture Notes in Statistics Vol. 139, Springer, New York.

Neumaier, A. (1990) *Interval Methods for Systems of Equations*. Cambridge University Press, Cambridge.

Nyberg, C.D. and Wallentinus, I. (2005) Can species traits be used to predict marine macroalgal introductions? *Biological Invasions* 7, 265–279.

Organisation for Economic Cooperation and Development (OECD) (2000) *Report of the Task Force for the Safety of Novel Foods and Feeds*. C(2000) 86/ADD1. OECD, Paris, France.

Palisade Corporation (1996) *Guide to Using @Risk*. Palisade Corporation, New York.

Pastorok, R.A., Bartell, S.M., Ferson, S. and Ginzburg, L.R. (eds) (2002) *Ecological Modeling in Risk Assessment*. Lewis, Boca Raton, Florida.

Puccia, C.J. and Levins, R. (1985) *Qualitative Modeling of Complex Systems: An Introduction to Loop Analysis and Time Averaging*. Harvard University Press, Cambridge, Massachusetts.

Ramsey, D. and Veltman, C. (2005) Predicting the effects of perturbations on ecological communities: what can qualitative models offer? *Journal of Animal Ecology* 74, 905–916.

Regal, P.J. (1994) Scientific principles for ecologically based risk assessment of transgenic organisms. *Molecular Ecology* 3, 5–13.

Regan, H.M., Colyran, M. and Burgman, M.A. (2000) A proposal for fuzzy International Union for the Conservation of Nature (IUCN) categories and criteria. *Biological Conservation* 92, 101–108.

Regan, H.M., Colyvan, M. and Burgman, M.A. (2002a) A taxonomy and treatment of uncertainty for ecology and conservation biology. *Ecological Applications* 12, 618–628.

Regan, H.M., Sample, B.E. and Ferson, S. (2002b) Deterministic and probabilistic ecological soil screening levels for wildlife. *Environmental Toxicology and Chemistry* 21, 882–890.

Regan, H.M., Hope, B.K. and Ferson, S. (2002c) Analysis and portrayal of uncertainty in a food web exposure model. *Human and Ecological Risk Assessment* 8, 1757–1777.

Regan, H.M., Akcakaya, H.R., Ferson, S., Root, K.V., Carroll, S. and Ginzburg, L.R. (2003) Treatments of uncertainty and variability in ecological risk assessment of single-species populations. *Human and Ecological Risk Assessment* 9, 889–906.

Regan, H.M., Ben-Haim, Y., Langford, B., Wilson, W.G., Lundberg, P. and Andelman, S.J. (2005) Robust decision making under severe uncertainty for conservation management. *Ecological Applications* 15, 1471–1477.

Seber, G.A.F. (1973) *The Estimation of Animal Abundance and Related Parameters*. Griffin, London.

Silverman, B.W. (1986) *Density Estimation for Statistics and Data Analysis*. Chapman & Hall, London.

Takewaki, I. and Ben-Haim, Y. (2005) Info-gap robust design with load and model uncertainties. *Journal of Sound and Vibration* 288, 551–570.

Van Dommelen, A. (1998) Useful models for biotechnology hazard identification: what is this thing called familiarity? In: Wheale, P., von Schomberg, R. and Glasner, P. (eds) *The Social Management of Genetic Engineering*. Ashgate, Aldershot, UK, pp. 219–236.

Vose, D. (2000) *Risk Analysis: A Quantitative Guide*. Wiley, Chichester, UK.

Williamson, R.C. and Downs, T. (1990) Probabilistic arithmetic I. Numerical methods for calculating convolutions and dependency bounds. *International Journal of Approximate Reasoning* 4, 89–158.

Wu, F.C. and Tsang, Y.P. (2004) Second order Monte Carlo uncertainty/variability analysis using correlated model parameters: application to salmonid embryo survival risk assessment. *Ecological Modelling* 177, 393–414.

Zadeh, L. (1978) Fuzzy sets as a basis for a theory of possibility. *Fuzzy Sets Systems* 1, 3–28.

8 Risk Management: Reducing Risk through Confinement of Transgenic Fish

G.C. Mair, Y.K. Nam and I.I. Solar

Introduction

A key way to manage risks associated with the use of transgenic fish in aquaculture is through the application of confinement measures designed to minimize the *likelihood* of transgenic fish causing harm to the environment (Kapuscinski, 2005). There are three primary aims of confinement strategies for transgenic fish:

1. Limiting the organism: confinement methods are used to prevent the whole fish (i.e. the individuals carrying the transgene) from entering and surviving in a receiving environment.
2. Limiting (trans)gene flow: confinement methods are targeted at preventing gene flow from the transgenic fish, either during aquaculture production or following its escape.
3. Limiting transgenic trait expression: confinement is targeted at limiting the expression of the traits encoded by the transgene, because it is likely that the trait – not the transgene itself – represents the risk to the receiving environment.

The objective of this chapter is to describe and review existing and potential methods of confinement, of which there are three primary types:

1. Physical confinement: confinement of the transgenic fish by preventing its entry into the receiving environment.
2. Geographical confinement: culturing a transgenic fish in a location where it cannot survive if it enters the surrounding environment.
3. Biological confinement: limiting (trans)gene flow into the receiving environment, primarily by limiting reproduction of the transgenic fish within the culture system, preventing reproduction of any transgenic fish once it enters the receiving environment, or preventing the expression of the transgene in the event of an escape.

 Environmental Risk Assessment of Genetically Modified Organisms: Vol. 3. Methodologies for Transgenic Fish (eds A.R. Kapuscinski *et al.*)

This chapter reviews the technical aspects of physical confinement measures that are necessary and appropriate to minimize the possibility of the entry of transgenic fish into a receiving environment. Geographic confinement is discussed, with reference to cases where it has been applied at a research level. The chapter also reviews a range of technologies that provide variable levels of biological confinement. These include producing single-sex (i.e. monosex) progeny, inducing sterility in transgenic fish using 'traditional' means, such as triploidy induction, and innovative use of genetic use restriction technologies (GURTs). A summary of the status and major issues related to the various confinement options is provided in Table 8.1, and each method is explored in greater detail in the chapter.

Preventing the entry of transgenic fish into the environment should always be a primary objective in aquaculture operations where they are used. Although reliable and complete physical confinement may be possible in most land-based systems, it is difficult to guarantee that escapes will not occur (see Chapter 5, this volume, for discussion of how to determine the probability of escape). This is especially true when factors such as human error or poaching are taken into account. The ideal confinement situation is thus one where redundancy is ensured by using a combination of physical, geographical and biological confinement measures. This reduces the probability of transgene flow in the receiving environment to a level deemed acceptable, i.e. a level of risk that meets criteria developed in advance by multiple stakeholders during the initial stage of a risk assessment framework (see Chapter 1, this volume, discussion of stage 'Identify, Define and Agree'). The degree and stringency of confinement methods ultimately utilized should reflect the level of risk management required to bring the estimated risk down to this acceptable level.

Physical Confinement

Physical confinement measures focus on preventing transgenic fish from entering a receiving environment by using physical barriers within culture systems, and they serve as the first line of defence in risk management. Although the issue of physical confinement for transgenic organisms, including aquatic species, has been considered in some detail (e.g. ABRAC, 1995; Scientist's Working Group on Biosafety, 1998), and several facilities with appropriate physical confinement measures do exist, there are no standardized recommendations for designing aquaculture facilities for transgenic fish. The only practical application of physical confinement of transgenic fish to date has been in research facilities (see Box 8.1), and these measures are insufficient to manage risks in commercial-scale operations (ABRAC, 1995). This section will discuss issues applicable to commercial systems for transgenic fish using recommendations developed primarily for research systems. These should be considered as minimum guidance for commercial operations.

Complete physical confinement measures that provide an acceptably low risk (as defined in the initial phase of risk assessment) of transgenic fish escapes

Table 8.1. Summary of status, strengths, weaknesses and related research and capacity needs of physical, geographical and biological confinement options.

Confinement approach	Status of development	Strengths	Weakness	Research or capacity needs
Physical	Various measures well established	• Acts as a first line of defence against escapes • Feasible in land-based systems • High level of confinement possible	• Costly • Training, monitoring and compliance capacities needed • Compromised by human activities (e.g. poaching, intentional release)	• Training on best practices, regulation and compliance
Geographic	Established for a few species (e.g. tilapia in regions with cold winters)	• Feasible for some species in some sites • Provides high level of confinement	• Only feasible with specific species and site combinations and species with narrow ranges of environmental tolerances	• Identification of appropriate species and site combinations
Biological				
Monosex populations	Well developed for some species (e.g. salmonids, tilapia)	• Complete monosex possible in some species (e.g. chinook salmon) • Useful where wild and feral stocks are absent	• Monosex not guaranteed in some species (e.g. tilapia and coho salmon) • Risk of broodstock contamination in hatchery – requires regular verification • Only applicable to few species • Does not prevent introgression of transgene to wild populations of reproductively compatible species	• Implementation of best practices in monosex induction for key species • Verification and protocols certification protocols

Continued

Table 8.1. *Continued*

Confinement approach	Status of development	Strengths	Weakness	Research or capacity needs
Triploid sterility	Technique developed for a range of species but needs to be further optimized in some The most feasible current biological confinement for several species (e.g. salmon, trout, grass carp)	• Technique developed in existing transgenic species (e.g. tilapia, carp, salmonids, mud loach)	• Risk of incomplete induction of triploidy • Fertility of some triploids (esp. males) • Costly to verify • Can result in reduced growth rate, survival and resistance to stress • Commercial scale induction not feasible in some species (e.g. tilapia)	• Research on how to optimize triploid yields and verify full sterility (especially males) • Efficient methods and systems of independent verification of triploidy in commercial settings • Biological reasons for failed cases of triploidy or sterility
Genetic Use Restriction Technologies (GURTs)	Initial stages of research	• If R&D successful, could provide robust methods of biological confinement	• Environmental risks associated with GURT transgenes themselves • Risk of transgene recombination affecting GURT function • Might be used for intellectual property protection, thus limiting access to benefits of fish	• Gain and maintain awareness of developments, risks and benefits

Box 8.1. A case study of physical confinement of finfish in a research facility at Auburn University, Alabama, USA. (From Rex Dunham, Auburn University, Auburn, Alabama, 2006, personal communication.)

A facility was established to hold fish resulting from research on transgenic common carp (*Cyprinus carpio*) and channel catfish (*Ictalurus punctatus*) in the early 1990s. At the time the facilities were planned, there were no formal policies or guidelines governing the design of confinement facilities for transgenic fish. The facility described below is the result of a 4-year consultation at institutional and federal levels. The then US Department of Agriculture's Office of Agriculture Biotechnology assumed regulatory authority, with oversight by a committee of relevant experts and stakeholders, and these actions resulted from an environmental risk assessment. The planning and development of this facility acted as a catalyst for formation of working groups to develop broader guidelines for environmental biosafety of transgenic fish and shellfish (i.e. ABRAC, 1995).

The main confinement features of the facility include:

- The site is located 35 ft above the 100-year flood plain.
- The earthen ponds have concrete walls extending 2 ft below the bottom edge of the pond to prevent burrowing and maintain pond integrity.
- The pond's site is surrounded by chain link fence with two strands of barbed wire on the top, and it is covered with bird netting. There is a solar-powered electric fence near the top of the perimeter fence, principally to ward off raccoons (see Fig. 8.2).
- The water level is maintained a few inches below the top of each pond's standpipe to greatly minimize any chance of water discharge except during harvest.
- The standpipe/discharge pipe is screened, and the discharge water passes through a double-screen filter box in a seep pond, and then through a gravel French drain into a slotted discharge pipe which exits into a drain ditch filled with aquatic predators (see example in Fig. 8.1). This ditch then empties into a 6.9 ha reservoir filled with aquatic predators. There are several other water bodies between the reservoir and the natural environment.
- A large drainage ditch cuts through the centre of the facility to further ensure no localized flooding.
- The ponds' dikes are designed so that, in the unlikely event of water build-up, they can only overflow to other ponds, then eventually into the seep pond and the French drain.
- Large amounts of rotenone are kept on hand so that, in the event of an emergency, the ponds can be poisoned.
- All personnel are trained in security and protocols of confinement, and the facility is kept locked.
- There are regular surprise inspections by Auburn University's Institutional Biosafety Committee.

are only feasible in land-based facilities, and thus these are considered separately from confinement measures appropriate for open aquaculture systems. Even advocates of transgenic fish in aquaculture maintain that 'it is essential that transgenic broodstock be maintained in secure, contained land-based facilities' (Fletcher *et al.*, 2002).

Confinement of transgenic fish in closed, land-based facilities

A closed, land-based facility is considered to be one that is isolated from its surrounding environment. It is not feasible to define a detailed series of universally applicable management measures when designing physical barriers for transgenic fish facilities; each system requires a case-by-case consideration of the species of concern, its biological properties, as well as the scale and geographic location of the aquaculture operation. However, some overarching general issues to be considered when designing effective physical barriers are discussed below.

Site selection

Ideally, the facility should be in a location where geographic confinement applies (see the next section). Criteria for site selection will differ between coastal marine or brackish-water sites and inland freshwater sites, as well as between sites within a particular area. Risk assessment strategies thus need to take into account the frequency and magnitude of regular and irregular events which might compromise confinement and establish site selection criteria based on consultation with stakeholders (Chapters 1 and 2, this volume). Clearly, one of the most significant risks related to site selection is flooding, and, for example, it might be prudent to site an inland facility above the 100-year flood level. If this is not known or is not feasible, then an alternative would be to manage surface water runoff and storm drains. Site selection in coastal areas needs to take into account potential rise in sea levels, and the risks and impacts of tsunamis.

Other factors to be considered when siting facilities are potential impacts of excessively high winds, snow loadings and seismic loadings. In such incidences, buildings housing the facility, the rearing units themselves and surrounding barrier infrastructure should be constructed in accordance with local construction standards (where these exist). The impacts of, for example, a seismic event can be mitigated by constructing a dike or other barrier around tanks to confine leaks and spills.

Potential escape pathways

It is necessary to review all possible escape pathways in the system when designing a facility for housing and culturing transgenic fish. Barriers should be installed to prevent escapes of all life stages of the transgenic fish. Special attention should be paid to minimizing the risk of release of gametes, particularly in situations where it might be possible for both the eggs and sperm to be released simultaneously. It is likely that the highest risk of escape will be during the early stages of the life cycle. The most likely escape pathways in land-based systems will be through influent and effluent waters, but they could also include waste slurries, cleaning wastes or equipment, and, in the case of larvae or gametes, even aerosols. Transport by predators, particularly by birds, dropping live fish some distance from the culture facility, is another potential way for fish to enter the receiving environment.

Physical and chemical barriers

Physical and chemical barriers involve the manipulation of physical or chemical factors to induce 100% mortality of all life stages of the transgenic fish species before they reach the receiving environment. Such barriers can include use of lethal temperatures (heating or freezing), lethal doses of radiation, biocidal or germicidal agents or lethal pH values. Effluent from a fish holding/rearing facility could either be treated as it flows from the facility or stored in isolation and treated prior to its release into the environment. It is necessary to ensure that any treatments used do not ultimately cause other environmental harm, and agents with this potential should be used with extreme caution. Although chemical treatment is effective in research facilities and could be feasible for commercial facilities primarily using recirculation, the volume of water involved in systems with any significant flow-through would render such treatments impractical or uneconomic.

Temperature or chlorine treatments are likely to be feasible chemical barriers. Upper or lower lethal treatments will vary between species. Doses of 15 mg/l of chlorine for 15 min or longer are sufficient to kill freshwater fish. Similarly, a 5.25% solution of hypochlorite (i.e. bleach) can be used to disinfect seawater (ABRAC, 1995). Mohanty *et al.* (1993) proposed a relatively affordable and practical solution of urea and bleaching powder (at 3 and 5 mg/l, respectively) as an effective piscicide. Chlorine treatments are relatively easy to ameliorate prior to releasing the effluent, by degassing over time or treatment with sodium thiosulphate. A common piscicide, rotenone, can also be used to treat effluent water, although this is expensive.

Mechanical barriers

Mechanical structures can be used to block the passage of one or more life stages of the fish. These could include various types of filters, such as gravel traps and drum or flat screens, perforated screens in drainage channels, floor screens and overflow or standpipe screens (Fig. 8.1). Such mechanical barriers are normally placed in a series, depending on the size and life cycle stage of the fish, and they should be positioned for optimal escape prevention. Redundant barriers are preferable in case one or more screens or filters fail, and barriers should be maintained regularly to avoid blockage or overflow. French drains, where effluent water seeps through several layers of gravel and fabric before being collected in perforated drainage pipes, can be designed to retain even the smallest fish size and are suitable for use in draining ponds (Fig. 8.1) (ABRAC,

(A)

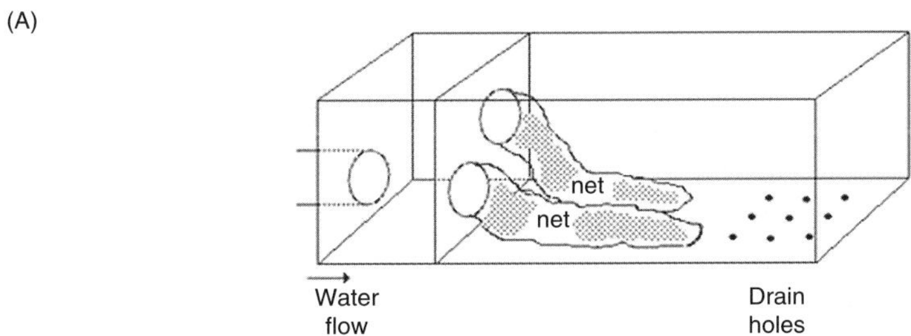

Fig. 8.1. Three examples of mechanical barriers that can be used to minimize the probability of escape of transgenic fish from aquaculture facilities via the drainage system. Full descriptions can be found in ABRAC (1995). (A) Sock filter trap with 0.3 mm screen for trapping eggs, embryos and larvae; (*Continued*)

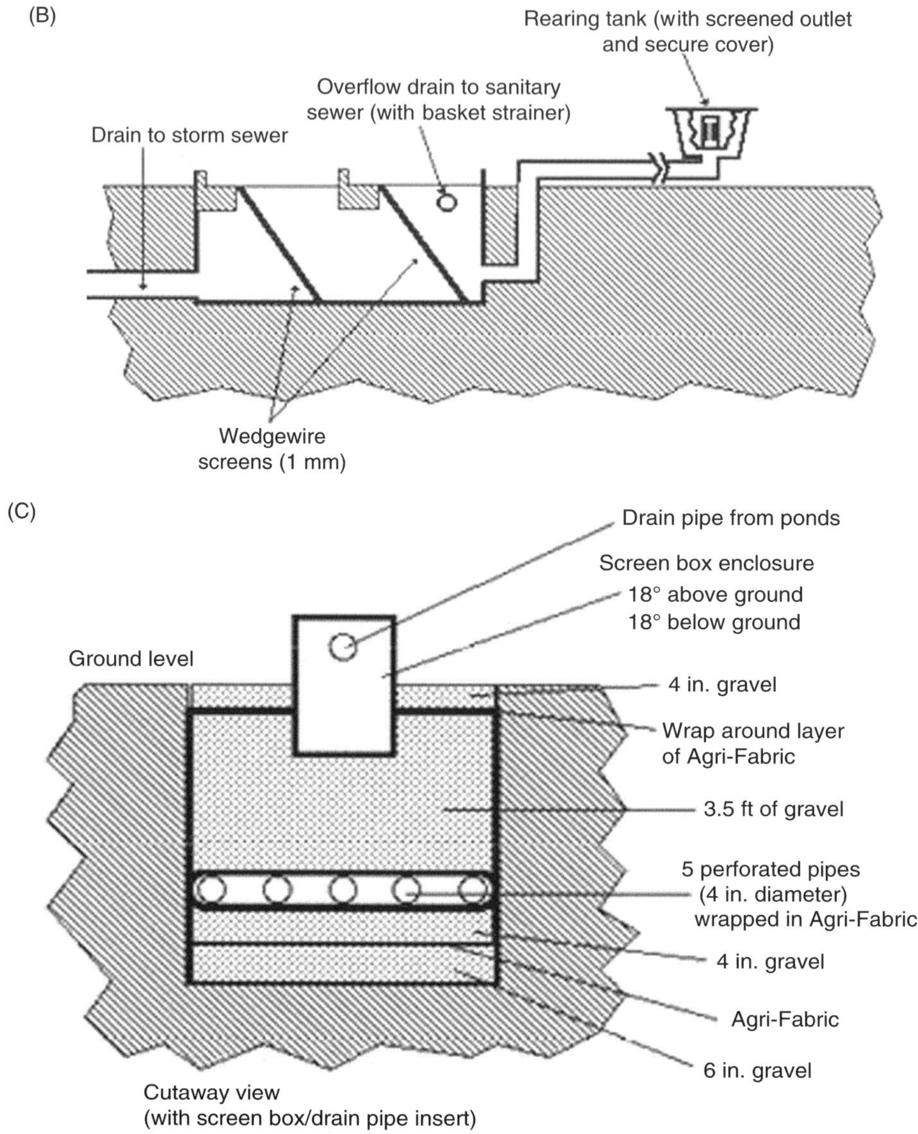

Fig. 8.1. *Continued.* (B) two stainless steel wedgewire screens used in a series act as final barriers for the common effluent from rearing units. Such screens can be used for larger fish (e.g. >2 cm total length); (C) schematic drawing of a French drain for screening effluent from outdoor fish ponds.

1995). Other mechanical barriers include pumps, which would kill most or all fish passing through them, nets and lids for fish tanks and buildings, as well as plastic or polythene tunnel house enclosures. Mechanical barriers should also be used to prevent predators from entering the facility and removing fish to the receiving environment. These types of barriers include bird nets, buried barriers against burrowing animals and electric fences (Fig. 8.2).

Fig. 8.2. Some of the physical security measures implemented at the Auburn University pond facility for transgenic fish research. Measures seen here include the perimeter fence with electrification, bird netting and concrete linings reaching the upper part of the pond dyke (From Rex Dunham, Auburn University, Auburn, Alabama, 2006, personal communication.)

Security measures for physical confinement

There are risks associated with deliberate removal of transgenic fish from production or breeding facilities, particularly given the potential value of transgenic broodfish. Risks of theft or poaching will vary according to the potential value of the fish, local socio-economic factors and the degree of knowledge and understanding of the value and risks of the fish by both the personnel at the facility and within the local community. Therefore, it is important to implement effective security measures in facilities holding or producing transgenic fish, and there should be agreed-upon operational plans and best management practices for confinement. An operational plan should include mechanisms by which confinement is ensured through normal operations, including full record keeping, particularly with regard to fish inventories and movements. All personnel should be trained in the implementation of the operational plan and be fully aware of the issues surrounding risk management of transgenic fish. Training programmes and operational plans should be developed in consultation with appropriate experts, be in accordance with regulations and environmental management plans and subjected to peer review. A more detailed summary of best management practices is provided in Box 8.2.

Confinement of transgenic fish in open systems

Physical confinement acts as the first line of defence for preventing escapes of transgenic fish, but it is not advisable as the sole confinement measure in open

Box 8.2. Best management practices for confinement of transgenic fish in aquaculture facilities.

Controlling and reducing the risks associated with the escape of transgenic fish, either from a research or a production facility, requires the development of best management practices and the application of stringent confinement measures. These measures, which should be regulated, certified and periodically inspected by the appropriate government agency, must include the following:

1. Restriction of access to authorized staff only. Ideally, background checks should be carried out on staff provided they comply with local privacy legislation. Any other persons entering confinement facilities must be on official business and be supervised while on site.
2. A system for accurate inventory of animals in the facility should be developed. This system should, where feasible, account for every transgenic individual entering or being generated within the facility. There should be accurate recording and justification for any change, loss or reduction in the exact number of transgenic fish within the facility. Fish too small to mark or count accurately should be tracked as groups until this is possible.
3. A documented security procedures manual must be written and followed. Copies of this document should be readily available for review by regular staff and used in the training of temporary or new personnel. This should promote awareness of the risks of escapes and ensure that such risks are monitored through regular meteorological checks and implementation of response measures for storms.
4. Personnel (including professional, technical, trainees, auxiliary and maintenance staff) must receive training on the potential risks associated with the work at the facility and the necessary precautions to prevent accidental release of confined transgenic organisms.
5. Personnel working in the confinement area must have general knowledge of the physical operation and design of the facility (e.g. confinement perimeter, staff traffic flow patterns, directional patterns of water flow, alarm signals for failure of life support, drainage traps and screening systems).
6. Entry and exit protocols for staff, visitors, fish, equipment, samples and waste must be written, posted and followed.
7. Emergency procedures should be designed, written, posted and followed for immediate response to failure of confinement measures. They must properly address fish escapes caused by natural disasters and other emergencies.

systems. Open systems include raceways linked to river systems, rice fields and aquaculture sites in irrigated or flood-prone areas, ponds, cages and pens. Given that open systems are contiguous with natural water systems, recovery of or limiting spread of escapees in the environment will be extremely difficult. It might be possible to isolate an inland lake or river to some extent (e.g. by closing a fish ladder), but it would be impossible in the case of escapes from sea cages. The majority of the world's aquaculture production is carried out in open systems; carp are farmed in ponds, rice fields and irrigation systems, tilapia are produced in inland ponds and cages, salmonids in coastal cages and shrimp in brackish-water ponds. The potential volume of transgenic fish production will be small if it is limited to closed systems, so it is important to determine how to minimize transgenic fish escapes from open systems. To do so, the previously mentioned physical confinement measures will need to be combined with highly efficient and reliable geographical or biological confinement mechanisms described below.

Raceways

In principle, raceway systems could incorporate several of the confinement measures described for closed land-based systems, but by definition they are sited near flowing water systems, either natural or artificial, significantly increasing the risk of escapes. Physical or chemical barriers will be difficult to deploy given the large water volumes, but mechanical barriers could be incorporated into the system. However, it will be difficult to screen for all stages of the life cycle; raceway systems will also need to rely upon geographical or biological confinement measures.

Ponds

Pond systems vary in their management and connectivity. Although the majority of current aquaculture ponds would fail to meet the siting criteria outlined for closed land-based facilities, it is possible to envision isolated pond systems where barriers to escape could be highly effective. Such systems would likely be inland, fed by groundwater or from raised inlets unreachable by fish, have no flood risk and be far removed and disconnected from natural water bodies. Physical, chemical and mechanical barriers that reduce the probability of escape could be developed for such pond systems. Ponds might also be geographically confined. Box 8.1 gives examples of measures that can be taken to minimize the possibility of escape of transgenic fish on a research scale, and these could be applied to commercial systems.

Cages and pens

Cages and pens are the most difficult culture facilities from which to prevent escapes. Indeed, there is considerable opposition to their use even for the culture of non-transgenic fish because escapes are common, generating concerns over impacts of escapees on wild stocks and on the environment (Silvert, 1992; Naylor *et al.*, 1998, 2005). Cage and pen facilities are prone to damage from storms, ice, boats, natural predators such as seals, crocodiles or predatory fish, or by poor maintenance and even poaching or sabotage, and the mesh can be breached and fish can escape. Careful site selection, double-layer netting, reinforced netting materials, predator avoidance devices, highly secure anchorages, security protocols, careful and regular maintenance programmes and covered cages or complete enclosures can help prevent such escapes. Many good management practices for land-based facilities, such as training of well-qualified staff, can also be applied to cage systems to minimize the chance of escapes.

Applying such methods to cages and pens may aid confinement, but they are unlikely to provide the required level of security to reduce the possibility of escape of transgenic fish to acceptably low levels. New water-based impoundment technologies are being developed to address concerns over environmental impacts of aquaculture (such as the semi-intensive floating tank system (SIFTS) under development and evaluation for inland saline aquaculture in Australia

(Partridge and Sarre, 2005)), and other systems are under development for marine aquaculture (GSA-CAAR, 2005). Bullen and Carlson (2003) reviewed non-physical fish barrier systems based on electrical fields, lighting or acoustics that might have some value in limiting escapes, but these require significant further research and demonstration of efficacy. Although these technologies are some distance from commercial application, they represent developments that could enable greater confinement of fish and reduce the possibility of transgenic fish escapes. These options also increase the cost of production, which may outweigh the economic benefit of culturing transgenic fish, and may not give the same level of secure confinement possible in land-based facilities.

Geographic Confinement

Siting aquaculture facilities in isolated geographic locations, where normal local or regional environmental parameters differ substantially from the range of conditions necessary for survival of the transgenic species, is one method to prevent escaped transgenic fish from surviving and reproducing in the wild. This approach can be in addition to physical confinement of cultured transgenic organisms in closed land-based facilities. Locating culture facilities away from other water resources will significantly reduce potential ecological effects of transgenic fish, but it could increase the cost of transport to processing facilities and markets. Additionally, local regulations restricting the use of pristine freshwater resources may prevent use of such isolated locations.

A promising alternative to geographic isolation is the culture of transgenic fish in areas where escaped animals will not be able to survive under local environmental conditions. Temperature and salinity are the main environmental limitations for aquatic organisms. Thus, it is possible to culture stenohaline freshwater species in areas surrounded by seawater, marine species in inland facilities using recirculated seawater (Hindar *et al.*, 1991) or tropical species in cold temperate regions.

Geographic confinement of, for example, tilapia using thermal restrictions will be more effective than salinity because the majority of tilapia species are euryhaline, or tolerant of a wide range of salinities (Chervinski, 1982). Initial plans to evaluate transgenic Nile tilapia (*O. niloticus*) in field trials in Thailand in 2001 included growing them in freshwater-fed tanks in a location surrounded by saline waters. Ultimately, these plans were abandoned because the surrounding waters had salinities in the range of 27–34 ppt, not high enough to guarantee failure of survival and reproduction of potential escapees (D. Penman, University of Stirling, UK, 2005, personal communication). Impacts could have included possible hybridization with feral *O. mossambicus*, which can tolerate such high salinities. Eventually trials of these transgenic tilapia were conducted using heated indoor systems in Hungary. The level of confinement was higher than the proposed saline barrier in Thailand because there was no risk of escapees surviving in surrounding coldwater environments during the winter (Rahman *et al.*, 2001). Thus, cultivation of transgenic tilapia, as well as tilapia species in regions where they are not native (i.e. the higher latitudes of northern Europe) using artificial or geothermal warm water, would lessen the risk of their survival outside the culture system. One important consideration is whether the temperature conditions remain restrictive year round.

Geographic confinement of cultured transgenic salmonids based on their thermosensitivity would be very difficult and costly in tropical or warm water areas due to the expense of maintaining low water temperatures in the culture facility. Geographic confinement of freshwater transgenic grass carp in facilities surrounded by sea or brackish water should succeed, given that Maceina and Shireman (1980) demonstrated suppressed growth and reproduction at salinities above 9 ppt. Kilambi and Zdaninak (1980) also showed that the species does not survive at salinities above 17 ppt. Temperature-based geographic confinement is not advisable because grass carp fry and fingerlings have been reported to tolerate water temperatures from 0°C to 40°C, while Chilton and Muoneke (1992) reported an upper lethal temperature range for fry of 33–41°C. Fingerlings in small ponds in Arkansas even survived for 5 months under heavy ice cover (Stevenson, 1965).

The Human Factor

Physical or geographical confinement of transgenic fish may be compromised by accidents caused by human error, negligence or even deliberate actions. These situations are more likely to occur during transport of live animals at some point in the production cycle; live sale of transgenic fish would be a particular concern. Failure to comply with established policies and procedures for secure handling, deficient record-keeping and inadequate maintenance of confinement facilities could result in release of transgenic fish to the receiving environment. High scientific or economic value of transgenic fish, particularly broodstock, could provoke unauthorized appropriation, vandalism or release of transgenic fish into the natural environment. These risks may be higher in resource-poor communities due to socio-economic conditions, and they emphasize the need for precise guidelines and regulations requiring certification, adequate training and compliance with best management practices (see Box 8.2) to achieve the highest level of security in facilities that maintain transgenic fish.

Biological Confinement

Biological confinement entails limiting or eliminating the reproductive potential of transgenic organisms. In fish, generation of single-sex progeny and induction of sterility through chromosome-set manipulation (deliberate change in haploid sets of chromosomes) are considered the most promising approaches to controlling reproductive potential (Devlin and Donaldson, 1992; Donaldson *et al.*, 1993). Research on various techniques for reducing or eliminating gonadal development, thereby restricting reproduction, includes surgical removal of gonads, hybridization, radiation or use of bioactive compounds. These techniques appear to have limited application because the procedures involved are highly labour-intensive, not fully reliable or the procedure or compounds used to induce sterility are not approved for use in fish destined for human consumption (Donaldson *et al*., 1993; NRC, 2004). However, there is potential for future development of one biochemical approach. This would involve preventing gonadal maturation in fish by inhibiting synthesis or release of gonadotropin using specific piscine gonadotropin-releasing hormone (GnRH) antagonists, preferably from the same species as the transgenic fish (Donaldson *et al*., 1993).

Single-sex populations

The generation of single-sex populations may provide an acceptably low level of confinement in instances where transgenic fish enter natural environments beyond their natural range and have no possibility of encountering conspecifics of the opposite sex. However, if conspecific stocks or closely related species are present in the receiving environment, successful reproduction between cultured transgenic fish and wild individuals may occur, thus enabling transgene flow into wild populations. This section briefly describes currently available methods for producing single-sex progeny, which are practical for large-scale applications, as would be needed to confine transgenic fish in commercial aquaculture.

In order for sex control to be an effective means of confinement, sex ratios must be absolutely monosex, especially where all-female populations are targeted. A single male in a seemingly all-female population could reproduce successfully and the species would no longer be biologically confined. This has been illustrated in tilapia, where virtually all the methods for producing single-sex populations (manual sexing, direct hormonal sex reversal, hybridization and breeding programmes for genetically monosex progeny) have failed to yield consistent 100% monosex populations; sex ratios typically range from 90% to 100% male. This failure can undermine biological confinement because presence of a few fertile females among the all-male escapees could lead to successful reproduction even when there are no wild relatives in the receiving environment.

Viable procedures for producing single-sex stocks are through direct sex reversal or genetic breeding programmes. The latter will depend on manipulating the species-specific mechanism of sex determination; research on several systems for fish will be discussed below. The major factor is whether sex is controlled by a monogenic sex-determining system with single dominant male or female chromosomes (e.g. XX/XY – female homogamety – or WZ/ZZ – male homogamety) or by a polygenic system with multiple sex-determining factors.

All-female stocks

The two main methods for generating monosex stocks in fish species, where the female is the homogametic sex, are by either direct or indirect feminization. Gynogenesis, a direct feminization method, consists of fertilizing normal ova with irradiated sperm, then restoring diploidy either through induced retention of the polar body or disruption of first mitosis in the fertilized egg. This restoration is done through the application of temperature or pressure shocks. Therefore, the activation of the ova takes place without the contribution of male-determining chromosomes from the paternal genome. Gynogenesis has two important drawbacks. First, induction rates can be very low, which would significantly increase broodstock and treatment requirements. Second, gynogenesis results in increased or complete homozygosity of the resulting gynogens. Increased homozygosity can reduce overall fitness of the fish, for instance, due to expression of recessive deleterious alleles.

Indirect feminization (Fig. 8.3, top) is achieved by fertilizing normal ova with only X-bearing sperm, derived from sex-reversed homogametic males (Donaldson

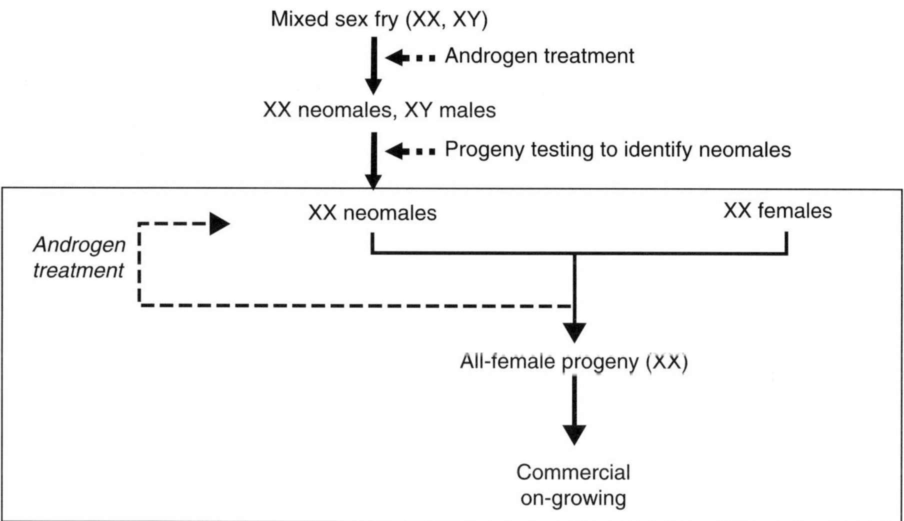

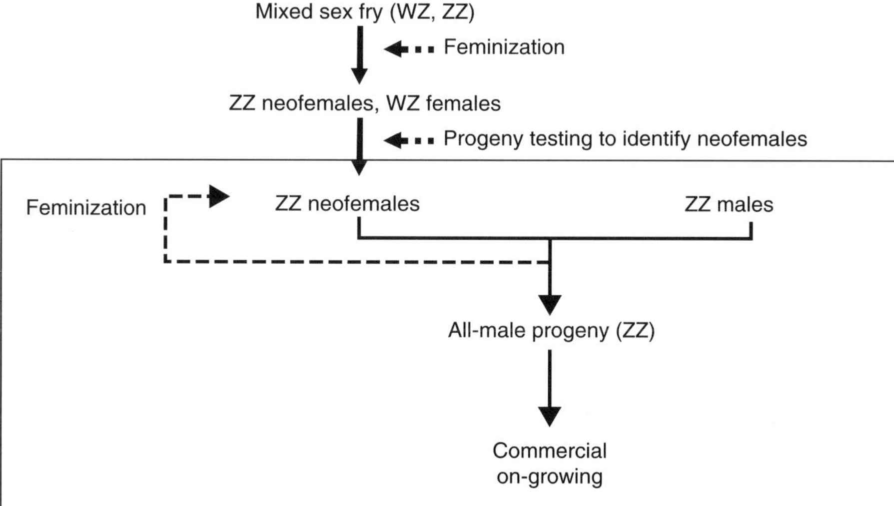

Fig. 8.3. Diagrams illustrating the breeding programmes for production of monosex females (top) in species with female homogamety (e.g. salmonids and silver barb). Programmes for monosex males are illustrated (bottom) for species with male homogamety (e.g. blue tilapia and *Macrobrachium*). The figures illustrate the initial treatment steps and the steps that need to be repeated in commercial production (denoted by the boxes around the process).

and Hunter, 1982; Solar *et al.*, 1987). These homogametic males, termed 'neomales' (Pongthana *et al.*, 1999), are genetically female but functionally male. They are generated by direct sex reversal using androgenic treatments of fry or embryos, administered either orally through food or through water immersion treatments. Identifying and segregating fish producing only X-bearing sperm may be done by progeny testing after treatment with androgens. Where available, male-specific genetic markers can also be used (Devlin *et al.*, 1994). Additional identification and segregation can be done by masculinization of known genetically female embryos or fry, which could be gynogens or progeny of previously confirmed neomales. Monosex female stocks have been successfully generated in several commercially important species, including salmonids (reviewed by Donaldson and Devlin, 1996), Nile tilapia (Abucay *et al.*, 1999) and Thai silver barb (*Barbodes gonionotus*) (Pongthana *et al.*, 1999).

All-male stocks

Direct masculinization can be achieved by overriding the effect of endogenous systems through the application of exogenous androgen (in solution or in the diet) during the period of sexual differentiation. However, indirect masculinization through a breeding programme to manipulate the genetic basis of sex may be more effective. Indirect masculinization can be done in species with female homogamety (XX) by generating novel YY 'super-males' and then crossing them with normal females. Similarly, species with male homogamety (ZZ) can produce monosex male progeny from novel sex-reversed ZZ 'neofemales' (Fig. 8.3, bottom).

Breeding programmes combining hormonal feminization and progeny testing can be used to generate the YY 'super-males' needed for indirect masculinization in female homogametic species. An additional method to produce 'super-males' is to use an androgenesis procedure. Androgenesis (similar to gynogenesis) involves fertilizing UV or gamma-irradiated ova with normal sperm. Surviving progeny will be either XX females or novel YY super-males. However, androgenesis is difficult to achieve in many species, and yields only a small number of fully homozygous progeny. Thus the breeding programme approach is more feasible for producing YY males. The breeding programme follows similar principles to those described for monosex female production but employs feminization to produce XY and YY neofemales and involves more steps. Monosex male production has been most widely researched and applied in female homogametic tilapia (*O. niloticus* and *O. mossambicus*) because all-male populations offer greater advantages for aquaculture (Mair *et al.*, 1991a, 1997). The major issues surrounding all-male production are reviewed by Beardmore *et al.* (2001).

All-male populations have also been produced in male homogametic species (Fig. 8.3, bottom), namely the blue tilapia (*O. aureus*) and the giant freshwater prawn (*Macrobrachium rosenbergii*). In both species, the male has better aquaculture production characteristics than the female (Sagi *et al.*, 1986; Mair *et al.*, 1991b). Large-scale production of monosex male progeny is feasible in blue tilapia using ZZ neofemales produced from a breeding programme combining feminization and progeny testing (Desprez *et al.*, 1995).

Androgenic hormones have not been identified in decapod crustaceans (e.g. freshwater prawns), and androgen treatments have not been effective in inducing their sex reversal. However, sex reversal may be induced by micro-surgical removal of the androgenic gland. This approach can induce ZZ males to differentiate as functional females, and these can be mated with normal males to generate all-male progenies (Sagi and Aflalo, 2005). Similar approaches may prove useful for restricting reproductive interactions in other aquatic organisms with similar systems of sex determination.

Triploidy and sterility

Fish embryos are amenable to various polyploid chromosome set manipulations, including the induction of triploidy and tetraploidy (Fig. 8.4). The induction of triploidy has generally involved the interruption of meiosis II by preventing the expulsion of the second polar body that contains a haploid set of maternal chromosomes. The resulting zygote then bears three sets of chromosomes; two maternal and one paternal (Benfey, 1999). Various methodologies have been used to induce triploidy in fish species. The second polar body can be retained by applying physical (heat, cold or extreme hydrostatic pressure) or chemical shocks to zygotes. Thermal shocks have most commonly been used, and their major advantage over hydrostatic pressure or chemical treatments is that no special equipment or toxic chemical compounds are needed, and large volumes of eggs can be treated at once (Felip *et al.*, 2001). However, effectiveness of induction is species-specific, particularly with regard to thermal shocks, and its parameters must be optimized for each species.

Treatments to induce triploidy should take into account the following three variables: (i) time (after fertilization) for initiating the shock; (ii) intensity of the shock; and (iii) shock duration. In many important aquaculture species like carps and salmonids, ovulated eggs are easy to obtain, rendering commercial-scale chromosome manipulations feasible. However, it is difficult to obtain ovulated eggs for *in vitro* fertilization in some species such as tilapia. In such species, direct induction of triploidy may be possible on a small scale (Mair, 1993), but commercial-scale production is not feasible.

Limitations of triploidy for sex control

Unfortunately, none of these currently employed methods can guarantee the induction of 100% triploidy. Although direct induction by application of physical shock has been used in many species, relatively few studies have systematically optimized shock parameters to consistently obtain 100% triploidy. Survival may be compromised with the use of more severe shocks required for maximum triploidy induction (Guoxiong *et al.*, 1989; Maclean and Laight, 2000). Even with optimized protocols, the efficacy of triploidy induction can be variable between species, strains and egg batches (for review, see Benfey, 1999; Felip *et al.*, 2001). Recent studies examining the efficacy of optimal pressure shock-induced triploidy for sterilization of growth hormone (GH) transgenic coho salmon (*Oncorhynchus kisutch*) have shown that rare diploid exceptions (from homozygous transgenic males crossed to wild-type females) can be either transgenic or non-transgenic. The former exceptions appear to

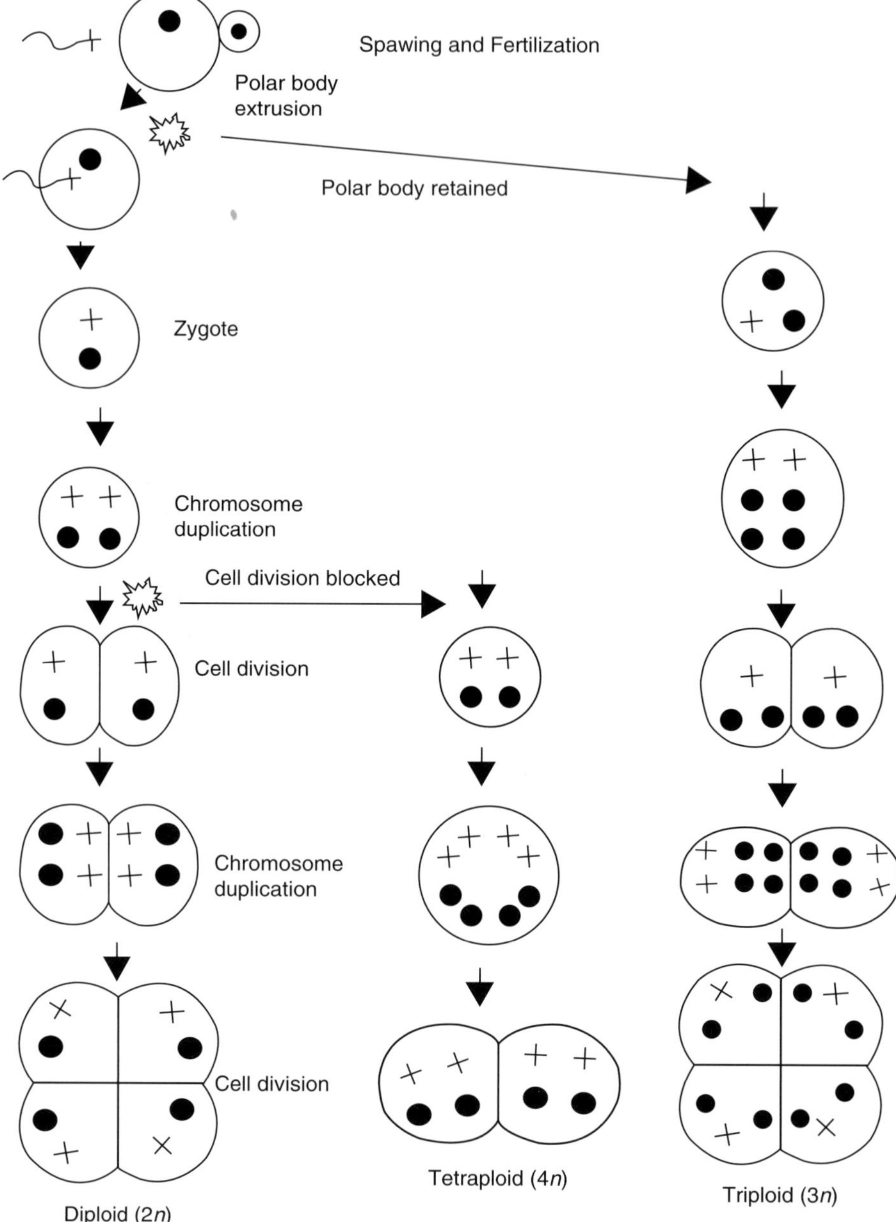

Fig. 8.4. Schematic representation of the options for induction of polyploidy in fish zygotes. The left schema indicates the initiation of normal development of diploid zygotes. The right schema shows how triploidy is induced by retention of the second polar body during meiosis II. The centre schema shows the induction of tetraploidy by disruption of metaphase in first mitosis. Symbols: denotes the point at which the physical shock is applied; • denotes one female-derived haploid chromosome set; and ✗ denotes one male-derived chromosome set. (Adapted from Donaldson and Benfey, 1987.)

arise from a failure to retain the second polar body, whereas the latter are from gynogens that lack paternal contributions of genetic material, including the transgene (Robert Devlin, Department of Fisheries and Oceans, Canada, 2006, personal communication). Thus, different and not fully understood genetic or cellular mechanisms can contribute to variation in confinement efficacy by triploidy.

Incomplete success in producing triploids may require individual verification of triploidy in order to identify and remove any residual diploids among the treated fish prior to their transfer to less secure grow-out facilities. Of the many methods used to identify triploid fish, flow cytometry is the most effective and reliable, with particle analysis also being useful (Wattendorf, 1986; Harrell *et al.*, 1998; Nell, 2002). These methods require very small quantities of blood or disaggregated tissue, permitting mass screening in a non-invasive manner. Participants in a risk assessment process (Chapter 1, this volume) could also consider the appropriateness of screening subsamples of embryos or juveniles and culling the entire lot if the error rate is above a pre-determined acceptance level. The need for absolute verification of 100% triploidy would depend on the probability of escape and the severity of its consequences, and thus depends on the properties of the transgenic fish in question, the status of potentially affected wild populations and other confinement measures being implemented (NRC, 2004).

Equipment for verifying that treated fish are indeed triploid would be too expensive for all but the best-funded hatcheries, although this work could be outsourced where the scale of production is significant. Kapuscinski (2002) estimated that the cost of flow cytometric screening of individual transgenic salmon would add US$0.02 to 0.04 per kg to the market cost of farmed Atlantic salmon. Recently, a new method of ploidy determination based on erythrocyte measurements, using computer-assisted analysis, has been reported as an inexpensive alternative to flow cytometry for identifying triploid rainbow trout (Espinosa *et al.*, 2005). However, this method requires verification over a broad range of species before wider application can be considered.

Triploid fish may have differing levels of transgene expression depending upon the transgene dosages per haploid genome, making their performance and physiology variable and unpredictable (Box 8.3). Depending on their maternal and paternal sources, transgenic triploids can possess the transgene on one, two or all three homologous chromosomes, and may display reduced levels of transgene expression when compared to their original diploid heterozygous transgenic parents. Lower-level increases of growth rate compared to diploids have already been observed in all GH-transgenic triploids of mud loach (*Misgurnus mizolepis*), Nile tilapia and coho salmon, which all harboured the GH-transgene on only one of the three homologous chromosomes (Razak *et al.*, 1999; Nam *et al.*, 2001a; Devlin *et al.*, 2004). Furthermore, triploidization itself may affect the physiology of fish, and there have been many reports claiming that the physiology of triploid fish is altered compared to normal diploids (Benfey, 1999). Such physiological factors can affect growth, stress response, hematology, immune function and susceptibility to diseases (Jhingan *et al.*, 2003). For this reason, careful examination of the performance of triploid transgenics should be made prior to practical application of this method of biological confinement.

Box 8.3. A case study of triploidy in mud loach (*Misgurnus* spp.) as a means of biological confinement for growth-enhanced transgenic fish.

The mud loach (*Misgurnus mizolespis*) is a small freshwater species, and it is an important food fish in Korea. This species has many advantages as a model system for studying transgenic fish and, in particular, is amenable to induced reproduction and techniques of chromosome set manipulation. Triploidy was induced in stable fast-growing transgenic mud loach lines incorporating the mud loach growth hormone (GH) gene, driven by the mud loach β-actin gene promoter, to help investigate the performance of transgenic triploid fish (Nam *et al.*, 2001a). Sperm obtained from heterozygous (hemizygous) transgenic diploid males was used to fertilize eggs from normal non-transgenic females. Triploidy was induced using protocols described by Kim *et al.* (1994). Under optimized induction conditions, over 90% of treated fish can become triploid.

Approximately 50% of triploid fish proved to be transgenic. This indicates successful germ line transmission from the males without adverse effects on the viability of transgenic fry, and it showed the absence of spontaneous gynogenesis. These triploids contained the transgene on only one of three homologous chromosomes and exhibited lower levels of growth stimulation than the original diploid transgenic fish, but they still grew significantly faster than non-transgenics. Four possible genotypes (diploid and triploid transgenics and non-transgenics) were reared communally for 6 months to investigate the relative effects of the transgene and triploidy on growth. Both diploid and triploid transgenics exhibited significantly faster growth than their non-transgenic counterparts. Diploid transgenic fish consistently exhibited the fastest growth throughout the trial, followed by triploid transgenics, triploid non-transgenics and finally diploid non-transgenics (Fig. 8.5). The lower transgene dosage in triploids (1/3) compared to diploids (1/2) appeared to explain the reduced growth rate in triploid transgenics. The transgenic triploids displayed food conversion and long-term viability comparable with that of the transgenic diploids.

Gonad development of transgenic triploid mud loaches was significantly retarded and depressed in both sexes, while diploids became sexually mature at 9 months of age (Fig. 8.5). There was no notable difference in maturation between non-transgenic and transgenic mud loaches. In triploids, only oogonia were present in ovaries of females, and then in very limited numbers. Triploid transgenic males showed poorly developed testes, and few sperm cells were observed. The potential fertilizing ability of these sperm has not been fully investigated and remains a potential concern with regard to biological confinement of transgenic males (Nam *et al.*, 2001a).

Additional research combined triploidization with interspecific hybridization. Triploid interspecific hybrids (allotriploids) were produced using females from the related species *M. anguillicaudatus* and autotransgenic male mud loach (Nam *et al.*, 2004). Triploid transgenic hybrids showed much more stringent sterility than autotriploids in both sexes. This was most notable in males, where their testes were underdeveloped and showed no signs of maturation at up to 2 years of age. Attempts to fertilize eggs using a testes preparation yielded no developing embryos. The production of allotriploids appears to have eliminated the possibility of germ line transmission of the transgene, but further research is needed to determine reproductive behaviour of these sterile fish.

The possibility of fertile triploids is one of the major shortfalls of triploidy as a biological confinement method for transgenic fish (Devlin and Donaldson, 1992; Kapuscinski, 2005). Although triploid fish appear to be effectively sterile, with significantly depressed or retarded gonad development, some studies report that triploid fish, especially males, might produce limited numbers of potentially functional gametes (Benfey, 1999; Dunham, 2004). Variable functional sterility of triploids suggests that triploidization alone cannot guarantee biological confinement of transgenic fish, especially in males. Furthermore, triploid oysters have shown that they can spontaneously revert to diploids (Allen *et al.*, 1996). Thus the use of triploids in finfish and shellfish cannot be a complete solution for confinement of genetically modified organisms; triploidization can only reduce the reproductive risk of transgenic organisms by varying degrees. The partial fertility of triploids may be resolved for some species by combining triploidization with monosex approaches, particularly where all females are produced (Donaldson and Devlin, 1996), or with cross-species hybridization. Examples include the production of all-female triploid salmonids (Galbreath and Thorgaard, 1995; Donaldson and Devlin, 1996) or hybridization of triploid loach (Nam *et al.*, 2004) (see Box 8.3).

Even fully effective triploid sterilization will not completely eliminate the risk of reproductive interactions between triploid transgenic fish and wild conspecifics, leading to possible disruption of the spawning of wild species. Such reproductive interference (or competitive interactions) may have negative impacts on recruitment in wild populations. This is a greater concern for triploid males than females because, in many fish species, males enter into courtship with multiple females. Thus, the production of all-female triploids combines the benefit of almost-guaranteed sterility of any escapees with the reduced risk of disruption of spawning in natural populations that might arise with triploid males.

Tetraploidy and the generation of triploid individuals

The use of tetraploid breeding lines of fish to generate all-triploid progeny is proposed as a way to overcome some limitations of direct induction of triploidy discussed above. The idea is to mate fertile tetraploids with diploids to produce 'interploid' triploids. Tetraploid genotypes can be induced in fish by blocking the first cleavage (first mitotic division) of fertilized eggs through physical or chemical shock treatments similar to those used to induce retention of the polar body in direct triploid induction (Fig. 8.5). However, the generation of viable tetraploids has been achieved in only a few species, and few true-breeding tetraploid lines are available outside of rainbow trout (Thorgaard, 1992) and mud loach (Nam *et al.*, 2001b).

Many issues, especially those involving the reproductive performance of tetraploids, need to be resolved before practical use of tetraploid technology. First, the cross between a tetraploid male and a diploid female may not be possible in some species (e.g. rainbow trout) because of incompatibility between the diameter of a diploid sperm head and the micropyle of a haploid egg (Chourrout *et al.*, 1986). Second, in rainbow trout and mud loach, it is not clear whether female tetraploid fish can successfully attain normal maturation and spawning.

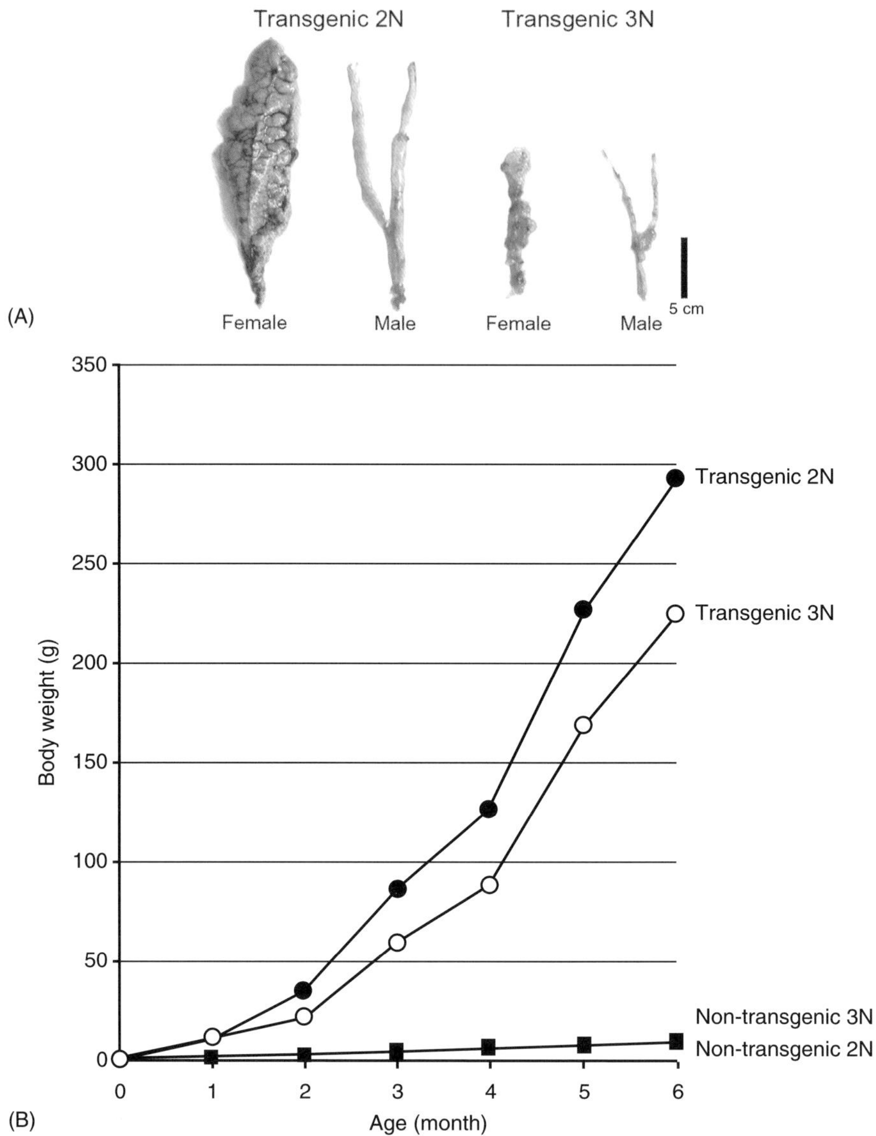

Fig. 8.5. Triploidy and transgenic mud loach (*Misgurnus mizolespis*). (A) Photographs showing the reduced gonad development in mature age triploid transgenics (right) compared to diploids (left). Vertical bar represents 5cm. (Photograph from Nam *et al.*, 2001a, reprinted with permission from World Aquaculture Society.) (B) Growth curves showing relative growth performance of diploid (2*n*) and triploid (3*n*) transgenic and non-transgenic loach, illustrating the slight negative impact of triploidy on the enhanced growth performance of transgenic fish. Triploid and diploid non-transgenics had very similar weights.

Third, not all somatic tetraploid fish can produce diploid gametes. As shown in mud loach, somatic tetraploid males (identified by flow cytometry of blood cells) may display variable reproductive performance, including depressed gonads and haploid, aneuploid and diploid sperm (Nam and Kim, 2004). This suggests that characterization of somatic cells with clearly tetraploid DNA content cannot always be used as an index for the production of diploid sperm (Nam and Kim, 2004). Poor survival and overall performance of tetraploid fish has discouraged mass production of interploid triploids. Tetraploid oysters, however, exhibit better performance, yielding high numbers of interploid triploids, and they have already been used by some commercial farms as a routine approach for producing triploids (Eudeline and Allen, 2000; Que *et al.*, 2003).

Research needs for triploid-based sterilization

Greater understanding of how to achieve triploid-based sterilization in commercially important fish species is critically needed. The properties of triploids and the methods used to induce them are well established for only a few aquaculture species, namely some salmonids, common carp and mud loach. If triploidy is to be adopted as the major method of biological confinement of transgenic fish, molluscs and crustaceans, then it is important to fully understand the properties of triploids (particularly their reproductive capacities) and to fully optimize methods for generating them.

Genetic use restriction technologies

Genetic use restriction technologies (GURTs) are a group of technologies defined as biotechnology-based switching mechanisms, and they are designed to restrict the use or inheritance of genetic material. Although originally developed by plant breeders to protect against unauthorized use of transgenic crop lines (Visser *et al.*, 2001a,b), they could also be applied for biological confinement of transgenic organisms (NRC, 2004). Research on agricultural applications of GURTs has been conducted for more than a decade, but their application to transgenic fish is in the very early stages of research and development. The discussion below addresses a few methods that are furthest developed.

'Sterile feral' approach

Rapid growth in knowledge about the functions of fish genes and inducible promoters provides the possibility to turn genes and their expression on or off at key stages of fish development. Such genetic switching may have applications for inducible disruption of reproduction or viability of fish, and such mechanisms have the potential to biologically confine transgenic fish once they enter the environment. For example, an Australian research group is developing a genetic switch to produce broodstock that are fertile in captivity but functionally sterile outside of hatchery conditions. This is done using a 'sterile feral' plasmid construct consisting of a temporally defined species-specific promoter coupled with a repressible gene element that drives the expression of a 'blocker gene'. This construct causes early mortality or reproductive failure in the fish by preventing the proper expression of essential endogenous genes (Grewe *et al.*, in press). Expression of this blocker gene, usually encoding an anti-sense RNA or double-stranded RNA-knockout sequence, should disrupt the function of an essential gene, inducing lethal effects.

A 'sterile feral' approach could be used for biological confinement of transgenic fish as follows. Transgenic fish treated with the repressor molecule during a critical period (usually early in the larval stage) would express the essential gene and survive. They would be transferred to a less secure facility to grow out. Any offspring produced from matings between two transgenic fish or between a transgenic fish and a wild conspecific would not be treated with the repressor molecule at the critical period, so the blocker gene would be expressed; assuming 100% efficiency of the blocker gene, this would cause mortality of all the progeny. The 'sterile feral' approach is still under development and not yet ready for application in aquaculture.

Disruption of reproduction through reversibly sterile fish

The reproductive ability of transgenic broodstock can be disrupted by 'knocking down' messages from genes responsible for successful reproduction. This approach involves the generation of a transgenic line expressing molecules that antagonize genetic functions necessary for fertility; the use of synthetic DNA sequences encoding an anti-sense RNA against a particular message has received the most attention to date. Transgenic fish expressing a blocker molecule should be sterile because the molecule prevents the reproduction-related message from being translated into protein. The fertility of the transgenic fish can be recovered by applying an exogenous compound (usually the final product of the target gene) through various routes, including feed administration or injection.

Several fish lines have been engineered with an anti-sense sequence that blocks expression of an essential reproductive hormone, GnRH. These lines include transgenic tilapia, carp and rainbow trout (Uzbekova *et al.*, 2000; Maclean *et al.*, 2002; Fu *et al.*, 2005). Completely reversible sterilization has yet to be achieved, but potential sterility, including depressed gonad development, has been demonstrated. For anti-GnRH transgenes to be useful for biological confinement of transgenic fish, expression of the target gene of interest needs to be completely silenced in 100% of the transgenic individuals of both sexes. The transgene construct also needs to be stably inherited without loss of its gene-silencing function. Furthermore, many species (such as salmonids) have at least two isoforms of GnRH DNA sequences (i.e. having slightly different DNA makeup), with some alternatively spliced mRNA forms. Careful optimization is needed to select the vital gene for maximizing the effectiveness of the anti-GnRH application (White and Fernald, 1998).

Concerns about GURTs

Although there may be potential opportunities to apply GURTs to biological confinement of transgenic fish, caution is advised because of the limitations and risks associated with GURTs themselves. Some are discussed below.

- First genetic switching may not guarantee 100% blockage of the expression of the target gene. There are also concerns about the stability of the GURT system over multiple generations. Over time, the expression of the blocker molecule or its promoter could be turned off by unwanted recombination, mutation or methylation, and the efficacy of the GURT's function could be diminished or become dissociated from the transgene(s) needing biological confinement. Although this may occur in only a very small fraction of fish, transgene rearrangement and gene silencing have already been reported in several transgenic fish lines (Iyengar *et al.*, 1996; Nam *et al.*, 1999).

- Another concern is that repressor molecules needed for normal reproductive development of engineered fish may also occur in the receiving environment. Pollution of many inland ecosystems or coastal areas from industrial and anthropogenic activities may increase the presence of repressor-like bioactive compounds (e.g. hormones, antibiotics, endocrine disruptors and other pharmaceuticals) and restore normal reproductive ability in the escapees.
- A final concern is that the genetic alteration of reproductive traits, creating the GURT itself, may pose potential risks to aquatic ecosystems, particularly because blocker-based approaches may contribute to reproductive interference with wild individuals. GURT-transgenic individuals might be able to participate in normal spawning events, where the blocker gene would function as a dominant allele, resulting in unviable offspring from matings between transgenics and wild individuals.
- GURTs are likely to be subject to many of the same concerns expressed by some stakeholders about the original transgene. Although the use of GURTs for biological confinement may assuage some environmental risk concerns, such use may actually increase other concerns regarding intellectual property rights, farmers' rights, human health impacts and ethics.

Integrated Confinement Systems

Confinement measures to reduce environmental risks posed by transgenic fish will be most effective if applied within an integrated confinement system. Policies would need to proscribe elements of integrated confinement systems, incorporating strict guidelines on physical confinement and providing for consideration of geographical and biological confinement options. Key elements of integrated confinement systems include (NRC, 2004; Kapuscinski, 2005): a commitment to confinement by managers and owners of transgenic fish research or production facilities for transgenic fish through the use of a written confinement plan, adoption of best management practices (as outlined in Box 8.2) and regular internal reviews and independent external audits of confinement, all conducted with transparency and appropriate public participation. Confinement options, their status of development and their physical and biological properties need to be well understood before being considered for use in a regulatory system governing the research and culture of transgenic fish. A significant component of such regulations should involve requiring that confinement measures be applicable to all stages of transgenic fish development: research, field trials and commercial aquaculture production.

Chapter Summary

There are three main methods for confining transgenic fish used in aquaculture: physical, geographic and biological (summarized in Table 8.1). Physical confinement is the most widely adopted measure, and it uses a series of physical, mechanical and chemical barriers to prevent fish from escaping from aquaculture facilities. It is difficult to physically contain fish in open systems, particularly in cages in open waters, and technologies do not yet exist for reducing the risk of escapes to acceptably low levels. Geographic confinement measures include culturing fish in areas removed from natural, contiguous bodies of water or in areas where the temperature or salinity of the surrounding waters prevents survival or reproduction. Biological confinement includes the generation of single-sex populations and sterile individuals. At present, the only technology that is adequately advanced and proven for the induction of sterility is triploidy. Further research is needed to improve the efficacy and practicality of biological confinement measures, particularly for species already being used to develop transgenic lines for aquaculture.

Another major area of research is the development of GURTs, which show potential for effective biological confinement of transgenic fish. These technologies are in the early stages of research and are thus briefly reviewed in this chapter. Methods furthest along in development for transgenic fish involve inserting another transgene designed to prevent reproduction or cause mortality at a certain life stage. These methods are subject to the same concerns as other biological methods regarding whether they are 100% effective, as well as additional societal concerns such as those over intellectual property rights, human health effects and ethics.

The use of redundant confinement measures and best management practices are important because no confinement method is 100% effective. Redundancy, for example, could be achieved by raising sterile fish in aquaculture systems with several mechanical barriers such as screens in water inlets and outlets. Best management practices for operations of an aquaculture facility are also important for minimizing transgenic fish escapes. Such plans detail who is allowed access to transgenic fish, training for staff members, how fish should be transported or destroyed and procedures to follow during emergencies. This chapter provides detailed discussion of critical components to include in a best management practice plan (see Box 8.2).

References

ABRAC (Agricultural Biotechnology Research Advisory Committee) (1995) *Performance Standards for Safely Conducting Research with Genetically Modified Fish and Shellfish. Parts I & II.* United States Department of Agriculture, Office of Agricultural Biotechnology. Document Nos. 95-04 and 95-05. Available at: http://www.isb.vt.edu/perfstands/

Abucay, J.S., Mair, G.C., Skibinski, D.O.F. and Beardmore, J.A.B. (1999) Environmental sex determination: the effect of temperature and salinity on sex ratio in *Oreochromis niloticus* L. *Aquaculture* 173, 219–234.

Allen, S.K. Jr., Guo, X., Burreson, G. and Mann, R. (1996) Heteroploid mosaics and reversion among triploid oysters, *Crassostrea gigas*: fact or artifact? *Journal of Shellfish Research* 18, 293.

Beardmore, J.A., Mair, G.C. and Lewis, R.I. (2001) Monosex male production in finfish as exemplified by tilapia: applications, problems and prospects. *Aquaculture* 197, 283–301.

Benfey, T.J. (1999) The physiology and behavior of triploid fishes. *Reviews in Fisheries Science* 7, 39–67.

Bullen, C.R. and Carlson, T.J. (2003) Non-physical barrier systems: their development and potential applications to marine ranching. *Reviews in Fish Biology and Fisheries* 12, 201–212.

Chervinski, J. (1982) Environmental physiology of tilapias. In: Pullin, R.S.V. and Lowe-McConnell, R.H. (eds) *The Biology and Culture of Tilapias (ICLARM Conference Proceedings 7)*. International Center for Living Aquatic Resources Management, Manila, The Philippines, pp. 119–128.

Chilton II, E.W. and Muoneke, M.I. (1992) Biology and management of grass carp (*Ctenopharyngodon idella*, Cyprinidae) for vegetation control: a North American perspective. *Reviews in Fish Biology and Fisheries* 2, 283–320.

Chourrout, D., Chevassus, B., Krieg, F., Happe, A., Burger, G. and Renard, P. (1986) Production of second generation triploid and tetraploid rainbow trout by mating tetraploid males and diploid females – potential of tetraploid fish. *Theoretical and Applied Genetics* 72, 193–206.

Desprez, D., Melard, C. and Philippart, J.C. (1995) Production of a high percentage of male offspring with 17α-ethynylestradiol sex-reversed *Oreochromis aureus*. 2. Comparative reproductive biology of females and F_2 pseudofemales and large-scale production of male progeny. *Aquaculture* 130, 35–41.

Devlin, R.H. and Donaldson, E.M. (1992) Confinement of genetically altered fish with emphasis on salmonids. In: Hew, C. and Fletcher, G. (eds) *Transgenic Fish*. World Scientific Publishing, Singapore, pp. 229–265.

Devlin, R.H., McNeil, B.K., Solar, I.I. and Donaldson, E.M. (1994) A rapid PCR-based test for Y-chromosomal DNA allows simple production of all-female strains of chinook salmon. *Aquaculture* 128, 211–220.

Devlin, R.H., Biagi, C.A. and Yesaki, T.Y. (2004) Growth, viability and genetic characteristics of GH transgenic coho salmon strains. *Aquaculture* 236, 607–632.

Donaldson, E.M. and Benfey, T.J. (1987) Current status of induced sex manipulation. In: Idler, D.R., Crim, L.W. and Walsh, J.M. (eds) *Proceedings of the Third International Symposium on the Reproductive Physiology of Fish*. Memorial University Press, St. John's, Newfoundland, Canada, pp. 108–119.

Donaldson, E.M. and Devlin, R.H. (1996) Uses of biotechnology to enhance production. In: Pennell, W. and Barton, B.A. (eds) *Principles of Salmonid Culture*. Developments in Aquaculture and Fisheries Science, No. 29. Elsevier, Amsterdam, The Netherlands, pp. 969–1020.

Donaldson, E.M. and Hunter, G.A. (1982) Sex control in fish with particular reference to salmonids. *Canadian Journal of Fisheries and Aquatic Sciences* 39, 99–110.

Donaldson, E.M., Devlin, R.H., Solar, I.I. and Piferrer, F. (1993) The reproductive confinement of genetically altered salmonids. In: Cloud, J.G. and Thorgaard, G.H. (eds) *Genetic Conservation of Salmonid Fishes*. Plenum Press, New York, pp. 113–129.

Dunham, R.A. (2004) *Aquaculture and Fisheries Biotechnology: Genetic Approaches*. CAB International, Wallingford, UK.

Espinosa, E., Josa, A., Gil, L. and Marti, J.I. (2005) Triploidy in rainbow trout determined by computer-assisted analysis. *Journal of Experimental Zoology* 303A, 1007–1012.

Eudeline, B. and Allen, S.K. (2000) Optimization of tetraploid induction in Pacific oysters, *Crassostrea gigas*, using first polar body as a natural indicator. *Aquaculture* 187, 73–84.

Felip, A., Zanuy, S., Carrillo, M. and Piferrer, F. (2001) Induction of triploidy and gynogenesis in teleost fish with emphasis on marine species. *Genetica* 111, 175–195.

Fletcher, G.L., Shears, M.A., King, M.J. and Goddard, S.V. (2002) Transgenic salmon for culture and consumption. In: Driedzic, W., McKinley, S. and MacKinlay, D. (eds) *Biochemical and*

Physiological Advances in Finfish Aquaculture, International Congress on the Biology of Fish. University of British Columbia, Vancouver, Canada, pp. 5–14.

Fu, C., Hu, W., Wang, Y. and Zhu, Z. (2005) Developments in transgenic fish in the People's Republic of China. *Revue Scientifique et Technique de l'Office International des Epizooties* 24, 299–307.

Galbreath, P.F. and Thorgaard, G.H. (1995) Saltwater performance of all-female triploid Atlantic salmon. *Aquaculture* 138, 77–85.

Grewe, P.M., Patil, J.G., McGoldrick, D.J., Rothlisberg, P., Whyard, S., Hinds, L.A., Hardy, C.M., Vignarajan, S. and Thresher, R.E. (2007) Preventing genetic pollution and the establishment of feral populations: a genetic solution. In: Bert, T.M. (ed.) *Ecological and Genetic Implications of Aquaculture Activities*. Springer, Dordrecht, The Netherlands, pp.103–114, (in press).

GSA-CAAR (Georgia Strait Alliance for the Coastal Alliance for Aquaculture Reform) (2005) New Aquaculture Technology: Closing in on Solutions, Briefing Notes. Available at: http://www.georgiastrait.org/Articles2005/salmon_newtech.php

Guoxiong, C., Solar, I.I. and Donaldson, E.M. (1989) *Comparison of Heat and Hydrostatic Pressure Shocks to Induce Triploidy in Steelhead Trout (Oncorhynchus mykiss)*. Canadian Technical Report of Fisheries and Aquatic Science 1719.

Harrell, R.M.,Van Heukelem, W. and Kerby, J.H. (1998) A comparison of triploid induction validation techniques. *The Progressive Fish-Culturist* 60, 221–226.

Hindar, K., Ryman, N. and Utter, F. (1991) Genetic effects of cultured fish on natural fish populations. *Canadian Journal of Fisheries and Aquatic Sciences* 48, 945–957.

Iyengar, A., Muller, F. and Maclean, N. (1996) Regulation and expression of transgenes in fish – a review. *Transgenic Research* 5, 147–166.

Jhingan, E., Devlin, R.H. and Iwama, G.K. (2003) Disease resistance, stress response and effects of triploidy in growth hormone transgenic coho salmon. *Journal of Fish Biology* 63, 806–823.

Kapuscinski, A.R. (2002) Controversies in designing useful ecological assessment of genetically engineered organisms. In: Leterouneau, D. and Burrows, B. (eds) *Genetically Engineered Organisms: Assessing Environmental and Human Health Effects*. CRC Press, Boca Raton, Florida, pp. 385–415.

Kapuscinski, A.R. (2005) Current scientific understanding of the environmental biosafety of transgenic fish and shellfish. *Revue Scientifique et Technique de l' Office International des Epizooties* 24, 309–322.

Kilambi, R.V. and Zdaninak, A. (1980) The effects of acclimation on the salinity tolerance of grass carp, *Ctenopharyngodon idella* (Cuv. and Val.). *Journal of Fish Biology* 16, 171–175.

Kim, D.S., Jo, J.Y. and Lee, T.Y. (1994) Induction of triploidy in mud loach (*Misgurnus mizolepis*) and its effect on gonadal development and growth. *Aquaculture* 120, 263–270.

Maceina, M.J. and Shireman, J.V. (1980) Effect of salinity on vegetation consumption and growth of grass carp. *Progressive Fish-Culturist* 42, 50–53.

Maclean, N. and Laight, R.J. (2000) Transgenic fish: an evaluation of benefits and risks. *Fish and Fisheries* 1, 146–172.

Maclean, N., Rahman, M.A., Sohm, F., Hwang, G., Iyengar, A., Ayad, H., Smith, A. and Farahmand, H. (2002) Transgenic tilapia and the tilapia genome. *Gene* 295, 265–277.

Mair, G.C. (1993) Chromosome-set manipulation in tilapia-techniques, problems and prospects. *Aquaculture* 111, 227–244.

Mair, G.C., Scott, A.G., Penman D.J., Beardmore, J.A.B. and Skibinski, D.O.F. (1991a) Sex determination in the genus *Oreochromis* 1. Sex reversal, gynogenesis and triploidy in *O. niloticus* (L). *Theoretical and Applied Genetics* 82, 144–152.

Mair, G.C., Scott, A.G., Penman, D.J., Skibinski, D.O.F. and Beardmore, J.A. (1991b) Sex determination in the genus *Oreochromis* 2. Sex reversal, hybridisation, gynogenesis and triploidy in *O. aureus* Steindachner. *Theoretical and Applied Genetics* 82, 153–160.

Mair, G.C., Abucay, J.S., Skibinski, D.O.F., Abella, T.A. and Beardmore, J.A. (1997) Genetic manipulation of sex ratio for the large scale production of all-male tilapia *Oreochromis niloticus* L. *Canadian Journal of Fisheries and Aquatic Sciences* 54, 396–404.

Mohanty, A.N., Chatterjee, D.K. and Giri, B.S. (1993) Effective combination of urea and bleaching powder as a piscicide in aquaculture operations. *Journal of Aquaculture in the Tropics* 8, 249–254.

Nam, Y.K. and Kim, D.S. (2004) Ploidy status of progeny from the crosses between tetraploid males and diploid females in mud loach (*Misgurnus mizolepis*). *Aquaculture* 236, 575–582.

Nam, Y.K., Noh, C.H. and Kim, D.S. (1999) Transmission and expression of an integrated reporter construct in three generations of transgenic mud loach *Misgurnus mizolepis*. *Aquaculture* 172, 229–245.

Nam, Y.K., Cho, H.J., Cho, Y.S., Noh, J.K., Kim, C.G. and Kim, D.S. (2001a) Accelerated growth, gigantism and likely sterility in autotransgenic triploid mud loach *Misgurnus mizolepis*. *Journal of World Aquaculture Society* 32, 353–363.

Nam, Y.K., Choi, G.C., Park, D.J. and Kim, D.S. (2001b) Survival and growth of induced tetraploid mud loach. *Aquaculture International* 9, 61–71.

Nam, Y.K., Park, I.S. and Kim, D.S. (2004) Triploid hybridization of fast-growing transgenic mud loach *Misgurnus mizolepis* male to cyprinid loach *Misgurnus anguillicaudatus* female: the first performance study on growth and reproduction of transgenic polyploid hybrid fish. *Aquaculture* 231, 559–572.

Naylor, R.L., Goldburg, R.J., Mooney, H., Beveridge, M., Clay, J., Folke, K., Kautsky, N., Lubchenco, J., Primavera, J. and Williams, M. (1998) Nature's subsidies to shrimp and salmon farming. *Science* 282, 883–884.

Naylor, R., Hindar, K., Fleming, I.A., Goldburg, R., Williams, S., Volpe, J., Whoriskey, F., Eagle, J., Kelso, D. and Mangel, M. (2005) Fugitive salmon: assessing the risks of escaped fish from net-pen aquaculture. *BioScience* 55, 427–437.

Nell, J. (2002) Farming triploid oysters. *Aquaculture* 210, 69–88.

NRC (National Research Council) (2004) *Biological Confinement of Genetically Engineered Organisms*. National Academies Press, Washington, DC.

Partridge, G. and Sarre, G. (2005) Paving the way for a viable inland saline aquaculture industry. *Austasia Aquaculture* February/March, 48–54.

Pongthana, N., Penman, D.J., Baoprasertkul, P., Hussain, M.G., Islam, M.S., Powell, S.F. and McAndrew, B.J. (1999) Monosex female production in the silver barb (*Puntius gonionotus* Bleeker). *Aquaculture* 173, 247–256.

Que, H., Zhang, G., Liu, X., Guo, X. and Zhang, F. (2003) All-triploids production by crossing male tetraploids with female diploids in pacific oyster, *Crassostrea gigas* Thunberg. *Oceanologia et Limnologia Sinica* 34, 656–662.

Rahman, M.A., Ronyai, A., Engidaw, B.Z., Jauncey, K., Hwang, G.L., Smith, A., Roderick, E.E., Penman, D.J., Varadi, L. and Maclean, N. (2001) Growth and nutritional trials on transgenic Nile tilapia containing an exogenous fish growth hormone gene. *Journal of Fish Biology* 59, 62–78.

Razak, S.A., Hwang, G.L., Rahman, M.A. and Maclean, N. (1999) Growth performance and gonadal development of growth enhanced transgenic tilapia *Oreochromis niloticus* (L.) following heat-shock-induced triploidy. *Marine Biotechnology* 1, 533–544.

Sagi, A. and Aflalo, E.D. (2005) The androgenic gland and monosex culture of freshwater prawn *Macrobrachium rosenbergii* (De Man): a biotechnological perspective. *Aquaculture Research* 36, 231–237.

Sagi, A., Ra'anan, Z., Cohen, D. and Wax, Y. (1986) Production of *Macrobrachium* in monosex populations: yield characteristics under intensive monoculture condition in cages. *Aquaculture* 51, 265–275.

Scientists' Working Group on Biosafety (1998) *Manual for Assessing Ecological and Human Health Effects of Genetically Engineered Organisms. Part 1: Introductory Materials and Supporting Text for Flowcharts*. The Edmonds Institute, Washington, DC.

Silvert, W. (1992) Assessing environmental impacts of finfish aquaculture in marine waters. *Aquaculture* 107, 67–79.

Solar, I.I., Baker, I.J. and Donaldson, E.M. (1987) Experimental use of female sperm in the production of monosex female stocks of chinook salmon (*O. tshawytscha*) at commercial fish farms. *Canadian Journal of Fisheries and Aquatic Sciences* 1552.

Stevenson, J.H. (1965) Observations on grass carp in Arkansas. *Progressive Fish-Culturist* 27, 203–206.

Thorgaard, G.H. (1992) Application of genetic technologies to rainbow trout. *Aquaculture* 100, 85–97.

Uzbekova, S., Chyb, J., Ferriere, F., Bailhache, T., Prunet, P., Alestrom, P. and Breton, B. (2000) Transgenic rainbow trout expressed sGnRH-antisense RNA under the control of sGnRH promoter of Atlantic salmon. *Journal of Molecular Endocrinology* 25, 337–350.

Visser, B., van der Meer, I., Louwaars, N., Beekwilder, J. and Eato, D. (2001a) The impact of 'terminator' technology. *Biotechnology and Development Monitor* 48, 9–12.

Visser, B., Eaton, D., Louwaars, N., van der Meer, I., Beekwilder, J. and van Tongeren, F. (2001b) Potential impacts of genetic use restriction technologies (GURTS) on agrobiodiversity and agricultural production systems. *FAO Background Study Paper* 15, Report of the Commission on Genetic Resources for Food and Agriculture – Ninth Regular Session, FAO, Rome, Italy.

Wattendorf, R.J. (1986) Rapid identification of triploid grass carp with a Coulter counter and channelyzer. *The Progressive Fish-Culturist* 48, 125–132.

White, R.B. and Fernald, R.D. (1998) Genomic structure and expression sites of three gonadotropin-releasing hormone genes in one species. *General and Comparative Endocrinology* 112, 17–25.

9 Risk Management: Post-approval Monitoring and Remediation

W. SENANAN, J.J. HARD, A. ALCIVAR-WARREN, J. TRISAK, M. ZAKARAIA-ISMAIL AND M. LORENZO HERNANDEZ

Introduction

One of the major concerns about commercial production of transgenic fish is that transgenic individuals could escape into the wild and trigger cascading ecological consequences (Kapuscinski, 2005; Chapters 5 and 6, this volume). The only way to detect such escapes and cascading ecological changes is through a well-designed monitoring programme. A monitoring programme could, therefore, serve two goals: (i) to detect escaped transgenic fish; and (ii) to provide evidence as to whether ecological changes occur due to the spread of transgenic fish. Detection of escapes could provide evidence showing the inadequacy of risk reduction measures (e.g. confinement measures) and allow initiation of necessary remedial responses at the earliest possible time. Even though this chapter only addresses a few measurable ecological end points, a monitoring programme could also be used to validate models for predictive environmental risk assessment, assuming there are adequate data and well-defined ecological parameters. A well-designed monitoring programme can also help maximize efficient use of limited resources in developing countries by reducing unnecessary sampling efforts.

This chapter addresses considerations for the design of a monitoring programme for detecting the presence of transgenic escapees and some of their subsequent ecological consequences (Fig. 9.1). It identifies monitoring end points and relevant scientific and statistical methodologies, as well as a range of remedial actions. It is important to plan for monitoring and remedial responses, on a case-by-case basis, *before* any transgenic fish are approved for use in commercial aquaculture. The plans should be responsive to new insights gained from monitoring and be consistent with the adaptive, risk-based management approach advocated in Chapter 1. A monitoring plan needs to be informed by existing data, especially those generated from empirical risk assessment research in laboratories and microcosms (Chapters 5 and 6, this volume).

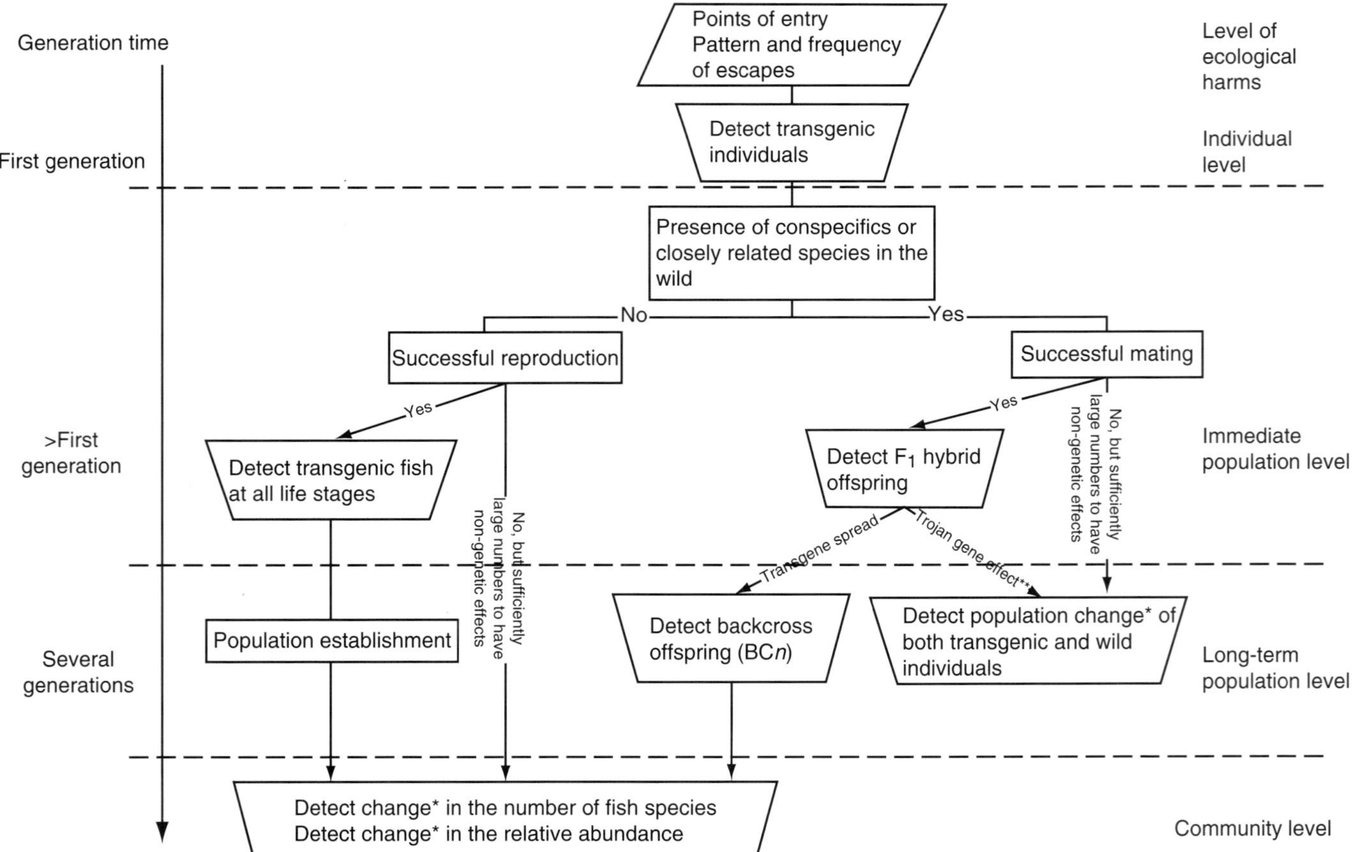

Fig. 9.1. Monitoring should focus on specific measurable end points (examples illustrated by the trapezoids) in the cascade of potential ecological changes that might result from escape and spread of transgenic individuals. Types of impacts are organized into individual, population and community levels. Generation time refers to the number of generations after the escapes. *indicates the need for baseline information prior to escape of transgenic fish. **Trojan gene effect is an outcome of gene flow (Chapter 5, this volume) that leads to the initial spread of transgenic genotypes and triggers a decline in both conspecific and transgenic populations.

General Approach to Monitoring

A monitoring programme needs clear objectives, both for scientific and practical reasons. Box 9.1 illustrates the objectives of a general approach to planning, conducting and evaluating a monitoring programme. Scientifically and statistically defensible monitoring strategies are used to verify changes predicted in the risk assessment end points (Chapters 1, 2 and 6, this volume). Collaboration between analysts and stakeholders is encouraged, through a deliberative risk assessment process, to identify, prioritize and select end points.

This chapter focuses on six measurable end points (Fig. 9.1, indicated by trapezoids). These end points are not exhaustive; this chapter focuses on them simply to illustrate the approach to take when designing a monitoring programme. There is usually a trade-off between the simplicity of the end point measurements and the direct linkage of the end points to stakeholders' ecological concerns. All of our proposed end points are measurable variables and, with two exceptions, arise solely from the presence and spread of transgenic fish. The two exceptions (change in either the number of fish species or in population abundance) may arise through a variety of natural and anthropogenic changes. The influence of these other factors makes it harder to identify the component of change, if any, in these end points that is induced by the presence and spread of transgenic fish. Measuring these end points may only be appropriate if there are enough existing baseline data to allow confident prediction and distinction between non-transgenic and transgenic-induced changes. Analysts may be advised to seek simpler end points if this information is not available.

The monitoring programme should be adequately sensitive[1] to detect impacts, if they exist, by using *inter alia* appropriate sampling designs, scientific tools and data analyses. Planning will help identify the number of samples to take, the time to initiate the sampling, as well as its intensity, frequency and location. All these factors depend on the characteristics of the transgenic species and prospective receiving ecosystems. Detecting no ecological effects at

Box 9.1. A basic iterative approach for planning, conducting and evaluating a monitoring programme requires interdisciplinary expertise.

General steps for planning, conducting and evaluating a monitoring programme:

1. Define the monitoring end points.
2. Define parameters and measurable variables.
3. Design sampling schemes and test for adequate statistical power of sampling.
4. Choose and employ appropriate sampling methods.
5. Analyse, interpret and store data.[a]
6. Refine the monitoring design based on lessons learned from data collection and analysis.
7. Inform risk management decisions.

[a]Stored data can inform future risk assessments.

[1] Having a high probability to detect changes that actually occur.

the end points could either reflect the real lack of effects from the transgenic organism or the lack of sensitivity of the sampling designs and methods. When designing a monitoring programme, careful consideration needs to be given to the statistical power of the sampling design; this will be detailed in the next section. Monitoring is a core component of any adaptive management framework (Chapter 1, this volume); additional insights gained from the monitoring effort should help refine the efficiency of the original monitoring plan and advise risk management decisions. Monitoring programmes often need refinement after initial surveys in order to improve the quality of data obtained and the adequacy of the sampling designs. Such refinement may necessitate multiple iterations of sampling and data analysis.

Relevant variables and baseline Information

Table 9.1 identifies six monitoring end points that reflect possible impacts of transgenic fish at various ecological levels, available tools to assess them and baseline information needed prior to escape of transgenic fish. We developed this set of end points based on possible ecological consequences of transgenic fish identified in the previous chapters. Ideally, selected end points should be sensitive to the disturbance (i.e. the presence and spread of transgenic individuals). Changes in most end points highlighted here are anticipated to occur over relatively short timeframes (e.g. a few generations or less). It is usually easier to distinguish the ecological changes mediated by the presence of transgenic fish from the ones mediated by other factors (e.g. chemical pollution) if the change is anticipated to occur quickly following the release or escape of transgenic fish. Changes anticipated to occur over longer timeframes require considerably more resources to measure, and it may be difficult to distinguish the reasons for end point changes.

The six measurable end points selected as examples for Fig. 9.1 include:

1. Presence of transgenic escapees;
2. Presence of F_1 hybrid offspring;
3. Presence of backcross hybrid offspring;
4. Presence of transgenic fish at all life stages;
5. Population change of both transgenic and wild individuals;
6. Changes in the numbers of local fish species and their relative abundance.

Monitoring of end points 1–4 should aim to determine the geographical spread of the transgene; results for these end points can guide selection of geographical areas in which to monitor end points 5 and 6. Furthermore, a monitoring effort may choose to focus on fewer end points depending on the target species and the geographic range of receiving environments. For example, if the approved transgenic species is not native to the area, only end points 1 and 4–6 would be relevant for monitoring.

The ability to detect escaped transgenic individuals (the first end point listed above) is critical because it affects the ability to monitor for all other end points. Detecting transgenic escapees depends on the stable inheritance and

Table 9.1. Examples of analytical tools and baseline information needed for measuring and inferring specific monitoring end points from Fig. 9.1. Tools outlined in the table are explored in detail in various sections of the chapter.

Monitoring end points	Possible associated ecological impacts	Tools for measuring the changes in proposed monitoring end points	Baseline information needed to make inferences
All		• Statistics • Sampling designs • Sampling techniques	• Life histories of target species • Habitat requirements and available habitats for transgenic fish in the receiving ecosystem • Sampling equipment for species and life stages
Individual level			
Presence of transgenic individuals	Prerequisite to other cascading ecological effects	• Molecular and morphological markers • Fishing gear appropriate for species	• Standardized sets of molecular and morphological markers unique to transgenic lines and potentially impacted populations and species
Immediate population level			
Presence of F_1 hybrid offspring	Gene flow to conspecifics and closely related species	• Molecular and morphological markers • Fishing gear appropriate for species	• Standardized sets of molecular and morphological markers unique to transgenic lines and potentially impacted populations and species
Presence of transgenic fish at all life stages (non-native species)	Population establishment	• Morphological markers • Fishing gear appropriate for species and life stages	• Morphology at various life stages of the non-transgenic species • Standard indexes for sexually mature adults (e.g. GSI, histology of gonads and body size)
Long-term population level			
Presence of backcross offspring	Persistent gene flow	• Molecular and morphological markers • Fishing gear appropriate for species	• Standardized sets of molecular and morphological markers unique to transgenic lines and potentially impacted populations and species
Change in numbers of transgenic individuals, their descendants and conspecifics	Genetic and non-genetic impacts on populations	• Population abundance estimates	• Abundance estimates of conspecific populations at control site(s) and potentially impacted site(s)
Community level			
• Increased number of transgenic individuals • Altered numbers of species and relative abundance in a fish community	Ecological interactions resulting in changes in fish species composition	• Fish assemblage estimates • Several types of fishing gear	• Typical characteristics of fish assemblage (richness and species-relative abundance) in control site(s) and potentially impacted ecosystems (with an emphasis on species potentially impacted by the proposed transgenic fish lines)

expression of the transgene construct. Thus, the approval process for commercialization of a transgenic line should ensure that the developers fully test for the stability of inheritance and expression of the transgene construct for many generations before filing the application. Wu *et al.* (2005) and Yaskowiak *et al.* (2006) are examples of transgenic lines with unstable and stable inheritance, respectively.

Scientific methodologies for end point assessment include analytical tools for detecting and estimating the abundance of transgenic escapees and their descendants. Morphological characters may be useful for detecting non-native or recently escaped transgenic individuals and molecular tools (Table 9.2) can be used to distinguish transgenic from conspecific individuals and a parental generation from F_1 and backcross or advanced generation offspring. Population abundance estimates can help detect changes in the numbers of transgenic individuals and their descendants in relation to conspecifics. Evaluation of fish species composition, namely species richness and relative abundance, will help indicate if there are potential impacts on local fish species from the spread of the transgene.

Successful reproduction of transgenic escapees and their descendants could occur either through gene flow or population establishment. Either of these events would be indicated by the presence of transgenic descendants (F_1 and backcross or advanced generation progeny), i.e. end points 2–4, or detectable changes in the numbers of transgenics relative to naturally reproducing conspecifics or closely related species (i.e. end point 5). Consequently, to reliably detect ecological changes, it is necessary to acquire baseline information on the characteristics of potential receiving populations or reference site(s). These data could be from: (i) ecosystems before escapes of transgenic fish; or (ii) similar environments that are isolated from escapes. Baseline information may be collected at the individual, community or ecosystem level (Table 9.1). Determining ecological changes in population abundance or fish community composition (end point 6) may be confounded by other natural and anthropogenic factors, such as major storms and capture fishing. These factors make it essential to have baseline data prior to and after escapes for both impacted sites and control sites, as well as appropriate statistical approaches, to accurately distinguish sources of end point changes (e.g. Underwood, 1997; Sitar *et al.*, 1999; Downes *et al.*, 2002).

Baseline information should be collected from appropriate target species and locations. For example, a combination of genetic markers (Table 9.2) unique to the transgenic line is needed to detect the presence of backcross offspring (end point 3). Relying solely on markers designed for the original transgene construct may prove inadequate because of potential mutations or rearrangements of the transgene sequences in some advanced generation progeny. Note also that some advanced generation individuals will no longer inherit the transgene. To ensure that genetic markers differentiate between transgenic and non-transgenic individuals, the markers need to be evaluated in both the transgenic line and potentially impacted populations. These markers may not be appropriate for monitoring other transgenic lines in other ecosystems.

Table 9.2. Types of molecular genetic markers, their characteristics and potential applications to differentiate non-transgenic and transgenic individuals. (Modified from Liu and Cordes, 2004 and Meehan-Meola *et al.*, 2007.)

							Potential application for producing baseline data for end points 1–4		
Marker type	Requires prior molecular information	Mode of inheritance	Type[a]	Locus under investigation	Likely number of alleles	Polymorphism or power	Population genetic diversity and differentiation of species	Differentiation of inter- and intraspecific hybrids	Differentiation of transgenic and non-transgenic counterpart
Allozyme	Yes	Mendelian, codominant	Type I	Single	2–6	Low	Yes	Yes[b]	Yes[b]
Mitochondrial DNA (mtDNA)	No	Maternal inheritance			Multiple haplotypes		Yes, maternal lineage	Yes[b]	Yes[b]
Restriction fragment length polymorphism (RFLP)	Yes	Mendelian, codominant	Type I or Type II	Single	2	Low	Yes	Yes[b]	Yes[b]
Random amplified polymorphic DNA (RAPD)	No	Mendelian, dominant	Type II	Multiple	N/A	Intermediate	Yes, population fingerprinting	Yes	Yes[b]
Amplified fragment length polymorphism (AFLP)	No	Mendelian, dominant	Type II	Multiple	N/A	High	Yes	Yes[b]	Yes[b]
Microsatellites or simple sequence repeats (SSR)	Yes	Mendelian, codominant	Primarily Type II[c]	Single	Multiple	High	Yes, paternity testing	Yes[b]	Yes[b]

Continued

Table 9.2. *Continued*

Marker type	Requires prior molecular information	Mode of inheritance	Type[a]	Locus under investigation	Likely number of alleles	Polymorphism or power	Potential application for producing baseline data for end points 1–4		
							Population genetic diversity and differentiation of species	Differentiation of inter- and intraspecific hybrids	Differentiation of transgenic and non-transgenic counterpart
Intersimple sequence repeat (ISSR)	Yes	Mendelian, dominant	Type II	Multiple	N/A	Moderate	Yes	Yes	?
Expressed sequence tags (EST)	Yes	Mendelian, codominant	Type I	Single	2	Low	Yes, physical, comp. mapping	Yes[b]	Yes[b]
Single nucleotide polymorphism (SNP)	Yes	Mendelian, codominant	Type I or Type II[d]	Single	2, but up to 4	High	Yes?	Yes[b]	Yes[b]
Insertions/ deletions (Indels)	Yes	Mendelian, codominant	Type I or Type II[e]	Single	2	Low	?	Yes[b]	Yes[b]

? = Needs further investigation to determine if these methods are applicable for collecting potential baseline information for detecting end points 1–4.
[a]Type I markers are associated with genes of known function, whereas Type II markers are associated with anonymous genomic segments.
[b]Can potentially be used with Type I or II markers and other genomics tools.
[c]Microsatellites are primarily Type II markers unless they are developed from expressed sequences or genes of known function.
[d]SNP markers are primarily Type II markers unless they are developed from expressed sequences (eSNP or cSNP).
[e]Indels can be Type I or Type II, depending on whether they are located in genes.

Very few existing databases provide quantitative and ecological data for fish. Instead, most databases provide brief species descriptions (e.g. FishBase, Froese and Pauly, 2007) or species diversity estimates at a large watershed scale (e.g. World Conservation Union reports; IUCNNR, 2006). For other environmental conditions, many developing countries may already have ongoing monitoring programmes for environmental quality. Countries may consider adapting the approaches and methods used to collect ecological data for some developed-country databases. Relevant examples include the SCECAP database of the South Carolina Estuarine and Coastal Assessment Program in the United States (2006), the National Water Quality Management Strategy (NWQMS) in Australia and the River Invertebrate Prediction and Classification System (RIVPACS; Wright *et al.*, 2000) in the UK. However, caution should be exercised when using established databases because their data may be generated from inconsistent protocols and analyses.

Planning a Monitoring Programme

Planning is important for a successful monitoring programme of any kind (e.g. Downes *et al.*, 2002). Planning steps include carefully identifying objectives, end points, sampling designs, sampling techniques and statistical analyses to be used. Because the design of the monitoring programme requires knowledge in many disciplines (e.g. statistics, genetics and ecology), scientists and statisticians with relevant and country-specific training should be consulted during the process. The sections below discuss these planning steps in more detail.

Statistical considerations

The sensitivity of a monitoring programme determines its ability to detect the presence and impacts of transgenic fish. High sensitivity can be achieved by using appropriate statistical tools, as well as being aware of and correcting for inferential problems and statistical powers. Relevant inferential issues include the likelihood of falsely concluding that transgene-mediated changes in an identified end point have occurred (Type I error) or falsely concluding that such changes have not occurred (Type II error). The latter issue is the core concern surrounding the use and possible escape of transgenic fish from aquaculture production facilities. These two inferential issues directly relate to the statistical power required to detect proposed ecological impacts, and therefore they influence the number of samples needed to detect ecological changes.

Statistical analyses

Types of statistical tools useful for detecting the presence and impacts of transgenic fish include:

1. Descriptive statistics, such as means and variances for counts (e.g. numbers of individuals, density) or measurements (e.g. catch per unit of effort (CPUE) and relative abundance);

2. Hypothesis testing for deviation from the null hypotheses (e.g. the presence/absence of transgenic individuals, changes in population abundance after the escapes and spread of transgenic fish) at specified probabilities;
3. Time-series analysis for long-term trends of some parameters (e.g. population abundance trends; Thomas, 1996).

Most monitoring end points highlighted in this chapter refer to the presence or absence of transgenic individuals and estimate their numbers in relation to conspecifics and other fish species at the time of sampling (Fig. 9.1). Detecting changes in the numbers of transgenic escapees and their descendants (as compared to conspecific populations) through time, but within a single location, will require further hypothesis testing or trend analysis. Detecting changes through time and across different locations may also require additional spatial statistical tools (e.g. Cressie, 1993). Furthermore, inferences about the changes will require appropriate baseline information (e.g. data prior to the escapes in potentially impacted and control sites, natural fluctuations and other anthropogenically induced changes in the end points of concern).

Inferential error and statistical power

Inferences from the empirical results of a monitoring effort are prone to two types of statistical errors. The first type, referred to as Type I error (α, see top cells of Table 9.3), occurs when an investigator rejects a true null hypothesis. The second type, known as Type II error (β, see lower cells of Table 9.3), occurs when an investigator fails to reject a false null hypothesis. Concerns about Type I error dominate most empirical biological studies because many researchers believe that Type I is more serious than Type II error (Cohen, 1988). This is not always true. For example, consider a sampling scheme designed to detect whether escaped transgenic individuals have successfully reproduced in the wild. Are the ecological consequences likely to be less serious if a researcher concludes that *escaped transgenics survive to reproduction in the wild when no transgenic individuals have in fact done so* (Type I error) than if the researcher concludes that *escaped transgenics did not survive to reproduction in the wild when, in fact, an appreciable number have survived to reproduce* (Type II error)? In the first case, risk management strategies may excessively restrict transgenic aquaculture operations; in the second, risk management strategies could allow uncontrolled escapes of transgenics, with resulting reduced viability

Table 9.3. Error hypothesis outcome table.

	Decision on null hypothesis H_0	
Status of null hypothesis H_0	Do not reject	Reject
H_0 true (H_1 false)	Correct decision made with $P(1 - \alpha)$	Type I error (α) made with $P(\alpha)$
H_0 false	Type II error (β) made with $P(\beta)$	Correct decision made with $P(1 - \beta)$

of wild populations. Consequences in both cases are serious, but they have distinctly different implications for the local aquaculture industry and for the genetic and ecological integrity of wild fish and aquatic communities. Consequently, the statistical evaluation of any monitoring programme should consider the competing consequences arising from different sources of inferential error. Cohen (1988), Peterman (1990) and Fox (2001) discuss this subject in detail.

The relationship between Type I and Type II error in evaluating statistical hypotheses is a fundamental aspect of scientific inference. In such an evaluation, the investigator wishes to correctly discriminate between the null and alternate hypotheses. Table 9.3 presents four possible outcomes relating conclusions to underlying reality (Taylor and Gerrodette, 1993). These outcomes reflect how conclusions from monitoring agree with reality, with two possible distinct types of errors. The probabilities of these outcomes reflect a complex relationship between several variables, only some of which are under the direct control of an investigator. Figure 9.2 shows the general relationship between the significance level (α), the probability of a Type II error (β), the effect size desired for detection, sample variability and statistical power ($1 - \beta$, the ability to detect the effect size if true). The effect size is the difference between the sample mean under the null hypothesis (μ_0) and the sample mean under the alternative hypothesis (μ_1). If the true results are described by the distribution defined by μ_0, then Type I error is shown by the darkly shaded area. It is important to note that the investigator can *a priori* set the magnitude of α. If the true results are described by the distribution defined by μ_1, then the magnitude of the Type II error is shown by the lightly shaded area.

An important question is how to minimize β, thereby maximizing statistical power. First, Fig. 9.2 indicates that for fixed n and α, β reduces as the effect size increases. For a given set of results and critical effect size, a trade-off arises between the levels of α and β, and, as a consequence, power declines with

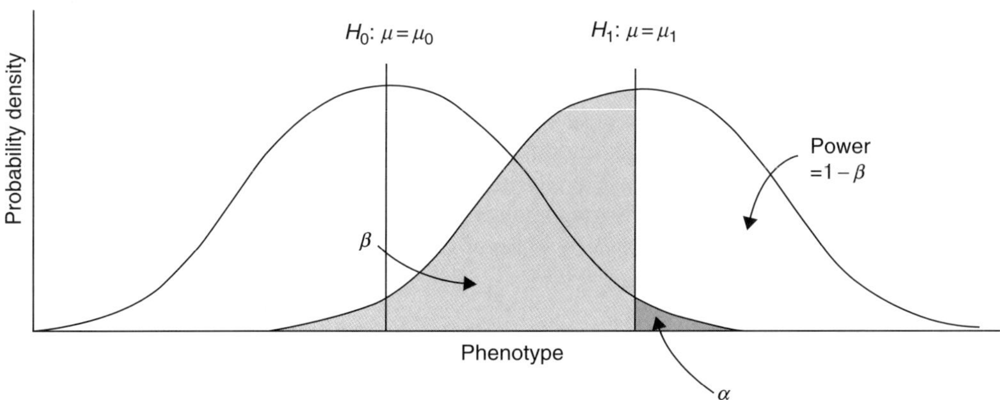

Fig. 9.2. Relationship between α, β, effect size, sample variability and power ($1 - \beta$). Effect size is the difference between μ_0 and μ_1. Type I error is shown by darkly shaded area and Type II error is shown by lightly shaded area.

greater protection against Type I error. Increasing the sample size, and thus the reliability of the estimated effect size, helps reduce the level of Type II error.

The investigator has the opportunity to control Type I error when designing a monitoring survey by setting a significance level to evaluate competing hypotheses. Although an α of 0.05 (equivalent to a 5% probability of rejecting a null hypothesis by chance) has achieved overriding importance as a design feature in much empirical science (Labovitz, 1968; Fox, 2001), it is often tempting to reduce the possibility of Type I error even further by reducing α. Unfortunately, this practice tends (other things being equal) to increase the likelihood that a Type II error (β) will be committed. That is, the ability to detect a small, but real, effect is reduced when α is too stringent, although this can be offset by increasing the sample size. Thus, the design of any monitoring programme, including one intended to detect transgenics or their effects, should take into account the relative costs of obtaining the data, as well as the monitoring goals (Cohen, 1988).

Analysts who design monitoring programmes must balance the statistical power of the programme against its costs. In ecological applications, monitoring programmes that are very sensitive (i.e. have high statistical power) are often very expensive. The usual ecological convention is to set power to 0.80, an arbitrary but well-accepted minimum for sample size calculations in biological sciences (Quinn and Keough, 2002). In the absence of specific cost information, we simply recommend that an investigator strive to maximize the power of their monitoring programme within their budget constraints. Investigators should note that the desired power will only be achieved approximately because effect size and variability are outside the investigator's control. Investigators should also recognize that, for fixed n and α, power is affected by whether a test is one- or two-tailed; power is greater for a one-tailed test (e.g. $A > B$; the presence of transgenic individuals is greater than zero) than for a two-tailed test (e.g. $A \neq B$; a change in population abundance is detected) because another alternative hypothesis ($A < B$) can be ignored.

Sample size requirements for detecting the presence of transgenic individuals and their hybrids and backcrosses

The detection of transgenic individuals and their descendants in the wild requires the ability to distinguish transgenic descendants from non-transgenic individuals either by morphological characteristics or molecular genetic markers. When using molecular genetic markers, the likelihood of detecting transgenic individuals and their offspring in the wild increases with both the frequency of hybrid progeny and sampling efforts. The sampling effort required to achieve a desired probability of detection can be estimated as follows (Kanda *et al.*, 2002):

$$\beta = (1-q)^{2nL} \tag{9.1}$$

where β is the acceptable level of not detecting a transgenic genotype where one exists (i.e. level of Type II error); q is the average frequency of transgenic alleles at loci diagnostic for transgenic individuals in the population (or level of introgression, Hitt *et al.*, 2003); n is the total number of individuals sampled; and L is the number of loci diagnostic for the transgenic lines. The sampling effort can be formidable when transgenic individuals are rare. For example, the ability to detect $q = 0.001$, with β set at 1% and only a single diagnostic marker ($L = 1$), would require a random sample of more than 2300 individuals. Increasing the number of diagnostic markers to 10 (at an average allele frequency of 0.1%) reduces the minimum n to 230 individuals. If the genetic contribution is higher (as manifested in a higher allele frequency, e.g. 0.5%), the sampling effort can be reduced further (e.g. $n = 45$).

Although these hypothetical scenarios are arbitrary, they serve to illustrate how logistical considerations and the minimum level of sampling effort vary with the number and frequency of diagnostic markers. Figure 9.3 shows a set

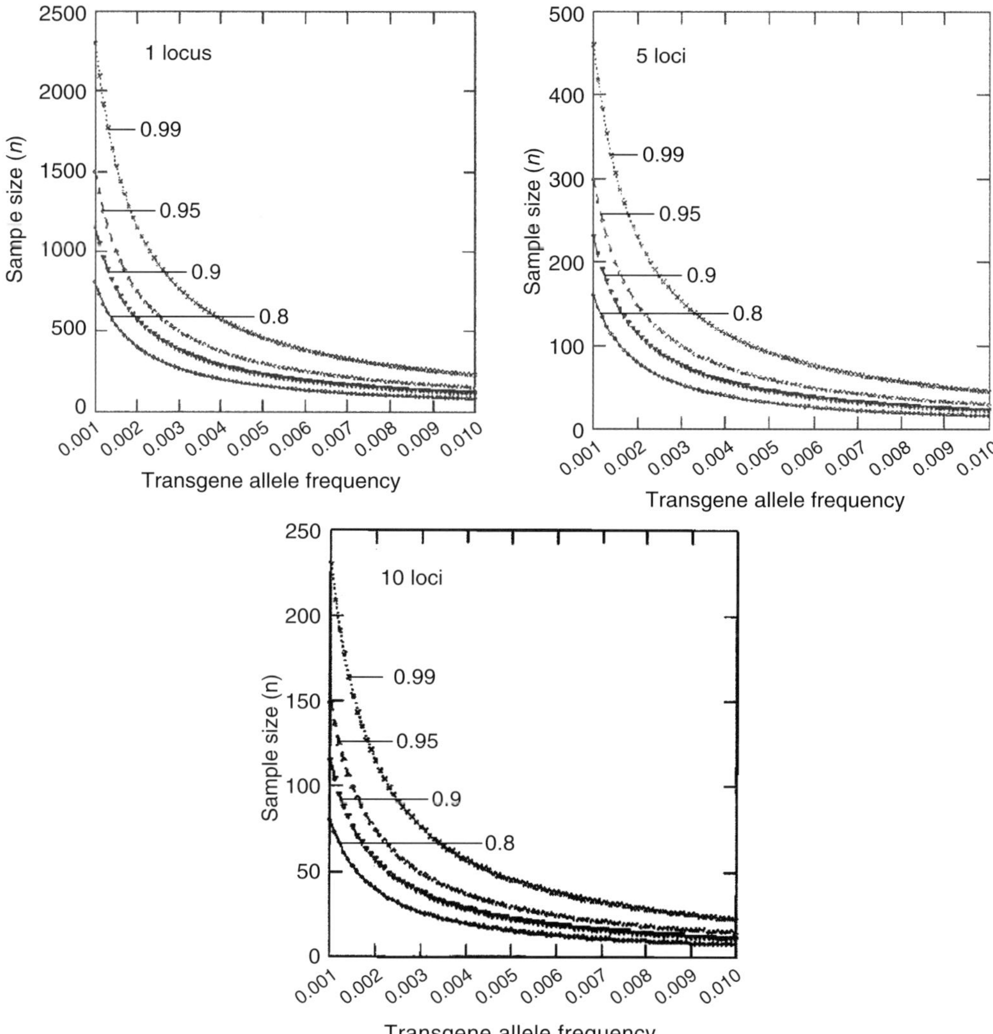

Fig. 9.3. Effect of transgene frequency (q) and number of individuals sampled (n) on the power to detect transgenes (β, values from 0.8 to 0.99) for various numbers of diagnostic loci evaluated (L = 1, 5 and 10).

of curves depicting the approximate relationship between n, q and β for various numbers of loci (L) sampled. In practice, the statistical power standard that the sampling strategy aims to achieve, and hence the tolerable level of Type II error, depends on sample costs, budget constraints and stakeholders' tolerance for risk. We therefore recommend addressing this issue during multi-stakeholder deliberations (Chapter 2, this volume).

When it is possible to use morphological characteristics to detect transgenic individuals, the required sampling effort can be estimated as follows (Burgman, 2005):

$$n = \frac{\log(1-p)}{\log(1-f)} \tag{9.2}$$

where n is the sample size, f indicates the level of scarcity (i.e. probability that transgenic individuals will be present in a randomly selected sampling unit at least once), with a probability p $(1 - \beta)$. For example, detecting the presence of transgenic fish, when their scarcity is 1 in a 1000 ($f = 0.001$), with a probability of 0.90 (p), would require sampling more than 2000 individuals. Burgman (2005) also suggests other applications for this equation relevant to monitoring: (i) the number of locations that should be sampled; (ii) the number of times to sample a site; and (iii) the likelihood that a species (either transgenic or conspecific) has become extinct. The main assumptions for this equation are that f is constant for all sampling units, and the probability of successfully detecting a transgenic individual is independent between sampling units.

Sample size requirements to estimate population abundance and changes in species composition

Monitoring end points 5 and 6 involves detecting a temporal change in fish population abundance and fish community composition and therefore requires accurate point estimates (means). Sample sizes required for point estimates depend on the number of transgenic fish released, the level of accuracy needed (i.e. a region around the true mean), the α level and the variability of the estimates (e.g. standard deviation, s; or coefficient of variation, CV). During the planning phase, these elements need a rough estimation based on other work, theories or experience. If large numbers of transgenic fish can be detected routinely in population surveys, analysts may estimate the sample sizes required to assess transgenic abundance using Eqs 9.3–9.5 below (Krebs, 1999). These equations are not applicable when numbers of transgenic fish are very small; in such situations, it is more efficient to focus monitoring efforts on end points 1–4 instead of 5 and 6. All equations with critical t values assume a normal distribution. For variables with other types of distribution, readers should refer to Hayek and Buzas (1997) and Krebs (1999).

For continuous variables, such as population abundance measurements, the sample size can be obtained using the equation:

$$n = \left(\frac{t_\alpha s}{d}\right)^2 \tag{9.3}$$

where n = sample size required for estimating the mean; t_α = student's t value for $n - 1$ degrees of freedom for the $1 - \alpha$ level of confidence; s = standard deviation of variables; and d = the region around the true mean.

For discrete variables, such as proportions of transgenics to non-transgenics or to other species, sample size can be estimated by using the equation:

$$n = \frac{t_\alpha^2 \hat{p}\hat{q}}{d^2} \tag{9.4}$$

where $\hat{p}$ = proportion of transgenic individuals; and $\hat{q}$ = proportion of conspecifics or other species $(1 - \hat{p})$.

Mark–recapture experiments are a common way to estimate population abundance (N), and the simplest mark–recapture method is the Petersen method (e.g. Krebs, 1999). In this method, fish are captured, marked, released and recaptured; the method only requires a single episode of marking individuals and a second single episode of recapturing individuals. Population abundance is then extrapolated from the proportion of marked individuals in the recapture trial. The sample size required for estimating the Petersen population abundance can be calculated as follows:

$$\mathrm{CV}(\hat{N}) = \frac{1}{\sqrt{R}} = \frac{1}{\sqrt{MC/N}} \tag{9.5}$$

where $\mathrm{CV}(\hat{N})$ = coefficient of variation of the Petersen population abundance estimate; R = expected number of marked fish to be caught in the recapture trial; M = number of fish marked and released in the first capture trial; C = total number of fish caught in the recapture trial; and N = the true population size (an educated guess is needed for a preliminary sample size calculation). For convenient use, Robson and Regier (1964; referenced in Seber, 1982; Krebs, 1999) provide charts of sample sizes given the desired accuracy (A, a percentage of the deviation between estimated population size and true population size in relation to the true population size) and number of fish present in a population.

Sample size required to detect temporal changes in ecological variables

In some instances, measurement end points may be temporal changes of an ecosystem attributed due to the spread of transgenic individuals (e.g. the abundance of transgenic individuals). The sample size required to detect temporal changes depends on the smallest true difference to be detected (i.e. effect size), variability of data and the desired level of statistical power. Assume that all population numbers are approximately normally distributed with the same underlying variance and that the following values are known: the standard deviation, the number of groups to be compared (>1; e.g. the proportions of transgenics to conspecifics at times 1 and 2) and the effect size. The numbers of individuals to be sampled can be calculated from the following equation (Tate *et al.*, 2003; Burgman, 2005):

$$n \geq 2\left(\frac{\sigma}{\delta}\right)^2 \left\{t_{\alpha[\upsilon]} + t_{2(1-power)[\upsilon]}\right\}^2 \quad (9.6)$$

where n = required number of replications per sample; σ = standard deviation (estimated prior to the analysis); δ = the smallest true difference desired to detect (or the magnitude of the ratio δ and σ); υ = degree of freedom (e.g. for two samples and n replicates in each sample, $\upsilon = 2[n - 1]$); α = significance level (Type I error rate); power = statistical power $(1 - \beta)$; $t_{\alpha[\upsilon]} + t_{2(1-\text{power})[\upsilon]}$ = critical values from a two-tailed t-table with υ degrees of freedom; and corresponding probabilities α and $(1 - \text{power})$. In the planning phase, the standard deviation of the samples needs to be estimated using information from preliminary estimates, data from previous work, theory or experience.

Sampling considerations

Sampling design

Sampling schemes, which are approaches for selecting observations/samples, are an important component of a sampling design. Different sampling schemes allow varying degrees of generalization of inferences about a (statistical) population from observations/samples. Examples of established sampling schemes used in ecology and fisheries science include random sampling, systematic sampling, stratified sampling, cluster sampling and adaptive sampling (e.g. Thompson, 1992; Brown and Austen, 1996; Hayek and Buzas, 1997; Krebs, 1999). Random sampling, where each sampling unit has an equal chance of being selected, *inter alia*, allows for generalized inferences about the (statistical) population from a sample. However, a simplified random sampling may not be practical for many aquatic species. Another common strategy is systematic sampling, where sampling units are selected in a systematic fashion (e.g. along equally divided transect lines). This is a common approach when collecting benthic invertebrates and water samples. However, for fish this approach may not be best because of their mobility and preference for certain kinds of habitats. If population distribution is heterogeneous, then stratified sampling is often used (Brown and Austen, 1996). This technique consists of dividing the site into strata and taking sampling units from each stratum. Future sampling can also be adapted using insights gained from early sampling efforts (Thompson, 1992).

In addition to sampling schemes, important considerations for designing a monitoring programme include geographic boundaries, sampling locations and sampling intensity (Box 9.2). These considerations depend first on knowing the pattern of escapes, life histories of target species, characteristics of the receiving ecosystems and efficiency of sampling equipment. For this chapter's end points, sampling efforts will also require knowledge about habitat requirements and mobility of the target species at specified life stages. Geographic boundaries for the survey area will be determined by the patterns of transgenic

Box 9.2. Factors to consider for efficient sampling design relevant to transgenic fish.

Sampling considerations	*Relevant questions*	*Factors to consider*
Geographic boundary	• Proximity to the point of entry into the environment? • How many sites within a watershed? • How many watersheds?	• Migratory pattern • Length of time since escape events • Pattern of escapes and efficiency of confinement measures • Size of approved sample areas • Connectivity of watersheds
Location	• Where to sample?	• Habitats of targeted species at various life stages • Appropriate sampling equipment and personnel
Intensity	• How often and how long to sample?	• End points to be monitored • The fluctuating nature of changes at end points • Number of transgenic individuals in the wild • Efficiency of sampling equipment • Statistical power of sampling

fish escapes and their migratory behaviour. Movements of different species, and even closely related fish species, can differ substantially. For example, in a lowland river in Australia, Crook (2004) found that radio-tagged common carp could travel up to 4 km within 90 days after release in a new habitat and that carp prefer slow-moving water over fast-moving water habitats. However, in another study in the upper Elbe River, Czech Republic, Slavik and Bartos (2004) found that radio-tagged Prussian carp (*Carassius auratus gibelio*) could travel up to 87 km downstream from a release site in 15 months.

The duration and frequency of sampling (i.e. the intensity) depends on the types of impacts to be monitored, the temporal and fluctuating nature of end point variables and the sensitivity of the sampling method employed. Some ecological impacts can be detected during the first generation (i.e. the presence of transgenic escapees), while others can only be detected if gene flow occurs or if the escaping transgenic fish establish a self-sustaining population. Sampling intensity is a very important consideration, especially for detecting fluctuating changes (e.g. seasonal fluctuation of population abundance); the periodicity of samples should be at least twice the periodicity of the fluctuating phenomena of interest (Legendre and Legendre, 1998). In some cases, a full cycle of the natural periodicity can take many years (e.g. the natural periodicity of fish population abundance). The sensitivity of the sampling method depends on the number of transgenic individuals present in the wild, the spatial and temporal distribution of the fish, the sampling efforts and the efficiency of sampling equipment.

Sampling equipment

All the end points detailed in this chapter can be monitored using sampling methods that involve capturing fish. However, if the results of the risk assessment process indicate the need to collect other organisms or monitor other environment variables, readers should consult standard sampling texts such as Greenberg *et al.* (1992) and Rabeni (1996).

The use of appropriate fishing gear is essential for detecting changes in selected end points. For example, for fish assemblage estimates, sampling equipment should capture the species of interest. Unfortunately, most fishing gear is either species or size selective, or both. The gear's efficiency can vary because of fish behaviour, habitat types and operator effects, as well as environmental factors such as light, water transparency and current (e.g. Arreguín-Sámchez and Pitcher, 1999; Ziegler *et al.*, 2003; Hayes *et al.*, 2005; Lindquist and Shaw, 2005). For example, Lindquist and Shaw (2005) found that the efficiency of light traps for collecting larval and juvenile fish reduces as current speed increases. Analysts planning monitoring strategies should be aware that the sample size needs to increase as sampling efficiency decreases in order to maintain a desired level of statistical power (e.g. Hayes *et al.*, 2005).

Both larval and adult fish can be sampled by using passive or active sampling gears, electricity and toxicants. Passive capture gears, such as nets and traps, capture fish by entanglement or entrapment (Lagler, 1978; Hubert, 1996; Kelso and Rutherford, 1996). Active fish sampling techniques, such as trawls and seines, capture fish by sieving them from the water medium (Hayes, 1983). Electrofishing uses electrical fishing devices powered by either a 12 or 24V battery or a generator. Rotenone is the most popular toxicant used in fisheries work in western countries (Lenon *et al.*, 1970); however, it is difficult to obtain permission to use it in some developing countries. Box 9.3 presents an overview of fishing gear typically available in developing countries for sampling adult fish.

Fishing gear typically used for adult fish may not be appropriate for fish larvae and fingerlings because of the differences in size, behaviour and distribution. Kelso and Rutherford (1996) suggest different types of fishing gear, including low-speed tows, high-speed tows, trawls, electrofishing and traps. The type and size of sampling gear should be based on the distribution of fry and fingerlings (e.g. benthic or pelagic), size of target species, habitat types and the availability of gear. Larval and fingerling samples should be appropriately preserved for further identification (e.g. using morphological and molecular genetic methods).

Another important consideration when selecting a sampling gear is the potential for conflict with local people. Conflicts may arise from the presence of researchers and use of fishing gear in ways that may be offensive to the local people (e.g. a trawl survey in their fishing grounds). It may be wise to consult and inform the local people before deciding what sampling gear to use, and if warranted, consider alternative sampling design and fishing gear. Involving local people in the participatory risk assessment process or the planning process may help alleviate tensions. Furthermore, local people may have specific knowledge about local conditions (e.g. fishing methods and fish habitat use), which can be used to improve a sampling design.

Box 9.3. Examples of fishing gear used in developing countries.

Many types of fishing gear are already used in commercial fisheries, such as gill nets or cast nets. They are usually species selective, but they can be applied to fish population and community surveys with specific objectives (i.e. a survey targeting a single species or a size class). Passive gears (e.g. gill nets) have been widely used to sample fish because they are cheap, readily available and require little specialized training to operate. They can be useful for studying the relative abundance of fish species in a lake or reservoir ecosystem. The size of the captured fish is determined by the mesh size. In lake habitats, gill nets can be set vertically throughout the entire depth of the shallow section of the reservoir or at a certain depth in a deeper portion of a lake. In a study of the ecology of a cyprinid fish, *Mystacoleucus marginatus*, in a reservoir in Malaysia, Zakaria-Ismail and Muhamad-Mazli (1995) used a vertical gill net suspended through the entire depth of the reservoir. If the study involves only a single species, and individual fish size is small, a 2.5cm stretch-mesh gill net is effective to catch the fish. The catch per unit of effort (CPUE) can be based on a 15-h sampling period.

A cheap and easily available active sampling gear is a cast net. Although this gear is very species specific, it can be used for studying the relative abundance of fish in shallow, slow-flowing riverine habitats. Martin-Smith (1998) used this sampling gear in the upper Segama River, Sabah, Malaysian Borneo. More than 6300 individuals, representing 21 species, were collected in the study. To reduce variability, the cast net should be of identical mesh size and used at permanent stations, which can be sampled over and over.

Electrofishing is the sampling method of choice in many stream or lake surveys (e.g. Zakaria-Ismail and Sabariah, 1994) because it does not have strong species selectivity compared to other methods. In a small stream, particularly in a remote area, a backpack shocker is extremely useful. The power source is either a 24V, deep-charge battery or a 115V AC generator such as MODEL 12 Smith-Root backpack electrofisher. The advantage of using electricity is that the specimens are alive, and can be released into the habitat once their biological parameters have been recorded. When used properly, mortality of fish from electrofishing is less than 3% (Zakaraia-Ismail, University of Malaya, Kuala Lumpur, 2006, personal communication). In a lake, a boat-mounted electroshocker such as Smith-Root MODEL SR-14H, powered by a 9.0 GPP generator, is suitable. Unfortunately, high cost, coupled with stringent government regulation, hinders the use of such sampling gear in many developing countries.

Methodologies and Approaches to Detect Key Ecological Changes

Presence of transgenic escapees and descendants

This section presents tools to help differentiate transgenic individuals (first generation[2]) and their descendants (advanced generations) from conspecific populations or closely related species. These tools include both phenotypic and molecular

[2] The generation indicated here refers to the first generation of escapes, not the number of generations the transgenic line is maintained in a laboratory. For example, escaped transgenic individuals can actually be a fourth-generation transgenic line.

approaches. For non-native species, it may be adequate to use morphological differentiation between the two species. For species that can hybridize with conspecifics or closely related species in the wild, additional genetic markers will be needed, especially when the natural reproduction of transgenic individuals is self-sustaining. This section will briefly mention morphological markers, focusing instead primarily on molecular approaches because they provide a clearer differentiation.

Morphological markers

If there are distinct morphological differences between transgenics and non-transgenics, detecting the presence of a transgenic species can be straightforward. Baseline information about morphological characteristics at all life stages will be necessary, and advice from taxonomists is valuable. For conspecifics or closely related species, it will be necessary to test the reliability of morphological characteristics to differentiate wild conspecifics, recently escaped transgenic individuals and advanced generation hybrids. Attempts to establish morphological baseline data for conspecifics are at very early stage. For example, Fonticiella and Arboleya (2003) have been developing a baseline for an index describing the length–weight relationships of feral tilapias in many lakes and rivers in Cuba. Distinctive index values between transgenic tilapia and feral tilapia could provide a useful morphological screening tool, at least for recently escaped transgenic individuals. Morphological characteristics may also be used in conjunction with molecular genetic markers.

Molecular genetic markers

Molecular genetic markers useful for detecting the presence of transgenic escapees and their descendents include the transgene construct itself, allozymes, mitochondrial DNA and other DNA markers. Multiple types of molecular markers are needed because offspring from the interbreeding of transgenic individuals and conspecifics (i.e. hybrids and backcross offspring) may not bear the original transgene construct, although first generation escapees and F_1 hybrid individuals should inherit at least one copy of the transgene construct.

Polymerase chain reaction (PCR) is the technique of choice for many types of molecular studies, especially ones requiring non-invasive sampling (e.g. Avise, 1994; Buitkamp and Epplen, 1996; Vignal *et al*., 2002; Liu and Cordes, 2004). This technique requires some knowledge about flanking nucleotide sequences of the target DNA (i.e. PCR primers), and the ability to store samples in preservatives other than formaldehyde, such as ethanol (e.g. Ferraris and Palumbi, 1996). For most commercially important fish species, the PCR primers are already available in many journal publications (e.g. Yue *et al*., 2004) and free databases (e.g. GenBank; NCBI, 2006; Blackwell Science's Molecular Ecology Notes[3]). A few examples of PCR primers useful for detecting transgenes in transgenic aquaculture species are illustrated in Table 9.4. Information about PCR primers that flank the transgene construct should be included in any application seeking the approval for commercial production of a transgenic line.

[3] Molecular Ecology Notes has searchable PCR primer information and can be accessed at: http://tomato.biol.trinity.edu/

Table 9.4 Examples of information needed for polymerase chain reaction (PCR) detection of transgene constructs in aquaculture species with potential for commercial development.[a]

Species	Promoter/gene used for plasmid construction	Forward (F) and Reverse (R) primer sequences	References
Oreochromis niloticus (Nile tilapia)	Ocean pout antifreeze protein (AFP) promoter co-ligated to a carp β-actin/lacZ reporter; chinook salmon growth hormone (GH) gene	F: 5'-TTGGCTCAGAAAATGTTCAATGA-3'[b,c] R: 5'-GGAATATCTTGTTCAGCTGTCTGC-3'	Rahman *et al.*, 1998; Caelers *et al.*, 2005
Cyprinus carpio (common carp)	Mouse metallothionein-1 promoter and regulation region/human GH mini-gene	F: 5'-GGTAAGCGCCCCTAAAATCC-3' R: 5'-TTGAAGATCTGCCCAGTCCG-3'	Wu *et al.*, 2005
Ictalurus punctatus (channel catfish)	Cecropin B gene from the moth *Hyalophora cecropia*, driven by the cytomegalovirus promoter	F: 5'-ATAAGCTTCACACACCATGTTATCTACAT-3' R: 5'-AATCTAGAGTCAGTCCCGACCATGCACA-3'	Dunham *et al.*, 2002
Oncorhynchus kisutch (coho salmon)	AFP; chinook salmon GH gene	F:5'-GTCAGGATCCAGCCTGGATGACAATGACTC-3' R:5'-GTCAGAATTCCTACAGAGTGCAGTTGGCCT-3'	Devlin *et al.*, 1995
Penaeus (Litopenaeus) vannamei (Pacific whiteleg shrimp)	Taura syndrome virus coat protein gene	RT-PCR with TSV-CP gene-specific primer pair A: F1: 5'-CTTAATTAATGCCTGCTAACCC-3' R2: 5'-ATTGATGTCTGCTTAGCATTCA-3' TSV-CP nested gene-specific primers 3 and 4: F: 5'-TGATACAACAACCAGTGGAGGAC-3' R: 5'-TGTCATCAGGTAGGGAAATTTC-3'	Lu and Sun, 2005

[a]Using PCR-RFLP, Southern blotting and fluorescence *in situ* hybridization (FISH) technologies. See also Chapter 3 for information on other transgenic species relevant for commercial aquaculture.
[b,c]PCR and Southern blotting techniques detailed in Rahman *et al.* (1998).

Table 9.2 summarizes the basic properties of various molecular genetics markers and their applications. They can be used to differentiate transgenic individuals from non-transgenic ones, inter- and intraspecific hybrids, and to estimate genetic diversity of transgenic lines and wild populations. Despite the usefulness of PCR, long-term detection of transgenic fish and their descendants, as well as determining how differences in genotypes and phenotypes of transgenic descendants may affect monitoring end points, would benefit from the combined usage of transgene constructs, relevant DNA marker types (Table 9.2) and recently developed genomics tools, such as microarrays and reverse transcriptase-PCR (RT-PCR). Such tools can quantify expression of the transgene and other genes in the same biochemical pathway. Although genomic tools are not required to detect this chapter's proposed end points, these tools could help identify the molecular mechanisms by which transgenics and their descendants respond to the environment; hence, they could also help identify relevant end points for future monitoring programmes.

Changes in population abundance

Population abundance of conspecifics

Estimating changes in population abundance of transgenic individuals and wild relatives over time could provide insights on the population-level impacts of transgenic fish. Increasing numbers or proportions of transgenic individuals in the samples would indicate the spread of transgenic individuals. Chapters 5 and 6 identify a number of scenarios for population responses due to genetic and non-genetic effects. For example, the Trojan gene effect[4] may trigger the decline in the numbers of both transgenic and conspecific individuals (Muir and Howard, 1999). To date, the predictions of changes in population abundance in response to spread of transgenic fish have been based solely on computer simulations, and still need empirical validation (Chapters 5 and 6, this volume).

Two approaches for estimating population abundance of transgenic individuals and wild relatives are: (i) measuring the ratio of two population estimates using a mark–recapture technique; and (ii) determining the abundance index, the CPUE. The first approach requires capturing, marking, releasing and recapturing individuals (both transgenics and conspecifics). The proportions of marked and unmarked fish allow for population abundance estimates. Many texts contain detailed discussion about different mark–recapture techniques (e.g. Otis *et al.*, 1978; Seber, 1982; White *et al.*, 1982; Williams *et al.*, 2002). The second approach estimates the numbers or weights of target species present in a specified fishing effort (e.g. a specific-size gill net set at specified hours or hours of electrofishing) at a one-time capture (e.g. Krebs, 1999; Paukert, 2004).

[4] This gene flow scenario refers to one possible outcome of interbreeding between transgenics and conspecifics: initially the transgene spreads, and then triggers a decline in the introgressed population. This occurs when the transgene has an antagonistic effect on different fitness traits. See Chapter 5 for more details.

The relative abundance of transgenic and non-transgenic fish within a mixed population can be estimated using the approach of Skalski *et al.* (1983), which is based on single mark–recapture estimates. This approach assumes closed populations (i.e. no immigration and emigration). See Box 9.4 for an overview of this methodology. Such a mark–recapture approach generally gives more

Box 9.4. Equations to use when calculating point estimates of the proportional abundance of transgenic and non-trangenic individuals, using a mark–recapture method. (From Skalski *et al.*, 1983.)

The proportional abundance (K), which is the ratio of the abundance of one group (N_1) to that of another group (N_2), can be obtained by conducting two independent mark–recapture experiments ($K = N_1/N_2$). The K estimates can be derived from either the absolute abundance of the two groups (K_1) or the number of individuals captured only once during the mark–recapture trials of the two groups (K_2). For K_2 to be an unbiased estimator, it is necessary to assume equal capture probability of two groups at both mark–recapture trials. When this assumption is met, the estimated variance of $\hat{K}_2$ is 2–20 times smaller than that of the estimated variance of $\hat{K}_1$.

The change in the proportional abundance over time (i.e. effect size) is estimated from the difference in the sequential estimates of K_s at times $t = 0$ and $t = t$.

Proportional abundance is obtained from absolute numbers of transgenics and conspecifics:

$$\hat{K}_1 = \frac{\dfrac{(n_{21}+1)(n_{22}+1)}{(m_2+1)} - 1}{\dfrac{n_{11}n_{12}}{m_1}}$$

where n = numbers of individuals collected for the capture and recapture trials and m = numbers of marked individuals.

The variance for $\hat{K}_1$ is estimated as

$$\hat{\mathrm{V}}\mathrm{ar}\left(\hat{K}_1/N_1, N_2\right) = \frac{m_1 n_{21} n_{22}\left([n_{22}-m_2][n_{21}-m_2][m_1 n_{11} n_{12}] + [n_{12}-m_1][n_{11}-m_1][m_2 n_{21} n_{22}]\right)}{m_2^{\,3} n_{11}^{\,3} n_{12}^{\,3}}$$

The first subscript denotes the group (group 1 or 2) and the second subscript denotes the mark–recapture sample (1 for an initial capture and mark trial and 2 for a recapture trial).

Proportional abundance obtained from the number of distinct individuals captured from the two trials of each group:

$$\hat{K}_2 = \frac{r_2}{r_1}, \text{ where } r_1 = n_{11} + n_{12} - m_1 \text{ and } r_2 = n_{21} + n_{22} - m_2$$

The variance for $\hat{K}_2$ is estimated as

$$\hat{\mathrm{V}}\mathrm{ar}\left(\hat{K}_2/N_1, N_2\right) = \frac{(n_2 - m.)(n_{.1} - m.)\left(\hat{K}_2 + \hat{K}_2^{\,2}\right)}{n_{.2} n_{.1} r_1}$$

where r_1 and r_2 are the number of distinct individuals captured from population 1 and 2, respectively. A dot (.) denotes summation across population.

precise estimates than CPUE estimates, but it may require more resources. Ways to minimize various technical limitations of mark–recapture approaches to estimating relative abundance are beyond the scope of this chapter. For example, a general recommendation for a typical mark–recapture experiment is to use different fishing gear types for the mark and the subsequent recapture in order to decrease any biases from gear selectivity (e.g. White *et al.*, 1982). Use of more gear types will increase the cost of sampling. The monitoring team should consider the trade-off between reducing potential bias versus increasing costs.

CPUE estimation provides an alternative method for determining population abundance. Data required to estimate CPUE include the catch data (e.g. numbers, weight or proportions of target species) and the amount of fishing effort. In the case of transgenic fish, the CPUE can be estimated for the relative abundance of transgenics and conspecifics, and changes in CPUE over time may indicate population-level impacts. For comparative purposes, CPUE estimates need to come from similar fishing efforts in order to minimize biases from sampling methods. The CPUE can be estimated for any fishing gear, such as gill nets or electrofishing. For gill nets, the CPUE could be set for the net size and the fishing time (e.g. a 20 m net, with 2.5 cm mesh, for 12–15 h; Zakaria-Ismail and Sabariah, 1994). For electrofishing, the CPUE can be standardized either to the area of shocking (grams or number of individuals/m^2/100 m transect) or the shocking time (grams or number of individuals/h). This approach has been used to monitor fish population trends, but it requires caution regarding possible biases due to sampling (e.g. Paukert, 2004 and references therein). When estimation of absolute abundance is not necessary, the CPUE approach is favoured over the mark–recapture approach because estimating the CPUE can be less time-consuming, less labour-intensive and more cost-efficient (Nielsen, 1983; Hall, 1986; Coble, 1992).

Weaknesses of this approach include the sensitivity of its key assumption, namely the CPUE being proportional to population abundance (CPUE = qN), as well as high variation around the CPUE estimates. To meet this critical assumption, there are several requirements: (i) catchability coefficient (q), the probability of each individual fish being caught in one unit of effort is constant under all sampling conditions for a given gear type and targeted species; (ii) q is independent of N; and (iii) CPUE is a linear predictor for N. In practice, these requirements may not be achieved easily. To ensure that these requirements are addressed, researchers may need to carefully evaluate a sampling design for its sensitivity for a given budget and technical constraints.

In natural settings, catchability coefficient (q) of a fishing gear type is not likely to be constant over time (i.e. non-stationary). The variation of q values can be attributed to changing gear efficiency, fish behaviour, habitat types and environmental conditions (e.g. Nielsen, 1983; Hall, 1986; Arreguín-Sámchez and Pitcher, 1999; Ziegler *et al.*, 2003). For example, fish schooling under certain environmental conditions can lead to increases in q and CPUE estimates, while N may not increase (e.g. Hilborn and Walters, 1992). Hence, changes in the CPUE could result from changes in catchability, rather than from actual changes in population abundance. To understand the pattern of change in q values for a gear type with factors mentioned above, fish stock assessment scientists usually employ rigorous experimental designs and appropriate statistical models (e.g. Kimura, 1988; McInerny and Degan, 1993; Maunder and Punt, 2004; Hanchet *et al.*, 2005).

In addition to variation in q values, unexpected fluctuation in N can also impact the relationship between the CPUE and N. There may be three types of relationships between the CPUE and N: (i) linear; (ii) 'hyperstable', where CPUE estimates decline at a slower rate than N (Hilborn and Walters, 1992); or (iii) 'hyperdepleted', where CPUE estimates decline at a faster rate than N (Harley *et al.*, 2001). Non-linear relationships between the CPUE and N can lead to inaccurate estimates of N. Harley *et al.* (2001) suggest a basic mathematical model to explain possible relationships between the CPUE and N:

$$U = q \times N^{\beta} \quad (9.7)$$

where U is the CPUE; q is the catchability coefficient; N is abundance or density; and β is a model parameter that determines the nature of the relationship between the CPUE and N. If $\beta = 1$, the relationship between the CPUE and N is linear; if $\beta > 1$, the relationship is 'hyperdepleted'; and if $\beta < 1$, the relationship is 'hyperstable'. Both of these non-linear cases will have very important implications on the interpretation of CPUE data. Examining these relationships will also require extensive CPUE data, abundance estimates from other methods, rigorous experimental designs and appropriate expertise in fish population dynamics (e.g. Worthington *et al.*, 1998; Quinn and Deriso, 1999).

In addition to problems associated with q and N, the CPUE data tend to have large variation (e.g. Paukert, 2004) and violate the assumption of normality, with a rightward skewed frequency distribution (e.g. Bannerot and Austin, 1985; Guy *et al.*, 1996), which could bias the CPUE estimates and hypothesis testing. Therefore, it is recommended to first evaluate the normality of the CPUE data and then decide whether to transform data or select other appropriate non-parametric methods (Miranda *et al.*, 1996).

Population establishment of non-native transgenic species

Monitoring for population establishment may involve surveying for the presence of all life stages of the transgenic species, as well as the numbers of sexually mature adults, in the wild. Baseline information on diagnostic morphological characteristics at all life stages and baseline measurements for sexual maturity, such as a gonadosomatic index (Peterson *et al.*, 2004) or a proportion of mature gametes in gonads (determined via visual observation, biopsy or histological analysis), will be helpful. Monitoring surveys should use appropriate sampling techniques for each life stage and habitat in order to obtain a good representation of all life stages. To detect increasing or decreasing numbers of transgenic fish over time, analysts could employ methods for estimating population abundance (previous section) and species-relative abundance (the next section), and then test for statistical changes in the parameters.

Changes in fish community composition

Ecologists usually describe fish community composition in terms of species richness and species-relative abundance (Hayek and Buzas, 1997; Krebs, 1999). Species richness refers to the number of species present in a specified sampling location. Estimating species richness assumes that clear taxonomic distinction among species is possible. Relative abundance is calculated as percentages or proportions of different species are sampled. The accuracy of richness and abundance estimates is sensitive to sampling design and techniques; using several sampling gears can reduce bias due to selectivity of each gear type.

Estimation of species-relative abundance is similar to estimation of the CPUE: the numbers or proportions of target species are standardized against the fishing gears. The sampling units – individuals versus clusters – will influence the calculation of means and variance (Box 9.5). Hayek and Buzas (1997) suggest two different approaches to estimate species-relative abundance and standard error of means, based on the type of sampling unit. Individuals can be randomly selected as independent sampling units when the sampling gear captures individual fish (e.g. electrofishing), or selected as a cluster when the gear captures a group of

Box 9.5. Equations for calculating point estimates of species-relative abundance and variance (Hayek and Buzas, 1997).

To detect the change in the species-relative abundance (p) over time (i.e. effect size), examine the difference in the sequential estimates of p at times $t = 0$ and $t = t$.

Sampling unit: individual

Observed proportion approximation (binomial distribution): $p = \hat{\mu} = \frac{1}{n}\sum_{i=1}^{n} x_i = \frac{1}{n}(k) = \frac{k}{n}$

Sampling variance: $\hat{\sigma}_p^2 = \frac{\hat{\sigma}^2}{n} = \frac{1}{n}\left(\frac{n}{n-1}\right)pq = \frac{pq}{n-1}$

where k = total number of individuals belonging to species i (e.g. transgenic species); p = the proportion of individuals belonging to species i; q = the probability that individual does not belong to the species i; and n = total number of all individuals collected.

Sampling unit: cluster

Observed proportion approximation (biased): $p = \frac{\sum_{i=1}^{n} a_i}{\sum_{i=1}^{n} m_i}$

Average number of individuals per replicate (cluster): $\hat{\mu}_m = \frac{\sum_{i=1}^{n} m_i}{n} = \frac{m}{n}$

The sample variance for the proportion: $\hat{\sigma}_{pclus}^2 = \frac{\sum_{i=1}^{n} m_i^2 (p - p_i)^2}{n(n-1)\hat{\mu}_m^2}$

where i = a specified cluster; n = numbers of replicates; a = individuals of interested species; and m = individuals of all species.

Standard errors (implying confidence limits) obtained from clustered sampling ($\hat{\sigma}_{pclus}$) are generally larger than those estimated for the independent individual sampling ($\hat{\sigma}_p$), except when $p < 1\%$ (i.e. the targeted species is rare).

fish (e.g. gill netting). Box 9.5 illustrates the calculations for obtaining point estimates, and their variance, of the species-relative abundance.

Remedial Responses

The detection of the chosen end points (e.g. presence of F_1 hybrids and backcross individuals) should trigger remedial responses. Therefore, it is important to simultaneously plan monitoring and remedial actions when conducting an environmental risk assessment of transgenic fish. Depending on the severity of the impacts, management options may include improving the efficiency of confinement measures in order to reduce the number of escapes, removing transgenic individuals from the affected environment or restricting aquaculture production of transgenic fish. Improving confinement can help if the escapees and descendants have not become widespread. For example, Wenner and Knott (1991) documented that regulations restricting aquaculture escapees of Pacific whiteleg shrimp (*Peneaus vannamei*) in South Carolina, USA, could reduce proportions of white shrimp in commercial catches from an average of 3% to 0.16% in as short a period as 1 year. However, highly successful invaders in local ecosystems could be very difficult to eradicate (e.g. common carp in Australia) (Koehn, 2004).

The analytic-deliberative process described in Chapters 1 and 2 will help identify appropriate mechanisms for monitoring and executing remedial responses. In some countries, there may already be established procedures for implementing monitoring programmes. For example, the Cuban National Center for Biological Safety (NCBS), an agency responsible for organizing, directing and executing the country's national system for biological safety, is establishing a procedure to review and evaluate monitoring and remedial action plans, submitted by the applicant (e.g. a biotechnology firm) on a case-by-case basis. If the plan is approved, the NCBS will assign third-party inspectors to conduct the monitoring plan.

To date, remedial measures have not been developed for any transgenic species. Potentially relevant measures are based on existing practices for fisheries management (e.g. Wenner and Knott, 1991). Generally, existing measures have not been very successful in removing invading aquatic species, such as carp or tilapia. Any risk management plan for transgenic fish should not rely solely on remedial measures.

Types of Research Needed to Address Knowledge Gaps for Monitoring

Developing countries generally have insufficient data to effectively plan or conduct a monitoring programme. Studies in the following areas could provide needed baseline information, as well as opportunities to evaluate sampling efficiency:

- Patterns of escapes and efficacy of confinement measures;
- Movement of the target species within the local habitats;

- Genetic diversity of transgenic lines and potentially impacted populations;
- Abundance of potential receiving populations and their temporal and spatial variation, which may require sampling of several generations (age-classes);
- Fish community composition and natural variation;
- Sampling efficiency for different locations, gear types and environmental conditions.

Research is also needed to improve methods for detection and removal of transgenic fish in the wild. Standardized molecular genetic or morphological methods are needed to distinguish transgenic fish from non-transgenic wild relatives. Also needed are feasible and cost-effective methods to eradicate or greatly reduce the abundance of transgenic fish that have established naturally reproducing populations in the wild.

Chapter Summary

A monitoring programme for detecting potential ecological consequences from the presence and spread of transgenic organisms is an important component of risk management in two ways. First, it provides evidence of transgenic fish escapes, reflecting the effectiveness of risk reduction measures (e.g. physical confinement measures). Second, it provides data on whether the ecological consequences predicted by experimental risk assessments and modelling actually occur. This chapter primarily addresses scientific considerations for a monitoring programme designed to serve the first purpose. The chapter's approaches can be used for the second purpose once empirical data from laboratories and microcosms become available (Chapters 5 and 6, this volume), and measurable end points can be defined adequately.

The chapter highlights six monitoring end points (Fig. 9.1), most of which are anticipated to occur over short timeframes (i.e. a few generations or less). Our approach represents an example of how to develop measurable variables for any monitoring programme. The proposed end points allow early detection of the impacts of transgenic fish, which is important for effective and timely remedial actions. The chapter, therefore, focuses on issues relevant to the sensitivity of tools to detect these end points.

The first consideration when designing a monitoring programme involves issues related to the sensitivity of sampling design and effort. The ability to detect the ecological end points depends on the presence of transgenic escapees and descendants in the wild, the efficiency of sampling design and equipment and the level of monitoring effort. A large amount of effort will be required to detect small ecological changes caused by the spread of transgenic individuals. If the impact is small (e.g. one transgenic individual in a wild population of 100,000 individuals), an affordable sampling design (typically one that samples only a few individuals using a small number of nets) will likely not be powerful enough to detect this impact.

If transgenic individuals are present at a detectable level, the second consideration deals with the sensitivity of methods used to estimate their population abundance and demographic contribution to the local fish community, as well as distinguishing transgenic individuals from their non-transgenic counterparts. We assume that the approved transgenic line has stable transmission of the transgene construct. Even though the scientific methods described in this chapter are common for aquatic ecology and fisheries research, their sensitivity and efficiency still needs to be evaluated for transgenic species, potentially impacted fish populations and communities, as well as potential receiving environments. Baseline information specific to transgenic lines and potential receiving ecosystems is also critical before making inferences about any ecological changes. In most developing countries, useful information is quite scattered. Scientists may need to search existing literature, as well as conduct new research, to collect baseline information and design more efficient protocols for sampling different fish species in habitats with different characteristics.

Possible remedial responses may range from improving confinement measures, removing transgenic fish from the wild and restricting the use of transgenics in aquaculture operations. It is important to note that the benefits of these actions are reduced once ecological impacts become widespread. Preventative approaches, such as highly effective and redundant confinements (Chapter 8, this volume), or early detection of impacts, could greatly aid effective risk management.

This chapter aims to provide planning and monitoring tools for a multidisciplinary team. Accurate field assessment of the monitoring end points discussed in this chapter is a very challenging task, with many potential pitfalls. Therefore, it is important that the monitoring team has the necessary breadth of expertise and recognizes important assumptions and technical requirements for each method. The relevant areas of expertise may include field ecology, population and quantitative genetics, molecular and genomics technologies, fish population dynamics and fish stock assessment, fisheries biology, biological statistics, geographic information systems (GIS), aquatic animal health and natural resource economics. The chapter also suggests types of research needed for gathering baseline data.

References

Arreguín-Sámchez, F. and Pitcher, T.J. (1999) Catchability estimates and their application to the red grouper (*Epinephelus morio*) fishery of the Campeche Bank, Mexico. *Fisheries Bulletin* 97, 746–757.

Avise, J.C. (1994) *Molecular Markers, Natural History and Evolution*. Chapman & Hall, New York.

Bannerot, S.P. and Austin, C.B. (1985) Using frequency distributions of catch per unit effort to measure fish-stock abundance. *Transactions of the American Fisheries Society* 112, 608–617.

Brown, M.L. and Austen, D.J. (1996) Data management and statistical techniques. In: Murphy, B.R. and Willis, D.W. (eds) *Fisheries Techniques*, 2nd edn. American Fisheries Society, Bethesda, Maryland, pp. 17–62.

Buitkamp, J. and Epplen, J.T. (1996) Modern genome research and DNA diagnostics in domestic animals in the light of classical breeding techniques. *Electrophoresis* 17, 1–11.

Burgman, M.A. (2005) *Risks and Decision for Conservation and Environmental Management*. Cambridge University Press, Cambridge.

Caelers, A., Maclean, N., Hwang, G., Eppler, E. and Reinecke, M. (2005) Expression of endogenous and exogenous growth hormone (GH) messenger (m)RNA in a GH-transgenic tilapia (*Oreochromis niloticus*). *Transgenic Research* 14, 95–104.

Coble, D.W. (1992) Predicting population density of largemouth bass from electrofishing catch per effort. *North American Journal of Fisheries Management* 12, 650–652.

Cohen, J. (1988) *Statistical Power Analysis for the Behavioral Sciences*, 2nd edn. Lawrence Erlbaum Associates, Hillsdale, New Jersey.

Cressie, N.A.C. (1993) *Statistics for Spatial Data*. Wiley, New York.

Crook, D.A. (2004) Movements associated with home-range establishment by two species of lowland river fish. *Canadian Journal of Fisheries and Aquatic Sciences* 61, 2183–2193.

Devlin, R.H., Yesaki, T.Y., Donaldson, E.M. and Hew, C.L. (1995) Transmission and phenotypic effects of an antifreeze/GH gene construct in coho salmon (*Oncorhynchus kisutch*). *Aquaculture* 137, 161–169.

Downes, B.J., Barmuta, L.A., Fairweather, P.G., Faith, D.P., Keough, M.J., Lake, P.S., Mapstone, B.D. and Quinn, G.P. (2002) *Monitoring Ecological Impacts: Concepts and Practice in Flowing Waters*. Cambridge University Press, Cambridge.

Dunham, R.A., Chatakondi, N., Nichols, A., Chen, T.T., Powers, D.A. and Kucuktas, H. (2002) Survival of F_2 transgenic common carp (*Cyprinus carpio*) containing pRSVrtGH1 complementary DNA when subjected to low dissolved oxygen. *Marine Biotechnology* 4, 323–327.

Ferraris, J.D. and Palumbi, S.R. (eds) (1996) *Molecular Zoology: Advances, Strategies, and Protocols*. Wiley-Liss, New York.

Fonticiella, D. and Arboleya, Z. (2003) *Peso estandar de la tilapia (Oreochromis aureus, Steindachner) en Cuba*. Ministerio de la Industria Pesqiera. Empresa Pesquera de Villa Clara 'PESCAVILLA' y Centro de Preparacion Acuicola de Mamposton 'CEPAM'. La Habana, Cuba.

Fox, D.R. (2001) Environment power analysis – a new perspective. *Environmetrics* 12, 437–449.

Froese, R. and Pauly, D. (eds) (2007) *FishBase*. World Wide Web electronic publication. Available at: www.fishbase.org, version (02/2007)

Greenberg, A.E., Clesceri, L.S. and Eaton, A.D. (1992) *Standard Methods for the Examination of Water and Wastewater*, 18th edn. American Public Health Association, Washington, DC.

Guy, C.S., Willis, D.W. and Schultz, R.D. (1996) Comparison of catch per unit effort and size structure of white crappies collected with trap nets ad gill nets. *North American Journal of Fisheries Management* 16, 947–951.

Hall, T.J. (1986) Electrofishing catch per hour as an indicator of largemouth bass density in Ohio impoundments. *North American Journal of Fisheries Management* 6, 397–400.

Hanchet, S.M., Blackwell, R.G. and Dunn, A. (2005) Development and evaluation of catch per unit effort indices for southern blue whiting (*Micromesistius australis*) on the Campbell Island Rise, New Zealand. *ICES Journal of Marine Science* 62, 1131–1138.

Harley, S.J., Myers, R.A. and Dunn, A. (2001) Is catch-per-unit-effort proportional to abundance? *Canadian Journal of Fisheries and Aquatic Sciences* 58, 1760–1772.

Hayek, L.A.C. and Buzas, M.A. (1997) *Surveying Natural Populations*. Columbia University Press, New York.

Hayes, K.R., Cannon, R., Neil, K. and Inglis, G. (2005) Sensitivity and cost considerations for the detection and eradication of marine pests in ports. *Marine Pollution Bulletin* 50, 823–834.

Hayes, M.L. (1983) Active fish capture methods. In: Nielsen, L.A. and Johnson, D.L. (eds) *Fisheries Techniques*. American Fisheries Society, Bethesda, Maryland, pp. 123–145.

Hilborn, R. and Walters, C.J. (1992) *Quantitative Fisheries Stock Assessment: Choice, Dynamics & Uncertainty*. Chapman & Hall, New York.

Hitt, N.P., Fissell, C.A., Muhlfeld, C.C. and Allendorf, F.W. (2003) Spread of hybridization between native westslope cutthroat trout, *Oncorhynchus clarki lewisi*, and nonnative rainbow trout, *Oncorhynchus mykiss. Canadian Journal of Fisheries and Aquatic Sciences* 60, 1440–1451.

Hubert W.A. (1996) Passive capture techniques. In: Murphy, B.R. and Willis, D.W. (eds) *Fisheries Techniques*, 2nd edn. American Fisheries Society, Bethesda, Maryland, pp. 157–192.

International Union for the Conservation of Nature and Natural Resources (IUCNNR) (2006) The World Conservation Union. Available at: http://www.iucn.org

Kanda, N., Leary, R.F., Spruell, P. and Allendorf, F.W. (2002) Molecular genetic markers identifying hybridization between the Colorado River-Greenback cutthroat trout complex and Yellowstone cutthroat trout or rainbow trout. *Transactions of the American Fisheries Society* 131, 312–319.

Kapuscinski, A.R. (2005) Current scientific understanding of the environmental biosafety of transgenic fish and shellfish. *Revue Scientifique et Technique de l'Office International des Epizooties* 24, 309–322.

Kelso, W.E. and Rutherford, D.A. (1996) Collection, preservation, and identification of fish eggs and larvae. In: Murphy, B.R. and Willis, D.W. (eds) *Fisheries Techniques*, 2nd edn. American Fisheries Society, Bethesda, Maryland, pp. 255–302.

Kimura, D.K. (1988) Analyzing relative abundance indices with log-linear models. *North American Journal of Fisheries Management* 8, 175–180.

Koehn, J.D. (2004) Carp (*Cyprinus carpio*) as a powerful invader in Australian waterways. *Freshwater Biology* 49, 882–894.

Krebs, C.J. (1999) *Ecological Methodology,* 2nd edn. Addison-Wesley/Longman, Menlo Park, California.

Labovitz, S. (1968) Criteria for selecting a significance level: a note on the sacredness of.05. *American Sociologist* 3, 200–222.

Lagler, K.F. (1978) Capture, sampling and examination of fishes. In: Bagenal, T. (ed.) *Methods for Assessment of Fish Production in Fresh Waters*. Blackwell Scientific, Oxford, UK, pp. 7–47.

Legendre, P. and Legendre, L. (1998) *Numerical Ecology*. Elsevier, Amsterdam, The Netherlands.

Lenon, R.E., Hunn, J.P., Schnick, R.A. and Burress, R.M. (1970) *Reclamation of Ponds, Lakes and Streams with Fish Toxicants: A Review*. Food and Agriculture Organization of the United Nations, Rome, Italy.

Lindquist, D.C. and Shaw, R.F. (2005) Effects of current speed and turbidity on stationary light-trap catches of larval and juvenile fishes. *Fisheries Bulletin* 103, 438–444.

Liu, Z. and Cordes, J.F. (2004) DNA marker technologies and their applications in aquaculture genetics. *Aquaculture* 238, 1–37.

Lu, Y. and Sun, P. (2005) Viral resistance in shrimp that express an antisense Taura syndrome virus coat protein gene. *Antiviral Research* 67, 141–146.

Martin-Smith, K.M. (1998) Temporal variation in fish communities from the upper Segama River, Sabah, Malaysian Borneo. *Polskie Archiwum Hydrobiologii* 45, 185–200.

Maunder, M.N. and Punt, A.E (2004) Standardizing catch and effort data: a review of recent approaches. *Fisheries Research* 70, 141–159.

McInerny, M.C. and Degan, D.J. (1993) Electrofishing catch rates as an index of largemouth bass population density in two large reservoirs. *North American Journal of Fisheries Management* 13, 223–228.

Meehan-Meola, D., Xu, Z. and Alcívar-Warren, A. (2007) Development of microsatellite genetic markers to assess genetic risks to wild penaeid populations caused by the accidental release of cultured or transgenic shrimp: preliminary results. *Journal of Shellfish Research* (accepted).

Miranda, L.E., Hubbard, W.D., Sangare, S. and Holman, T. (1996) Optimizing electrofishing sample duration for estimating relative abundance of largemouth bass in reservoirs. *North American Journal of Fisheries Management* 16, 324–331.

Muir, W.M. and Howard, R.D. (1999) Possible ecological risks of transgenic organism release when transgenes affect mating success: sexual selection and the Trojan gene hypothesis. *Proceedings of the National Academy of Sciences USA* 96, 13853–13856.

National Center for Biotechnology Information (NCBI) (2006) *GenBank*. U.S. National Library of Medicine, Bethesda, Maryland. Available at: http://www.ncbi.nih.gov/Genbank/index.html

Nielsen, L.A. (1983) Variation in the catchability of yellow perch in an otter trawl. *Transactions of the American Fisheries Society* 112, 53–59.

Otis, D.L., Burnham, K.P., White, C.C. and Anderson, D.R. (1978) Statistical inference from capture data on closed animal populations. *Wildlife Monographs* 62, 1–135.

Paukert, C.P. (2004) Comparison of electrofishing and trammel netting variability for sampling native fishes. *Journal of Fish Biology* 65, 1643–1652.

Peterman, R.M. (1990) Statistical power analysis can improve fisheries research and management. *Canadian Journal of Fisheries and Aquatic Sciences* 47, 2–15.

Peterson, M.S., Slack, W.T., Brown-Perterson, N.J. and McDonald, J.L. (2004) Reproduction in nonnative environments: establishment of Nile tilapia, *Oreochromis niloticus*, in coastal Mississippi watersheds. *Copeia* 4, 842–849.

Quinn II, T.J. and Deriso, R.B. (1999) *Quantitative Fish Dynamics*. Oxford University Press, New York.

Quinn, G.P. and Keough, M.J. (2002) *Experimental Design and Data Analysis for Biologists*. Cambridge University Press, Cambridge.

Rabeni, C.F. (1996) Invertebrates. In: Murphy, B.R. and Willis, D.W. (eds) *Fisheries Techniques*, 2nd edn. American Fisheries Society, Bethesda, Maryland, pp. 335–351.

Rahman M.A., Mak, R., Ayad, H., Smith, A. and Maclean, N. (1998) Expression of novel piscine growth hormone gene results in enhancement of growth in transgenic tilapia (*Oreochromis niloticus*). *Transgenic Research* 7, 357–369.

Seber, G.A.F. (1982) *The Estimation of Animal Abundance and Related Parameters*, 2nd edn. Macmillan, New York.

Sitar, S.P., Bence, J.R., Johnson, J.E., Ebener, M.P. and Taylor, W.W. (1999) Lake trout mortality and abundance in southern Lake Huron. *North American Journal of Fisheries Management* 19, 881–900.

Skalski, J.R., Robson, D.S. and Simmons, M.A. (1983) Comparative census procedures using single mark-recapture methods. *Ecology* 64, 752–760.

Slavik, O. and Bartos, L. (2004) What are the reasons for the Prussian carp expansion the upper Elbe River, Czech Republic? *Journal of Fish Biology* 65 (Suppl. A), 240–253.

South Carolina Estuarine and Coastal Assessment Program (2006) South Carolina Department of Natural Resources (SCDNR) and South Carolina Department of Health and Environmental Control. Available at: http://www.dnr.sc.gov/marine/scecap/index.htm

Tate, W.B., Allen, M.S., Myers, R.A. and Estes, J.R. (2003) Comparison of electrofishing and rotenone for sampling largemouth bass in vegetated areas of two Florida lakes. *North American Journal of Fisheries Management* 23, 181–188.

Taylor, B.L. and Gerrodette, T. (1993) The uses of statistical power in conservation biology: the Vaquita and northern spotted owl. *Conservation Biology* 7, 489–500.

Thomas, L. (1996) Monitoring long-term population change: why are there so many analysis methods? *Ecology* 77, 49–58.

Thompson, S.K. (1992) *Sampling*. Wiley, New York.

Underwood, A.J. (1997) *Experiments in Ecology: Their Logical Design and Interpretation Using Analysis of Variance*. Cambridge University Press, Cambridge.

Vignal, A., Milan, D., San Cristobal, M. and Eggen, A. (2002) A review on SNP and other types of molecular markers and their use in animal genetics. *Genetics Selection Evolution* 34, 275–305.

Wenner, E.L. and Knott, D.M. (1991) Occurrence of Pacific white shrimp, *Penaeus vannamei*, in coastal waters of South Carolina. In: *Proceedings of the Conference and Workshop, Introduction and Transfers of Marine Species: Achieving a Balance Between Economics and Development and Resources Protection*. South Carolina, pp. 173–181.

White, G.C., Anderson, D.R., Burnham, K.P. and Otist, D.L. (1982) *Capture–Recapture and Removal Methods for Sampling Closed Populations*. Los Alamos National Laboratory Publication, LA-8787-NERP, Los Alamos, New Mexico.

Williams, B.K., Nichols, J.D. and Conroy, M.J. (2002) *Analysis and Management of Animal Populations*. Academic Press, New York.

Worthington, D.G., Andrew, N.L. and Bentley, N. (1998) Improved indices of catch rate in the fishery for blacklip abalone, *Haliotis rubra*, in New South Wales, Australia. *Fisheries Research* 36, 87–97.

Wright, J.F., Sutcliffe, D.W. and Furse, M.T. (eds) (2000) *RIVPACS and Other Techniques*. Freshwater Biological Association, Ambleside, UK.

Wu, B., Sun, Y.H., Wang, Y.W., Wang, Y.P. and Zhu, Z.Y. (2005) Characterization of transgene integration pattern in F_4 hGH-transgenic common carp (*Cyprinus carpio* L.). *Cell Research* 15, 447–454.

Yaskowiak, E.S., Shears, M.A., Agarwal Mawal, A. and Fletcher, G.L. (2006) Characterization and multi-generational stability of the growth hormone transgene (EO-1alpha) responsible for enhanced growth rates in Atlantic salmon. *Transgenic Research* 15, 465–480.

Yue, G.H., Ho, M.Y., Orban, L. and Komen, J. (2004) Microsatellites within genes and ESTs of common carp and their applicability in silver crucian carp. *Aquaculture* 234, 85–98.

Zakaria-Ismail, M. and Muhamad-Mazli, A. (1995) Autecological study of *Mystacoleucus marginatus* (Teleostei: Cyprinidae) in the Klang Gates reservoir, Selangor, Malaysia. *Malaysian Journal of Science* 16A, 7–11.

Zakaria-Ismail, M. and Sabariah, B. (1994) Ecological study of fishes in a small tropical stream (Sungai Kanching, Selangor, Peninsular Malaysia) and its tributaries. *Malaysian Journal of Science* 15A, 3–7.

Ziegler, P.E., Frusher, S.D. and Johnson, C.R. (2003) Space-time variation in catchability of southern rock lobster *Jasus edwardsii* in Tasmania explained by environmental, physiological and density-dependent processes. *Fisheries Research* 61, 107–123.

10 Risk Assessment of Transgenic Fish: Synthesis and Conclusions

A.R. Kapuscinski, G. Dana, K.R. Hayes, S. Li, K.C. Nelson, Y.K. Nam, Z. Gong, R.H. Devlin, G.C. Mair and W. Senanan

Introduction

This book presents a comprehensive risk assessment framework and science-based methodologies for assessing and managing potential environmental risks pertaining to aquaculture and other uses of transgenic fish. The framework and methodologies are also useful for addressing potential environmental risks associated with other kinds of genetically improved aquatic organisms, such as selectively bred lines. It represents a global collaboration among scientists to synthesize and refine the most recent scientific knowledge and methodologies needed to inform environmental risk assessments of transgenic fish. Given that transgenic finfish and shellfish are likely to be the first commercialized transgenic animals produced on a large scale, this book provides a timely, science-based tool for informing the development and implementation of biosafety policy.

There is an urgent need for rigorous environmental risk assessment as biotechnologists further develop different lines of transgenic fish and shellfish (Tables 3.1 and 3.2). Conducting such assessments on a case-by-case basis, as recommended throughout this book, may seem daunting, particularly when the necessary scientific capacity, confined testing facilities and funds are limited. The previous chapters in this book provide much needed guidance on how to structure and conduct an environmental risk assessment process using a systematic approach of science-based risk assessment methods combined with multi-stakeholder deliberations to address the potential environmental effects of transgenic fish. Previous chapters also provide methodologies for managing identified environmental risks, including confinement and post-approval monitoring. This concluding chapter summarizes the main messages of the book, identifies opportunities to improve existing knowledge, suggests ways to improve capacity for conducting environmental risk assessment and management and proposes a proactive approach for addressing environmental biosafety from the initial stages of development of a transgenic fish line.

Major Messages from the Book

The major messages contained in the preceding chapters are given below.

1. Risk assessment should be a participatory and transparent process. Risk assessment should involve all interested and affected parties (the stakeholders), incorporating their perspectives and knowledge, to ensure that the process produces a socially, as well as scientifically acceptable, outcome (NRC, 1996). This approach requires that all participating parties believe that risk assessment is an appropriate decision support tool in this context. Stakeholder participation can enhance legitimacy and public trust of the risk assessment conclusions and improve the quality of the assessment. This is because people with diverse experiences can provide information and insights that a technically oriented team of scientists and risk analysts simply cannot have (Gibbons, 1999; Sagar *et al.*, 2000; Susskind *et al.*, 2000; Wondolleck and Yaffee, 2000). Each country will need to determine who its stakeholders are and how to involve them in the process of risk assessment and management (Chapter 2, this volume). Clearly, the precise composition of stakeholder groups will vary from case to case, but participants will likely include developers of the transgenic fish, risk analysts from the range of required scientific fields, aquaculturists who might produce transgenic fish, any affected groups that ultimately bear the risks associated with the release of transgenic fish (e.g. aquaculturists, fish importers and trade organizations, fishermen), fisheries, aquatic and environmental managers, as well as parties with an interest in and working knowledge of the environment into which transgenic fish may be released. Observers of risk assessment deliberations may include additional groups of stakeholders (Chapter 2, this volume).

It is important that a diverse and representative assemblage of stakeholders share a common understanding of the societal problem (e.g. insufficient production of nutritious fish protein) that would be addressed by use of the proposed technology (i.e. transgenic fish). The problem formulation and options assessment (PFOA) methodology is designed to enable a group of stakeholders to consider and develop a shared understanding of future alternatives that serve as potential options for meeting a societal need, to evaluate each option's attributes, potential system changes and adverse effects if they are used (Chapter 2, this volume). One option might be to farm transgenic fish with specific traits that boost yields per unit of input; other ways to achieve the same goal could include changing the management of live and formulated feeds in the aquaculture system, shifting from a monoculture to a polyculture system or introducing selectively bred strains of fish. Arriving at such a shared understanding of the problem and potential options at the outset builds the foundation for subsequent steps in the risk assessment and management process.

This book espouses a highly interactive risk assessment framework with stakeholders involved at key stages, using the PFOA process to structure their involvement. Each country will need to determine an appropriate level of stakeholder participation that fits with its society and available resources.

Stakeholder deliberation can be particularly helpful at the following steps: defining the scope and boundaries of the risk assessment, selecting end points of environmental changes to be estimated, identifying criteria for acceptable and unacceptable risks, identifying environmental hazards to be considered, bringing forward information that will assist analysts to reduce or characterize uncertainty in the risk calculations and determining risk management strategies.

2. Risk assessment should link analysis and deliberation. An effective risk assessment requires scientific analysis and multi-stakeholder deliberation (NRC, 1996; Hilbeck and Andow, 2004; Hilbeck *et al.*, 2006). This book recommends quantitative and qualitative methods to gather and analyse relevant data on the hazards and risks associated with transgenic fish. Chapter 5, for example, presents approaches for estimating the probability of gene flow to wild relatives (upon escape of transgenics), while Chapters 4, 5 and 6 present methods to assess the potential effects of this hazard at molecular, organismal, population and ecosystem levels. Scientifically defensible risk analysis requires that the types and sources of uncertainty associated with empirical data and risk assessment methods are identified and described (Chapters 5 and 6, this volume) and, as far as possible, honestly propagated through the risk assessment (Chapter 7, this volume). An informed deliberation process should ensure that stakeholders are aware of the availability of data, their limitations and the different sources of uncertainty in a risk assessment process. Risk assessments that honestly acknowledge uncertainty will usually provide intervals that extend above and below the most likely or plausible risk estimate. Careful deliberation and consideration of the probable or plausible range of a risk estimate is important in order to make decisions about whether the risk is acceptable or not.

3. Risk assessment as an iterative process. The process of risk assessment and deliberation should allow for iteration of key steps in light of new information. Aspects of the risk assessment process, such as data gathering, risk estimation, uncertainty analysis and risk management evaluation, may need to be repeated as data and insights generated within one part of the assessment process reveal the need to revise another part (Chapter 1, this volume). For example, if an experiment conducted to evaluate competition between transgenic fish and a native species (Chapter 6, this volume) also uncovers a genotype-by-environment interaction that was not included in the prior estimation of the probability of gene flow (Chapter 5, this volume), it would be essential to determine whether this interaction could affect the probability and effects of gene flow, and if so, to repeat the gene flow assessment. Likewise, iterative deliberation would be needed to allow the multiple stakeholders to consider major new findings of this sort (Chapter 1, this volume), and also to integrate these findings with broader societal considerations (Chapter 2, this volume).

4. Carefully define and prioritize the problem addressed by the risk assessment. Defining the risk assessment problem using the input of multiple stakeholders will ensure that the subsequent analysis addresses the most important questions and issues. In this context, defining the problem refers not to the larger societal problem identified by multiple stakeholders at the start of a PFOA (Chapter 2, this volume), but rather to the scope, boundaries, end points and hazards associated with the technology option(s) that stakeholders have agreed should undergo a complete risk assessment. Careful problem definition is also important when constructing a conceptual model of the societal and environmental system that might be affected by application of the technology (Chapters 1 and 2, this volume).

The definition of a conceptual model and efficient hazard analysis requires a certain level of abstraction and aggregation that reduces the complexity of the real world to a manageable level without losing sight of critical interactions (Chapter 6, this volume). For instance, multi-stakeholder deliberations on potential adoption of transgenic tilapia in a specific area of East Africa could start defining the system of concern by hand-drawing a visual map that shows the area's tilapia fish farms, their connections to natural wetlands and lakes and then listing the species of organisms and capture fishing occurring in these natural waters. The species of fish, aquatic invertebrates, aquatic plants and birds that co-occur in natural habitats of native tilapia might then be aggregated into functionally equivalent groups, such as competitors, prey and predators, for the purposes of the risk assessment. Poor problem definition, abstraction, aggregation and conceptualization can lead the subsequent hazard analysis astray, possibly leading to prioritization of the wrong hazards and a misguided risk assessment that does not answer the most critical concerns of affected parties. For instance, the existing trade of live tilapia in East Africa could be a major pathway for future spread of transgenic fish; thus it could be a serious mistake to omit such trade in a conceptual model of the system. The information needed to adequately describe the system of concern for this hypothetical example underscores the importance of involving the appropriate mix of stakeholders, in addition to biologists and risk specialists. Much of the information mentioned above will have to come from farmers who presently raise tilapia, fishermen, merchants, traders or others intimately familiar with relevant farming systems and trade practices.

5. Focus on measurable end points. Great care must be taken to identify traits of the transgenic fish and components of the ecosystem which are feasible to measure, as well as being good scientific indicators of whether a specific environmental harm will occur. The potential environmental hazards associated with escape or intentional release of a specific line of transgenic fish range from changes occurring at the molecular level to the ecosystem level. The decline of a native fish species caused by gene flow from transgenics to wild relatives is one example of a potential end point of a hazard event chain; this event chain can be broken down into discrete stages such as escape of the transgenic fish into the wild, survival of escapees to sexual maturation, successful mating in the wild with native fish and subsequent negative impacts of expression of the transgene on the fitness of the hybrid offspring (Chapter 5, this volume). It is impossible to conduct prospective analysis or post-approval monitoring of gene flow without specifying a measurable end point (Chapters 5 and 9, this volume), such as introgression of the transgene into the wild population at the first backcross generation (Fig. 5.1). However, in general a risk assessment's ability to provide honest and accurate predictions diminishes as the length of hazard event chains increases because of parallel increases in its complexity. The risk assessment should, therefore, carefully choose assessment end points that seek to minimize the complexity of the assessment process while maintaining relevance to stakeholders. In the context of the gene flow example, the assessment may be well advised to seek measurement end points earlier in the event chain, such as escape of transgenic fish into the wild or survival of escaped transgenic fish to sexual maturity, depending on the details of the case in question and the level of scientific uncertainty.

6. Risk assessment and management should be done on a case-by-case basis. The myriad of considerations that inform the environmental risk assessment process necessitates a case-by-case approach to assessing and managing risks of transgenic fish. Each proposed use of a specific line of transgenic fish is different, being influenced by a variety of processes at the molecular, organismal, population and ecosystem levels (Chapters 4–6, this volume). The opportunities, challenges and priorities for risk management also vary case-by-case, depending on the characteristics of the transgenic fish line, the aquaculture systems and the natural aquatic ecosystems (Chapters 8 and 9, this volume). Furthermore, each country's regulatory and policy environment is different; its societal problems and its stakeholders will differ in each context. The fact that each country operates in a different policy environment is most vividly displayed in Chapter 2 in the country-specific diagrams of potential policy frameworks into which the multi-stakeholder methodologies of the PFOA could be incorporated. Clearly, the risk assessment process must be flexible enough to accommodate each country's specific situation. The range of potential hazards and definition of harms will often be different in each situation, and different prioritizations and levels of acceptable risk will arise in each deliberation. It is impossible to make overarching statements about levels of risk and its acceptability, making it critically important to conduct a risk assessment and plan and implement risk management measures on a case-by-case basis. Nevertheless, many of the elements of scientific analysis undertaken for a risk assessment in one context may be applicable in others.

7. The methodologies in this book apply to genetically improved aquatic organisms other than transgenic fish. The methodologies compiled in this book are applicable to various kinds of genetically improved lines of aquatic species, not only transgenic ones. Other kinds of genetically improved fish and shellfish, such as those generated through selective breeding or interspecific hybridization, can also pose environmental risks when the improved lines escape from aquaculture systems or are purposefully released to enhance a capture fishery (see reviews in Kapuscinski and Brister, 2001; Gupta *et al.*, 2004; Bartley and Kapuscinski, in press). Indeed, PFOA (Chapter 2, this volume) provides a structured process for considering various aquaculture technology options for meeting a broad societal problem (e.g. how to increase the production of nutritious fish protein). It is possible that a multi-stakeholder group would identify selectively bred and other genetically improved lines, along with a line of transgenic fish, as possible options when considering how to increase aquaculture production.

Although a substantial amount of research is currently devoted to transgenic strains (Chapter 3, this volume), this technology has yet to be proven in commercial production. Therefore, affected parties may also wish to explore alternative options for meeting their specified aquaculture goals. For example, one option for increasing productivity of tilapia aquaculture is to farm a selectively bred 'genetically improved farmed tilapia' (GIFT) strain (Asian Development Bank, 2005). The use of the GIFT strain may raise environmental biosafety concerns similar to those of transgenic tilapia, such as the consequences of gene flow to native wild tilapia populations in Africa or to feral tilapia elsewhere; and the methodologies presented in this book can be equally applied to address these concerns. The scientific methodologies discussed in Chapters 5–9 are also applicable for risk assessment and management of selectively bred lines of any fish or shellfish species.

The methodologies for scientific analysis and multi-stakeholder deliberation presented in this book are also well suited for assessing and managing the environmental risks posed by transgenic shellfish. Most of the chapter discussions have focused on transgenic finfish because they are further developed and because there is more publicly available, relevant information on them than on transgenic shellfish.

8. Capacity building is necessary to improve risk assessment methods and their use. Increased capacities (e.g. trained scientists, risk assessment specialists, trained conveners of the PFOA process, laboratory facilities and data-sharing tools) for conducting environmental risk assessment of transgenic fish, and other options such as selectively bred lines, are urgently needed, particularly in developing countries. A lack of baseline data, trained risk analysts, laboratory and research facilities and policies that enable these gaps to be filled greatly impedes a nation's ability to conduct effective environmental risk assessment for transgenic fish. These deficiencies in turn limit a country's ability to craft and implement effective science-based biosafety policies and regulations. Capacity needs are discussed in more detail later in the chapter.

9. Risk management strategies and practices must be based on reliable risk analysis. Risk management strategies such as physical and biological confinement and ecological monitoring programmes should be based on the results of a scientifically defensible and socially acceptable risk assessment. Multi-stakeholder deliberations should guide decisions about appropriate confinement measures, and stakeholders should help choose adequate monitoring end points that represent a step in the event chain leading to the ecological effect (harm) that the stakeholders have agreed they wish to reduce or avoid (Chapters 8 and 9, this volume). Using a participatory process to decide management strategies helps to ensure that they focus on the most critical risks as identified by affected and interested parties, enhancing the legitimacy and public trust of management decisions.

Implementation of risk management strategies does not replace the need to conduct a science-based and socially accepted risk assessment process on transgenic fish. Many of the confinement methodologies applicable to transgenic fish are still in their infancy and have not been proven 100% effective, particularly in large-scale operations (Chapter 8, this volume). Research on biological confinement methods has been limited to a few aquatic species, and the risk of reproduction by transgenic fish can be reduced only by varying degrees, not completely eliminated. Physical confinement measures are more developed because many have already been used in aquaculture production. However, there is a need to better quantify their reliability (or failure rates) under different local environments and across different aquaculture operations.

Monitoring programmes designed specifically to detect indicators of ecological change caused by the escape or reproduction of transgenic fish in receiving environments do not exist yet. Therefore, monitoring methodologies and remedial actions discussed in this book primarily involve adapting existing practices from fisheries management. Limitations of monitoring (Chapter 9, this volume), coupled with variable reliability of biological confinement measures (Chapter 8, this volume), require that the potential risks of each transgenic fish line are anticipated early in its research and development process in order to find ways to mitigate them far in advance of commercial use. The limitations and variable reliability of management strategies also necessitate conducting a thorough science-based, multi-stakeholder risk assessment to guide risk management decisions.

10. Redundancy of confinement measures is essential when confinement is a major component of risk management. Currently used confinement measures are not 100% effective; therefore, it is important to build redundancy into confinement systems. Three levels of confinement can be used to prevent or minimize the entry of transgenic fish or transgenes into the environment: physical, geographic and biological. A combination of the three can be achieved, for example, by culturing monosex fish (biological confinement) in land-based systems (geographic confinement) which incorporate screens and filters (physical confinement). Best management practices incorporating awareness and training programmes for staff, stringent protocols for maintaining inventories and transporting fish and emergency action plans (Box 8.2) are also important for maintaining securely confined aquaculture operations culturing transgenic fish.

Capacity Needs for Conducting Environmental Risk Assessment

A country's human and institutional capacity to conduct a variety of scientific analyses influences the extent to which interested parties can apply the methodologies presented in this book. During this book's international book-writing workshop, capacity gaps surfaced repeatedly when participants discussed the difficulties associated with developing and implementing methodologies for environmental risk assessment and management of aquatic transgenic organisms. Resources necessary for conducting an effective environmental risk assessment include trained scientists, adequate and confined laboratory facilities, enabling policies and monetary support for research on and implementation of environmental risk assessment and management, trained facilitators for the PFOA and risk assessment process and avenues for collaboration between scientists and policy makers. This list is very similar to the major capacity needs identified by an Ad Hoc Technical Expert Group on Risk Assessment convened at the request of the Parties to the Cartagena Protocol on Biosafety (CBD, 2005).

Improving scientific capacities necessary for risk assessment of transgenic organisms, particularly in developing countries, is recognized by key international organizations such as the partners of the Global Environment Facility (GEF). Article 22 of the Cartagena Protocol on Biosafety urges parties to the Protocol to cooperate in 'the development and strengthening of human resources and institutional capacities in biosafety' including scientific and technical training in 'the use of risk assessment and risk management for biosafety, and the enhancement of technological and institutional capacities in biosafety' (CBD, 2000). Furthermore, the Ad Hoc Technical Expert Group on Risk Assessment identified 'a need for further development of guidance, research on ecological effects, and capacity-building' for genetically modified animals used in aquaculture (CBD, 2005). Other international documents, such as the Dhaka Declaration on Ecological Risk Assessment of Genetically Improved Fish and the Nairobi Declaration for the Conservation of Aquatic Biodiversity and Use of Genetically Improved and Alien Species for Aquaculture in Africa, call for increased capacity building for environmental risk assessment of genetically improved fish of all types (ICLARM – The WorldFish Center *et al.*, 2002; WorldFish Center, 2003).

Capacity needs

Baseline ecological data: collection and access

One of the most pressing scientific capacity issues is the absence of baseline data about critical components of natural fish communities and aquatic environments that transgenic fish could potentially alter. Risk assessment can still proceed in the absence of specific baseline data by using robust methods for reducing the complexity of the real world (Chapter 1, this volume) and for addressing uncertainty (Chapter 7, this volume). However, having key baseline data will improve the analysis and deliberation in a risk assessment, as well as the design of the most appropriate risk management strategy.

Environmental risk assessment depends on the ability to determine how wild aquatic species and other ecological components will be affected by transgenic organisms entering and persisting in the ecosystem. Many scientific methodologies for environmental risk assessment and management require knowledge of, for example, specific DNA sequences of the unmodified organism, population genetic structure and genetic diversity of wild populations with which transgenic fish can mate or otherwise interact, the species with which transgenic fish can interact and their key interspecific interactions in natural aquatic environments (Chapters 4, 5, 6 and 9, this volume). Data on existing patterns of fish escapes from aquaculture facilities will improve the ability to predict the probability of gene flow from future escapes of transgenic fish (Chapter 5, this volume) and to design and implement effective physical confinement (Chapter 8, this volume). Aquatic monitoring systems, a potential component of transgenic fish risk management, will be more effective if they have ecological baseline data as reference points for detecting changes (e.g. population decline of wild relatives) occurring after transgenic fish enter and spread in a natural ecosystem. Gathering these and other kinds of baseline data (as mentioned in Chapters 5, 6, 8 and 9, this volume) can begin far ahead of any proposal to introduce transgenic fish. Moreover, they can provide great value for assessing and managing environmental risks imposed by ongoing escapes of non-transgenic fish and shellfish from existing aquaculture production systems. They will also be directly relevant to assessing risks posed by other kinds of introduced species (e.g. a Latin American species of shrimp, *Penaeus vannamei*, introduced to shrimp farms in South-east Asia) and introduced genotypes (e.g. proposals to farm the selectively bred GIFT in Africa's centre of origin of tilapia).

The authors of this book identify the steps needed to prioritize, collect and collate appropriate baseline data to inform a particular risk assessment. Chapter 5 identifies the types of confined experiments and field studies useful for collecting baseline data necessary to estimate gene flow from transgenic fish to wild relatives. Figure 5.3 indicates options for prioritizing baseline data needs on a case-by-case basis. Table 5.1 summarizes the types of studies, data and scientific expertise needed to assess specific events in the introgression fault tree (Fig. 5.1). For example, data on feral populations of the same or related species as the transgenic fish that resulted from prior escapes of unmodified farmed fish can greatly assist predictions about the probability of escape and establishment of transgenics in the same environment. Case studies presented from Thailand and Uganda on fish farm escapes and subsequent establishment of feral populations (Box 5.1) provide examples of needed baseline information about escapes that can be collated and made easily accessible.

Chapter 6 discusses the kinds of baseline data about biotic and abiotic components of the receiving ecosystem needed before one can begin assessing ecological risks of transgenic fish (e.g. Table 6.1, column 4). Chapter 9 discusses baseline data needs, and appropriate data collection and analysis methods, for detecting changes in potential measurement end points (Fig. 9.1). These include estimates of abundance of potential receiving populations and their temporal and spatial variation, fish community composition and natural variation and sampling efficiency for different locations, gear types and environmental conditions.

Baseline ecological data need to be made readily available to the scientific community, risk analysts and stakeholders. Publicly accessible databases can be helpful in this regard. For example, the global database FishBase houses information on the distribution, life history and ecology of over 29,000 fish species, and scientists are able to submit new and updated information easily (Froese and Pauly, 2007). This information can be accessed using common or scientific names, and includes information on habitat and worldwide distribution, including introductions beyond the species' natural range. Information is also available on feeding and reproductive biology, life history traits such as maximum age and size, length–weight relationship and growth information. Other countries, including some developing countries, are building other databases specific to their fish species. For example, the Indian National Bureau of Fish Genetic Resources has developed a database on the fish biodiversity of India with information on systematics, habitat and distribution of 2118 finfishes found in India (Ponniah and Gopalakrishnan, 2000; Kapoor *et al.*, 2002). These kinds of broad ecological databases provide a helpful starting point, but they usually lack the depth and specificity of baseline data needed for a specific risk assessment. Targeted collection of priority baseline data for specific cases is still required (see discussion of necessary baseline data in Chapters 5, 6, and 9, this volume).

Broad, public databases nevertheless provide several benefits for the process of risk assessment and risk management if they are well managed and publicized. They should provide easy access, either via online or through low price or free compact discs. The data can be made available in multiple languages, easing the language barrier often encountered in scientific publications. Many people can contribute information, and with good quality control, this can greatly increase the breadth of information available about global fish species. Electronic media also offer the easiest way to disseminate baseline data around the world.

A few organizing principles should guide the construction of new databases and expansions of existing ones relevant to aquaculture. For instance, it would be advantageous to generate life history and ecology template sheets, which would guide researchers to record relevant data on the specific life history and ecology of a population being studied (see examples in FishBase; Froese and Pauly, 2007). Such data sheets would greatly facilitate the numerous steps involved in assessing and monitoring ecological effects of transgenic fish on natural populations (Chapters 5, 6 and 9, this volume). Ideally, databases should be at the population level because existing global and national databases often do not contain biological information at the intraspecies level, reducing their utility for assessing gene flow from transgenic fish to wild populations in potentially affected water bodies. This is particularly true when there is interest in detecting unique genetic diversity within certain native fish populations. Population-level databases are being built by projects like the Fish DNA Bank and Genomics Laboratory, Inland Fisheries Resources and Research Institute and the Department of Fisheries, Bangkok, Thailand. The facilities consist of a genetic analysis laboratory and facilities to store DNA from randomly sampled tissue collections of indigenous fish from both wild and aquaculture populations. A printed barcode is also assigned to each sample, and each record is organized and controlled by a computer system (W. Kamonrat, Department of Fisheries, Thailand, 2005, personal communication).

Baseline geographical data are also sorely needed to guide effective and efficient application of physical confinement measures for transgenic fish.

A useful approach would be to compile and maintain GIS maps with different layers for: (i) aquaculture facility locations; (ii) their location in relation to water bodies; (iii) possible connections between these water bodies and aquaculture facilities; and (iv) native species and wild relative population distributions. Such information would be invaluable in guiding risk management approaches that include physical confinement measures.

Research facilities

Research drawing upon a wide variety of disciplines is necessary when collecting baseline data and when conducting confined research or tests to obtain necessary empirical data pertaining to environmental risk assessment of a specific line of transgenic fish. Examples of important fields to draw upon are freshwater and marine community ecology, population and quantitative genetics, population dynamics, fish behaviour and behavioural ecology, molecular ecology and physiology, ecosystem modelling and fisheries management. Ideally, a country involved in developing or considering the adoption of transgenic fish should have appropriate research facilities for the range of methodologies discussed in this book (Chapters 4, 5, 6, 8 and 9, this volume).

Countries should have a fish molecular genetics laboratory run by geneticists trained in molecular genetic diagnostic methods and who are experienced in working with fish tissues. Such a laboratory should be equipped to conduct PCR-based analyses and appropriate statistical methods to detect transgenes, determine genetic structure and genetic diversity of wild populations vulnerable to gene flow from transgenic fish and discriminate between species that can hybridize with the transgenic's parental species.

Support should also be provided for establishing new or improving existing fish research facilities, with appropriate confinement systems, for conducting whole organism, breeding, ecological risk assessment and biological confinement experiments needed to fill critical data gaps (Chapters 5, 6 and 8, this volume). Provision of semi-natural conditions in such research facilities would be especially important, as stressed in Chapters 5 and 6. These facilities need to be well confined to prevent escape or unauthorized removal of transgenic fish (Chapter 8, this volume). Properly equipped facilities for this range of risk assessment studies will be expensive to build and operate. One way around this obstacle would be to establish international cooperative testing facilities, ideally in several ecologically distinct regions. National and international organizations could play a major role in supporting the establishment of such regional facilities.

Training programmes for key risk assessment participants

Conducting an environmental risk assessment and designing appropriate risk management measures is also contingent upon having adequately trained scientists, risk specialists, biosafety policy decision makers and facilitators for the deliberative and iterative processes. Countries need their own adequately trained professionals in order to conduct their own environmental risk assessment and management of transgenic fish. Exchange of information and personnel is critical not only for collaborations between developing and developed countries, but also for developing international partnerships in order to achieve widespread dissemination of expertise in research and risk assessment and management. At the international workshop convened to start writing this book, participants identified this as a critical need, as have other relevant organizations and meetings (e.g. CBD, 2005), and a few possible options for addressing it are presented below.

Training doctoral and postdoctoral students and selected working professionals in risk assessment and management methodologies, particularly from countries with direct experience in aquatic biotechnology, and with well-developed lines that could have commercial application (e.g. growth-enhanced transgenic carp in China, Chapter 3, this volume), is an effective way to increase scientific capacity for environmental risk assessment and management of transgenic fish. It is important to train biologists from the range of fields drawn upon when assessing and managing environmental risks of transgenic fish – from molecular biology to aquatic ecology and statistics and modelling (Chapters 4–9, this volume). Across this diversity of fields, scientists need to experience (ideally through hands-on and interdisciplinary projects) how to apply their current field of expertise to answer major questions in environmental risk assessment and management of transgenic aquatic organisms. The interdisciplinary collaboration among junior and senior scientists that occurred during the writing of this book is one example of this kind of hands-on learning, and such experiences need to be increased. Training efforts should also strive to establish valuable international collaborations among scientists. Such international collaborations will lead to increased quality and quantity of scientific methods and data worldwide for conducting environmental risk assessment and risk management for all kinds of introduced aquatic organisms, from transgenic lines to alien species of fish, molluscs and crustaceans.

Training in risk assessment and risk management methodologies

Persons interested in carrying out the risk assessment and risk management steps described in this book need various levels of formal training and expertise in multiple disciplines. A thorough and systematic uncertainty analysis, the development and analysis of conceptual models, formal hazard identification and prioritization, application of biological and ecological models to risk assessment, risk calculation and the design and implementation of statistically defensible monitoring regimes are all steps that require at least college or university training. Risk practitioners will undoubtedly benefit from the variety of training courses provided by universities and professional societies (e.g. Society for Risk Analysis[1]). International teams of scientists and risk specialists, from both developed and developing countries, who focus on biosafety of transgenic fish, can also help capacity building and training in multiple countries as they formulate their own country-specific risk assessment methodologies. Good examples of these actions were the GMO Environmental Risk Assessment workshops in Kenya, Brazil, and Vietnam (Nelson *et al*., 2004; Capalbo *et al*., 2006; Nguyen van Uyen *et al*., Wallingford, UK, 2007, unpublished data) and the international workshop convened to begin writing this book.

[1] Available at: http://www/sra/org

Persons interested in convening multi-stakeholder deliberations also need training in relevant methods. The *PFOA Handbook* (Nelson and Banker, 2007) is designed as a support tool for guiding individual countries through the construction and application of a customized PFOA. It introduces the PFOA process, guides users through techniques and resources that can assist them in formulating the process and makes suggestions for integrating it into environmental risk assessment procedures for transgenic organisms. Although this handbook is not intended to be a training guide for the general public's participation in a PFOA, per se, it is written for a general audience to provide information that could help participants prepare for and become trained on how to participate in a PFOA more effectively.

Methods for defining, analysing and mitigating uncertainty in a risk assessment are critical because of the complexity of interactions between biotic and abiotic components of the receiving environment and the genetic background of the transgenic organism. This book, together with a number of comprehensive texts (Morgan and Henrion, 1990; Burgman, 2005), and the references therein, provides wide-ranging and detailed guidance for managers, analysts and stakeholders about qualitative and quantitative risk assessment methods and approaches to identify and analyse main sources of uncertainty in ecological risk assessments. To date, we are unaware of any published examples that apply these uncertainty analysis methods in an environmental risk assessment of transgenic fish. There are, however, many examples that are truly generic (i.e. can be applied to all forms of environmental stress) or sufficiently similar to transgenic fish (e.g. escape of farmed fish, spread of invasive species, etc.), and they provide analysts with a rich repository of both theory and application (Chapter 7, this volume).

It is also important to build awareness and capacity with regard to understanding the benefits and limitations of confinement approaches. This is particularly important in the developing world, where pressure on regulators and policy makers to enable production of transgenic fish may be greater due to a perceived lower ratio of environmental risk to economic benefit where food security is a pressing concern. Those empowered with setting policies on environmental biosafety and regulating aquaculture and other uses of transgenic fish in developing countries may lack critical knowledge on which to base such policies. It is thus important to raise awareness of the key confinement issues through different means, including training programmes for government-certified biotechnology safety specialists and through widespread circulation of critical knowledge (e.g. Chapter 8, this volume).

Cooperative efforts between research institutions, governments and regional and international organizations can help reduce resource burdens for increasing the capacity of each country to conduct its own environmental risk assessment and management of transgenic aquatic organisms. A variety of international partnerships have begun small steps in this direction. For example, the WorldFish Center and FAO, along with other partners, have organized international consultations and workshops exploring options for genetic improvement (including transgenesis) in aquaculture species in developing countries, while also outlining some relevant environmental biosafety questions and research and capacity needs (Pullin *et al.*, 1999; Gupta *et al.*, 2004).

These types of partnerships should continue and be expanded to support projects dedicated to environmental biosafety research.

'Safety First' Approach

Many of the chapters in this book identify knowledge gaps and analytical uncertainties associated with our current abilities to conduct environmental risk assessments of transgenic fish. However, the fact that transgenic aquaculture technology is still in its infancy provides an excellent opportunity to focus on environmental biosafety at the beginning stages of research and development, in order to generate transgenic lines that will be safer from the outset. A proactive approach would be a departure from the prevailing approach, in which risk assessment and management questions are primarily raised and addressed at the tail end of research and development, when proponents have already spent years and substantial money to develop a line of transgenic organisms and start applying for regulatory approval for its dissemination and use. A proactive approach to biosafety is encapsulated in the 'safety first' approach, which recognizes the value of interacting with multiple stakeholders, starting at early stages of research and development, to identify and assess potential risks of the technology (Kapuscinski, 2003). Results from this early, prospective risk assessment can guide planning and implementation of specific proactive measures to reduce or prevent the likely risks as the development of a transgenic line proceeds. Experience with the development of other complex technologies (e.g. improving the safety of aircraft, starting with the design of their parts) indicates that applying such proactive safety planning and design can make the technology safer, alleviate acrimonious debates and consumer resistance and increase durable public trust of a new technology (Kapuscinski *et al.*, 2003 and references therein).

Many of the risk assessment and management methodologies discussed in this book could be applied in the context of this proactive safety-first approach. Indeed, the approaches for engaging a variety of stakeholders to define and frame the problem, determine effective and measurable risk assessment end points and identify and prioritize possible hazards (Chapters 1 and 2, this volume) should ideally begin early in the development of a stable line of transgenic fish, rather than waiting until a public or private entity is seeking regulatory approval to disseminate the line.

Practical application of the safety-first approach could begin at the molecular level, following the recommendations in Chapter 4 for designing safer transgene constructs. Developers should strive for better control over the copy number, site integration and tissue-specific expression of inserted transgenes, leading to more uniform and predictable engineered traits (Chapters 3 and 4, this volume). These actions, along with careful screening and evaluation of transgenic lines for stability of transgene transmission and expression, can assist in the assessment of gene flow and ecological effects (Chapters 5 and 6, this volume) by partly reducing the variability of transgene expression and incertitude about whole-organism effects of transgene expression. This helps to characterize and reduce two known sources of scientific uncertainty. Developing inducible expression of transgenic constructs could help control the expression of novel traits if the transgenic fish were to escape into an environment where it is not intended (Chapter 4, this volume). Other improvements in biological confinement methods (Chapter 8, this volume) would provide better tools for building environmental safety into the transgenic fish lines. Additional research on and refinement of scientific models for assessing the consequences of gene flow to wild relatives, such as the net fitness methodology (Muir and Howard, 2001) and ecosystem models (e.g. Ecopath[2]), would allow earlier predictions of a transgenic fish's possible impacts on wild relatives and natural fish communities (Chapters 5 and 6, this volume).

Conclusion

The development of transgenic organisms for aquaculture is one possible option for increasing the productivity of aquaculture systems around the world. Aquaculture is critical to the economic and food security of many countries, particularly in developing nations, and transgenic research with finfish and shellfish will continue. Meanwhile, many farmers have yet to fully realize productivity gains offered by other technology options (e.g. Dey *et al.*, 2005a). For instance, advances in selective breeding (e.g. Dey *et al.*, 2000; Asian Development Bank, 2005), fish farming techniques (e.g. Brummett and Costa-Pierce, 2002; Dey *et al.*, 2005b) and integrated aquaculture–agriculture (Dey *et al.*, 2006) have increased productivity and farmer incomes. Ongoing research on these options is likely to generate further large improvements (Costa-Pierce, 2002; Edwards *et al.*, 2002; Li and Cai, 2003; Basiao *et al.*, 2005; Costa-Pierce *et al.*, 2005; Ponzoni *et al.*, 2005; Li *et al.*, 2006). Focusing on aquaculture technology options *alone* will not achieve widespread improvements in food security, incomes of resource-poor farmers, and environmentally responsible aquaculture. It is crucially important to: (i) establish widespread and effective policies that support socially and environmentally sustainable aquaculture; (ii) increase human capacity to practise sustainable aquaculture (via appropriate and effective training and extension); (iii) secure land tenure and water resource access for the poorest farmers; and (iv) improve access to financial capital and improve essential infrastructure (e.g. roads, aquaculture seed supplies) (Delgado *et al.*, 2003; Asian Development Bank, 2004; Dey *et al.*, 2005c).

For transgenic aquaculture research to come to fruition and for its potential applications to be of best use for society, its risks must be honestly and accurately analysed and understood by society. A transparent, flexible, participatory and scientifically sound process of risk assessment and management is crucial to achieving this task. Ideally, this process would begin proactively using a safety-first approach and proceed on a case-by-case basis. A proactive approach would allow more time for the science-driven analysis and deliberation among stakeholders necessary for an effective risk assessment process, thereby

[2] Available at: http://www.ecopath.org

decreasing acrimonious debate and the probability of unanticipated and unacceptable risks of the technology, facilitating its utilization by society.

This book presents a range of analytical and deliberative methodologies that countries can apply to conduct their own risk assessments and develop their own risk management measures for proposed uses of transgenic fish on a case-by-case basis. In presenting and discussing the utility and limitations of these different methodologies, each chapter also indicates key references that provide more detailed guidance on relevant knowledge and methods. These methodologies are envisioned to be flexible and useful tools in multiple settings. They are relevant for countries actively developing their own lines of transgenic aquatic organisms, as well as those considering exporting or importing transgenic aquatic organisms. Finally, using this book's chapters for guidance, countries can begin to approach the task of creating effective, scientifically sound and socially responsive biosafety policies for transgenic fish and other aquatic organisms.

References

Asian Development Bank (2004) *Special Evaluation Study on Small-Scale Freshwater Rural Aquaculture Development for Poverty Reduction – Case Studies*. Operations Evaluation Department, Asian Development Bank, Metro Manila, The Philippines. Available at: http://www.adb.org/Documents/Reports/Evaluation/sst-reg-2004-07/case-studies.asp

Asian Development Bank (2005) *An Impact Evaluation of the Development of Genetically Improved Farmed Tilapia and Their Dissemination in Selected Countries*. Operations Evaluation Department, Asian Development Bank, Metro Manila, The Philippines. Available at: http://www.adb.org/Documents/Books/Tilapia-Dissemination/IES-Tilapia-Dissemination.pdf

Bartley, D. and Kapuscinski, A.R. What makes fishery enhancements responsible? In: Nielsen, J., Dodson, J., Friedland, K., Hamon, T., Hughes, N., Musick, J. and Verspoor, E. (eds) *Proceedings of the Fourth World Fisheries Congress: Reconciling Fisheries with Conservation*. American Fisheries Society, Symposium 49, American Fisheries Society, Bethesda, Maryland (in press).

Basiao, Z., Arago, A.I. and Doyle, R.W. (2005) A farmer-oriented Nile tilapia, *Oreochromis niloticus* L., breed improvement in the Philippines. *Aquaculture Research* 36, 113–119.

Brummett, R.E. and Costa-Pierce, B.A. (2002) Village-based aquaculture ecosystems as a model for sustainable aquaculture development in Sub-Saharan Africa. In: Costa-Pierce, B. (ed.) *Ecological Aquaculture: Evolution of the Blue Revolution*. Blackwell Science, Oxford, UK, pp. 145–160.

Burgman, M.A. (2005) *Risks and Decisions for Conservation and Environmental Management*. Cambridge University Press, Cambridge.

Capalbo, D., Simon, M.F., Nodari, R.O., Valle, S., Dos Santos, R.F., Coradin, L., De O. Duarte, Miranda, J.E., Dias, E.P.F., Quyen, L.Q., Underwood, E. and Nelson, K.C. (2006) Consideration of Problem Formulation and Option Assessment for Bt-cotton in Brazil. In: Hilbeck, A., Andow, D.A. and Fontes, E.M.G. (eds) *Environmental Risk Assessment of Genetically Modified Organisms, Volume 2: Methodologies for Assessing Bt Cotton in Brazil*. CAB International, Wallingford, UK, pp. 67–92.

Convention on Biological Diversity (CBD) (2000) Cartagena Protocol in Biosafety to the Convention on Biological Diversity: Text and Annexes. Montreal: Secretariat of the Convention on Biological Diversity. Available at: http://www.biodiv.org/doc/legal/cartagena-protocol-en.pdf

Convention on Biological Diversity (CBD) (2005) Report of the Ad Hoc Technical Expert Group on Risk Assessment, UNEP/CBD/BS/COP-MOP/3/INF/1, 6 December 2005. Montreal: CBD Secretariat. Available at: http://www.biodiv.org/doc/meeting.aspx?mtg = MOP-03&tab = 1

Costa-Pierce, B. (ed.) (2002) *Ecological Aquaculture: Evolution of the Blue Revolution.* Blackwell Science, Oxford, UK, pp. 145–160.

Costa-Pierce, B., Desbonnet, A., Edwards, P. and Baker, D. (2005) *Urban Aquaculture.* CAB International, Wallingford, UK.

Delgado, C.L., Wada, N., Rosegrant, M.W., Meijer, S. and Ahmed, M. (2003) *Fish to 2020: Supply and Demand in Global Markets.* WorldFish Center Technical Report 62, IFPRI (International Food Policy) and WorldFish Center, Penang, Malaysia. Available at: http://www.lib.noaa.gov/docaqua/noaa_matrix_program_reports/03_siwa_msangi.pdf

Dey, M.M., Eknath, A.E., Li, S. Hussain, M.G., Tran, T.M., Pongthana, M., Nguyen, V.H. and Paraguas, F.J. (2000) Performance and nature of genetically improved farmed tilapia: a bioeconomic analysis. *Aquaculture Economics and Management* 4, 83–106.

Dey, M.M., Rab, M.A., Paraguas, F.J., Bhatta, R., Alam, Md.F., Koeshendrajana, S. and Ahmed, M. (2005a) Status and economics of freshwater aquaculture in selected countries of Asia. *Aquaculture Economics and Management* 9, 11–37.

Dey, M.M., Prein, M., Mahfuzul Haque, A.B.M., Sultana, P., Cong Dan, N. and Van Hao, N. (2005b) Economic feasibility of community-based fish culture in seasonally flooded rice fields in Bangladesh and Vietnam. *Aquaculture Economics and Management* 9, 65–88.

Dey, M.M., Paraguas, F.J., Srichantuk, N., Xinhua, Y., Bhatta, R. and Dung, L.T.C. (2005c) Technical efficiency of freshwater pond polyculture production in selected Asian countries: estimation and implication. *Aquaculture Economics and Management* 9, 39–63.

Dey, M.M., Kambewa, P., Prein, M., Jamu, D., Paraguas, F.J., Pemsl, D. and Briones, R.M. (2006) Impact of development and dissemination of integrated aquaculture–agriculture (IAA) technologies in Malawi. *NAGA, WorldFish Center Quarterly* 29(1 & 2), 28–35.

Edwards, P., Little, D.C. and Demaine, H. (eds) (2002) *Rural Aquaculture.* CAB International, Wallingford, UK.

Froese, R. and Pauly, D. (eds) (2007) *FishBase.* World Wide Web electronic publication. Available at: www.fishbase.org, version (02/2007)

Gibbons, M. (1999) Science's new social contract with society. *Nature* 402, C81–C84.

Gupta, M.V., Bartley, D.M. and Acosta, B.O. (eds) (2004) *Use of Genetically Improved and Alien Species for Aquaculture and Conservation of Aquatic Biodiversity in Africa.* WorldFish Center Conference Proceedings 68, Penang, Malaysia.

Hilbeck, A. and Andow, D.A. (eds) (2004). *Environmental Risk Assessment of Genetically Modified Organisms, Volume 1: A Case Study of Bt Maize in Kenya.* CAB International, Wallingford, UK.

Hilbeck, A., Andow, D.A. and Fontes, E.M.G. (eds.) (2006) *Environmental Risk Assessment of Genetically Modified Organisms, Volume 2: Methodologies for Assesing Bt Cotton in Brazil.* CAB International, Wallingford, UK.

ICLARM – The WorldFish Center, Technical Center for Agricultural and Rural Cooperation (CTA), Food and Agriculture Organization of the United Nations (FAO), The World Conservation Union (IUCN), United Nations Environment Program (UNEP) and Convention on Biological Diversity (CBD) (2002) *Nairobi Declaration: Conservation of Aquatic Biodiversity and Use of Genetically Improved and Alien Species for Aquaculture in Africa.* WorldFish Center, Penang, Malaysia. Available at: http://www.cta.int/pubs/nairobi/declaration.pdf

Kapoor, D., Dayal, R. and Ponniah, A.G. (2002) *Fish Biodiversity of India.* National Bureau of Fish Genetic Resources (NBFGR), Lucknow, Uttar Pradesh, India.

Kapuscinski, A.R. (2003) From reactive to pro-active biosafety: science, technology and capacity needs. In: Sandlund, O.T. and Schei, P.J. (eds) *Proceedings of the Norway/UN Conference on Technology Transfer and Capacity Building, Trondheim, 23–27 June 2003.* Available at: http://english.dirnat.no/wbch3.exe?ce = 18026

Kapuscinski, A.R. and Brister, D.J. (2001) Genetic impacts of aquaculture. In: Black, K.D. (ed.) *Environmental Impacts of Aquaculture*. Sheffield Academic Press, Sheffield, UK, pp. 128–153.

Kapuscinski, A.R., Goodman, R.M., Hann, S.D., Jacobs, L.R., Pullins, E.E., Johnson, C.S., Kinsey, J.D., Krall, R.L., La Vina, A., Mellon, M. and Ruttan, V. (2003) Making 'safety first' a reality for biotechnology products. *Nature Biotechnology* 21, 599–601.

Li, S.F. and Cai, W.Q. (2003) Genetic improvement of the herbivorous blunt snout bream (*Megalobrama amblycephala*). *NAGA, WorldFish Center Quarterly* 26, 20–23.

Li, S.F., He, X.J., Hu, G.C., Cai, W.Q., Deng, X.W. and Zhou, P.Y. (2006) Improving of growth performance and stripe pattern on caudal fin in F_6–F_8 generations of GIFT Nile tilapia (*Oreochromis niloticus*) by mass selection. *Aquaculture Research* (accepted).

Morgan, M.G. and Henrion, M. (1990) *Uncertainty: A Guide to Dealing with Uncertainty in Quantitative Risk and Policy Analysis*. Cambridge University Press, Cambridge.

Muir, W.M. and Howard, R.D. (2001) Assessment of possible ecological risks and hazards of transgenic fish with implications for other sexually reproducing organisms. *Transgenic Research* 11, 101–114.

NRC (1996) *Understanding Risk: Informing Decisions in a Democratic Society*. National Academy Press, Washington, DC.

Nelson, K.C. and Banker, M. (2007) *Problem Formulation and Options Assessment Handbook: A Guide to the PFOA Process and How to Integrate it into Environmental Risk Assessment (ERA) of Genetically Modified Organisms (GMOS)*. GMO-ERA Project (in press). Available at: www.gmo-guidelines.info (in preparation)

Nelson, K.C., Kibata, G., Muhammad, L., Okuro, J.O., Muyekho, F., Odindo, M., Ely, A. and Waquil, J.M. (2004) Problem Formulation and Options Assessment (PFOA) for genetically modified organisms: The Kenya case study. In: Hilbeck, A. and Andow, D.A. (eds) *Environmental Risk Assessment of Genetically Modified Organisms*, Volume 1: *A Case Study of Bt Maize in Kenya*. CAB International, Wallingford, UK, pp. 57–82.

Ponniah, A.G. and Gopalakrishnan, A. (eds) (2000) *Endemic Fish Diversity of the Western Ghats*. NBFGR-NATP (Publ.1), National Bureau of Fish Genetic Resources (NBFGR), Lucknow, Uttar Pradesh, India.

Ponzoni, R.W., Hamzah, A., Tan, S. and Kamaruzzaman, N. (2005) Genetic parameters and response to selection for live weight in the GIFT strain of Nile tilapia (*Oreochromis niloticus*). *Aquaculture* 247, 203–210.

Pullin, R.S.V., Bartley, D.M. and Kooiman, J. (eds) (1999) *Towards Policies for the Conservation and Sustainable Use of Aquatic Genetic Resources*. ICLARM Conference Proceedings, Manila, The Philippines.

Sagar, A., Daemmrich, A. and Ashiya, M. (2000) The tragedy of the commoners: biotechnology and its publics. *Nature Biotechnology* 18, 2–4.

Susskind, L., Levy, P.F. and Thomas-Larmer, J. (2000) *Negotiating Environmental Agreements: How to Avoid Escalating Confrontation, Needless Costs, and Unnecessary Litigation*. Island Press, Washington, DC.

Wondolleck, J.M. and Yaffee, S.L. (2000) *Making Collaboration Work: Lessons from Innovation in Natural Resource Management*. Island Press, Washington, DC.

WorldFish Center (2003) *Dhaka Declaration on Ecological Risk Assessment of Genetically Improved Fish*. WorldFish Center, Penang, Malaysia. Available at: http://www.worldfishcenter.org/Pubs/Dhaka%20booklet/Dhaka_booklet.pdf

Glossary

Acceptance criteria: The criteria used to distinguish acceptable from unacceptable changes in risk, such as a 50% chance of a 30% decline in Shannon's index of native fish biodiversity per year. *See* **Assessment end point** and **Measurement end point**.

Allozyme: An enzyme produced by one allele at a locus; allelic form of an enzyme that can be distinguished by electrophoresis.

Androgenesis: Production of offspring with all chromosomes and genes obtained from the father.

Androgens: Any compound with male sex hormone activity, such as testosterone. Usually synthesized in the testis.

Aneuploid: Bearing a number of chromosomes that is not an exact multiple of the haploid number typical for the species.

Antisense: A nucleic acid sequence (usually RNA) complementary to all or part of a functional mRNA molecule, to which it binds, blocking its translation.

Assessment end point: A formal expression of the environmental values to be protected, such as biodiversity of native fish species.

Bayesian analysis: Bayesian analysis is a way of rationally updating beliefs in the face of accumulating evidence using Bayes theorem. It is a method of statistical inference that uses evidence or observations to update or infer the probability that a hypothesis may be true.

cDNA: Complementary DNA; a DNA strand synthesized *in vitro* from a mature RNA template using reverse transcriptase. DNA polymerase is then used to create a double-stranded molecule.

Chromosome set manipulation (or chromosome manipulation): Intentional change in the number of haploid sets of chromosomes in a cell or organism.

Cloned gene: Refers to the synthesis of multiple copies of a chosen DNA sequence using a bacterial cell or another organism as a host.

Copula: A mathematical function that joins univariate distribution functions to form multivariate distribution functions.

Diagnostic markers: A marker (often genetic-based) that uniquely distinguishes an individual organism or group of organisms from among a larger population.

Diploid: A cell, tissue or organism having two complete sets of chromosomes.

Ectopic expression: The transcription (and possibly also the translation) of a transgene in tissues where it would not normally occur.

Embryonic stem cell: Cells of the early embryo that can give rise to all differentiated cells, including germ line cells.

Feral: Pertaining to formerly domesticated animals now living in a wild state.

Fitness: The success in survival and reproduction of an individual organism, a population or a species, relative to other individuals, populations or species; an individual's number of offspring that survive to reproduce.

Founder generation: In genetic engineering, the first generation of organisms into which DNA (a transgene) has been transferred.

Future alternatives: Options for addressing a particular problem, including options that are in use now (*status quo*) and continued into the future, options that exist but could be more widely implemented or new options such as aquaculture of a transgenic fish.

Gene construct: Consists of a structural gene linked to promoter, enhancer and regulatory sequences, and sometimes marker gene sequences. Often, but not always, part or all of the gene construct has been derived from a different species than that of the recipient.

Gene flow: The exchange of genes between different populations of the same or closely related species.

Genetic drift: Random changes in allelic frequencies due to natural sampling errors that occur in each generation. The smaller the population, the greater is the genetic drift, with the result that some alleles are lost, and genetic diversity is reduced.

Genetically modified organism (GMO): In this book, synonymous with transgenic organism and living modified organism. *See* **Transgenic organism**.

Genotype-by-environment interaction (G × E interaction): When a specific difference of the environment may have a greater effect on one genotype than on another. A more dramatic form of interaction involves a change in the relative ranking of genotypes when the same traits are measured under different environments. *See also* **Phenotypic plasticity.**

Gonadosomatic index: The ratio of gonadal tissue mass to total body mass, often expressed as a percentage.

Gonadotropin releasing hormone (GnRH): Produced in the hypothalamus and influences synthesis and secretion of gonadotropin molecules from the pituitary gland, which, in turn, influence development of ovaries and testes in vertebrates.

Harm: Undesirable consequences to humans and the things that they value; the social construction of and shared agreement on potential loss based on social values.

Hazard: An act or phenomenon that has the potential to produce harm or other undesirable consequences to what humans value.

Hemizygous: Having genes present only once in the genotype and not in pairs.

Heterogametic: The sex with differently shaped sex chromosomes (e.g. X and Y).

Homogametic: The sex with similarly shaped sex chromosomes (e.g. XX), producing only one kind of gamete in regard to the sex chromosomes.

Homologous chromosomes: Chromosomes that pair during meiosis and contain the same linear gene sequences.

Homozygous: Having the same allele at a particular gene (or DNA sequence) on the two homologous chromosomes.

Keystone species: A species having a dominating influence on the composition of a community, which may be revealed if the keystone species is removed.

Measurement end point: The metric used to measure changes in the assessment end point, such as Shannon's index of biodiversity. *See* **Assessment end point**.

Mosiacism: In transgenesis, refers to organisms in which only a fraction of cells have integrated copies of the transgene. In chromosome set manipulations, refers to the presence, within one individual, of cells with different chromosome numbers.

Open reading frame (ORF): Sequence of messenger RNA, or corresponding region of DNA, that can be translated to produce a protein.

Phenotypic plasticity: The general responsiveness of phenotypes to environmental conditions.

Plasmid: A circular, self-replicating, non-chromosomal DNA molecule found in many bacteria, although many artificial ones have been made. Often used as vectors for genetic engineering.

Pleiotropy: The phenomenon in which a single gene affects more than one phenotypic characteristic.

Ploidy: A multiple of the full set of homologous chromosomes in a cell. Haploidy indicates a single set and diploidy a double set of chromosomes. *See* **Diploid, Triploid** and **Tetraploid**.

Polymerase chain reaction (PCR): A molecular genetic technique used to amplify the number of copies of a target DNA sequence.

Polyploidy: Organism, tissues or cells having more than two complete sets of chromosomes.

Population: A group of organisms that belong to the same species and freely interbreed.

Promoter: Region of DNA in front of a gene that binds RNA polymerase and initiates transcription, thereby promoting gene expression.

Recombination: Exchange of alleles between homologous chromosomes due to crossing over during meiosis.

Regulatory sequence: A DNA sequence involved in regulating the expression of a gene, e.g. a promoter or operator region (in the DNA molecule).

Retroviral vector: A gene construct containing sequences from a retrovirus, which facilitate integration of the transgene into the host's genome.

Retrovirus: A class of RNA viruses that, by using reverse transcription, can form double-stranded DNA copies of their genomes, which can integrate into the chromosomes of an infected cell.

Risk: The likelihood of harm occurring from a specified hazard (or set of hazards), or as a result of some behaviour or action (including no action).

Stakeholder: Anyone who has an interest in an issue, or anyone who shares the burden of the risks resulting from a particular decision. In a social context, an individual or a representative of a group affected by or affecting the issues in question.

Structural gene: A gene that codes for a protein.

Tetraploid: A cell, tissue or organism having four complete sets of chromosomes.

Transgene: A sequence of DNA, usually a gene construct, transferred into the cell(s) of an organism and often inserted into the host genome.

Transgenesis: The process of transferring DNA into an organism by artificial means. *See* **Transgenic organism**.

Transgenic organism: An organism whose genetic composition has been altered to include selected genes or DNA sequences from other organisms of the same or different species by methods other than those used in traditional breeding.

Transposon: Small DNA molecules that can move in and out of specific positions within or between chromosomes or plasmids. In moving, they may or may not leave a copy of their DNA sequence behind; sometimes referred to as 'jumping genes' or 'selfish DNA'.

Triploid: A cell, tissue or organism having three complete sets of chromosomes.

Vector: A small DNA molecule (plasmid, virus, bacteriophage, artificial or cut DNA molecule) used to deliver DNA into a cell; and must be capable of being replicated and contain cloning sites for the introduction of foreign DNA.

Index

Note: page numbers in *italics* refer to figures, tables and boxes